MW01105422

Bioinformatics in Cancer and Cancer Therapy

Cancer Drug Discovery and Development

Beverly A. Teicher, *Series editor*

For other titles published in this series, go to
www.springer.com/series/7625

Bioinformatics in Cancer and Cancer Therapy

Edited by

GAVIN J. GORDON, PhD
Brigham and Women's Hospital, Harvard Medical School,
Division of Thoracic Surgery, Boston, MA, USA

Editor
Gavin J. Gordon
Brigham and Women's Hospital
Division of Thoracic Surgery
Boston, MA
USA

Series Editor
Beverly A. Teicher
Genzyme Corporation
Framington, MA
USA

ISBN: 978-1-58829-753-2 e-ISBN: 978-1-59745-576-3
DOI: 10.1007/978-1-59745-576-3

Library of Congress Control Number: 2008931368

Printed on acid-free paper

9 8 7 6 5 4 3 2 1

springer.com

Preface

Bioinformatics can be loosely defined as the collection, classification, storage, and analysis of biochemical and biological information using computers and mathematical algorithms. Although no single person or group started the field wholly on their own, Temple Smith, Ph.D., a professor at Boston University, is generally credited with coining the term. Bioinformatics represents a combination of biology, medicine, computer science, physics, and mathematics, fields of study that have historically existed as mutually exclusive disciplines.

In the past twenty years we have witnessed an explosion of interest in computer-assisted bioinformatics-based analysis of cancer. Although this approach to experimental science is not new, it has recently gained traction among a diverse set of academic and business professionals from varied backgrounds. Concurrently, bioinformatics has vaulted into the public's eye in newspapers and magazines, most notably in the area of (personalized) DNA sequencing. The combined result is that bioinformatics is being heralded as a panacea to the current limitations in the clinical management of cancer. While certainly over-optimistic in some regards, this designation is not without promise, particularly in the area of cancer diagnosis and prognosis.

The focus of this book is to: (1) provide a historical and technical perspective of the analytical techniques, methodologies, and platforms used in bioinformatics experiments, (2) show how a bioinformatics approach has been used to characterize various cancer-related processes, and (3) demonstrate how the bioinformatics approach is being used as a bridge between basic science and the clinical arena to positively impact patient care and management.

Boston, MA ***Gavin J. Gordon***

Contents

Contributors

EGISTO BOSCHETTI • *Bio-Rad Laboratories, Hercules, CA, USA*
LARS BULLINGER • *Department of Internal Medicine III, University of Ulm, Ulm, Germany*
GEORGE A. CALIN • *Comprehensive Cancer Center, Ohio State University, Columbus, OH, USA*
STEVEN W. COLE • *Division of Hematology-Oncology, Department of Medicine, David Geffen School of Medicine, University of California, Los Angeles, CA, USA Molecular Biology Institute, Jonsson Comprehensive Cancer Center, University of California, Los Angeles, CA, USA*
CARLO M. CROCE • *Comprehensive Cancer Center, Ohio State University, Columbus, OH, USA*
KEVIN K. DOBBIN • *Biometric Research Branch, National Cancer Institute, Bethesda, MD, USA*
HARTMUT DOHNER • *Department of Internal Medicine III, University of Ulm, Ulm, Germany*
ANTHONY EL-KHOUEIRY • *Kenneth Norris Comprehensive Cancer Center, Keck School of Medicine, University of Southern California, Los Angeles, CA, USA*
STEVEN A. ENKEMANN • *H. Lee Moffitt Cancer Center & Research Institute, 12902 Magnolia Drive, Tampa, FL 33612, USA*
MANUELA FERRACIN • *University of Ferrara, 44100 Ferrara, Italy*
JEFFREY E. GREEN • *Laboratory of Cell Regulation and Carcinogenesis, National Cancer Institute, National Institutes of Health, Bethesda, MD, USA*
MARK J. HOENERHOFF • *Laboratory of Cell Regulation and Carcinogenesis, National Cancer Institute, National Institutes of Health, Bethesda, MD, USA*
ERIC H. JENSEN • *H. Lee Moffitt Cancer Center & Research Institute, 12902 Magnolia Drive, Tampa, FL 33612, USA*
HEINZ JOSEF LENZ • *Keck School of Medicine, Kenneth Norris Comprehensive Cancer Center, University of Southern California, Los Angeles, CA, USA*
CHANG-GONG LIU • *Comprehensive Cancer Center, Ohio State University, Columbus, OH, USA*
JAMES M. MCLOUGHLIN • *H. Lee Moffitt Cancer Center & Research Institute, 12902 Magnolia Drive, Tampa, FL 33612, USA*
ALEKSANDRA M. MICHALOWSKI • *Laboratory of Cell Regulation and Carcinogenesis, National Cancer Institute, National Institutes of Health, Bethesda, MD, USA*

Massimo Negrini • *Comprehensive Cancer Center, Ohio State University, Columbus, OH, USA University of Ferrara, 44100 Ferrara, Italy*

Carla Oliveira • *Institute of Pathology and Molecular Immunology, University of Porto (IPATIMUP), Porto, Portugal*

Christos A. Ouzounis • *Computational Genomics Unit, Institute of Agrobiotechnology, Center for Research & Technology (CERTH), Thessalonica, 57001, Greece*

Jonathan R. Pollack • *Department of Pathology, Stanford University, Stanford, CA, USA*

Ting-Hu Qiu • *Laboratory of Cell Regulation and Carcinogenesis, National Cancer Institute, National Institutes of Health, Bethesda, MD, USA*

Nagesh P. Rao • *Department of Pathology & Laboratory Medicine, David Geffen School of Medicine, University of California, Los Angeles, CA, USA*

Raquel Seruca • *Institute of Pathology and Molecular Immunology, University of Porto (IPATIMUP), Porto, Portugal*

Richard M. Simon • *Biometric Research Branch, National Cancer Institute, Bethesda, MD, USA*

Sérgia Velho • *Institute of Pathology and Molecular Immunology, University of Porto (IPATIMUP), Porto, Portugal*

Stefano Volinia • *Comprehensive Cancer Center, Ohio State University, Columbus, OH, USA University of Ferrara, 44100 Ferrara, Italy*

Scot Weinberger • *GenNext Technologies, Inc., Montara, CA, USA*

David T. Wong • *School of Dentistry, Dental Research Institute, University of California, Los Angeles, CA, USA Molecular Biology Institute, Jonsson Comprehensive Cancer Center, University of California, Los Angeles, CA, USA*

Timothy J. Yeatman • *H. Lee Moffitt Cancer Center & Research Institute, 12902 Magnolia Drive, Tampa, FL 33612, USA*

Xiaofeng Zhou • *Center for Molecular Biology of Oral Diseases, College of Dentistry, Graduate College, UIC Cancer Center, University of Illinois, Chicago, IL, USA Guanghua School & Research Institute of Stomatology, Sun Yat-sen University, Guangzhou, China*

1 The Emergence of Bioinformatics: Historical Perspective, Quick Overview and Future Trends

Christos A. Ouzounis

ABSTRACT

We provide a concise overview of the history of bioinformatics, its current status and some possible future trends, with a specific emphasis on cancer research.

Key Words: Bioinformatics, Computational biology, Genome research, Cancer genomics, historical view, Future trends

1 INTRODUCTION

Attempting to condense the emergence of the field of bioinformatics into a few pages can be a huge challenge. Any such endeavor invariably entails the past, present, and future: a historical narrative, a status report, and possible future trends, as the very title suggests. In the context of this entire book, and the role of bioinformatics in cancer research, this undertaking is all the more demanding, because it has to translate achievements from an abstract field of genome research into applications for human health and welfare. These applications are eloquently described further in the following chapters, providing a valuable synopsis of the methodologies and platforms used in bioinformatics experiments, the characterization of various cancer-related processes, and the demonstration of how this field has been used to bridge basic science and the clinical arena to positively impact patient care and management. The humble role of this introductory chapter is thus to provide the historical and practical context in which all these activities are being conducted. Citations to previous work will be used sparsely, so that pointers to relevant reviews can act more as entry points to subareas of bioinformatics and not necessarily to original research, with the exception of few genuine classics. By exploring its past history and providing a quick overview of the present status of this field, it is possible to identify certain future trends that are likely to occur, at the fringes of bioinformatics with other disciplines, this time closer to hardware than software.

G.J. Gordon (ed.), *Cancer Drug Discovery and Development: Bioinformatics in Cancer and Cancer Therapy*, DOI: 10.1007/978-1-59745-576-3_1,

2 A HISTORICAL PERSPECTIVE: THE THEORY, FUSION OF MOLECULAR GENETICS AND COMPUTER SCIENCE

Bioinformatics can be narrowly defined as a field at the crossroads of biology and computer engineering, responsible for the storage, distribution, and analysis of biological information. This field has emerged as an independent discipline during the 1990s, consolidating previously dispersed activities across different departments and faculties. The formation of graduate programs in bioinformatics, the founding of societies and regional groups, the creation of research groups and programs, the targeted calls for proposals by funding agencies, substantial investments by private industries, the development of technological platforms, and finally the publication of specialized journals all contributed toward the recognition of this multidisciplinary activity as a stand-alone area (Boguski 1994). In fact, all of the above contain the hallmarks of a scientific revolution in the making, in the sense that this frenetic activity was driven both by research and business opportunities in biology, medicine, and engineering in the broadest sense.

Arguably, bioinformatics might be one of the most rapidly evolving fields of science ever, having gone from a peripheral, almost obscure area o f biological sciences, into a most mainstream technology field, often under the limelight. The discovery of this field by the media and press less than a decade ago, however, does not necessarily mean that this discipline did not have a long, and sometimes highly under-appreciated, history (Ouzounis 2000). The development of bioinformatics in the 1990s is a case study of a scientific explosion, where the needs and requirements of a field grow suddenly into a different direction, heavily influenced by data availability and technology in DNA sequencing (Fields 1996). When this happens, there is a social change, with industry pressing for more courses in this field, academia instigating coordinated activities, the job market experiences a shortage which has to be supplemented by short-term improvisation and medium-term multidisciplinary actions, and finally the funding agencies scrabbling for relocation of resources toward the accomplishment of some of the above objectives.

Now that the dust has settled, it is hard to believe that 15 years ago, publishing computational analyses of biological information was a challenge. Molecular biologists, in particular, were not accustomed in computationally derived results, and viewed experimental observations as the only way of making progress. Today, it is almost inconceivable that a high-impact research publication in biology does not contain some elements of computing, at different levels, from instrumentation to interpretation. Thus, it can be argued that there has been a paradigm shift, where biology was transformed from a purely experimental science to a hybrid field of computation and experiment (Ouzounis 2002).

Yet, 50 years ago[1], some of the most fundamental problems in molecular biology presented some formidable algorithmic issues, including the structure of DNA and the encoding of genetic information (Ouzounis and Valencia 2003). A list of classic papers during the 1960s on the evolution of genes and proteins, the structural properties of polypeptide chains, the informational content of DNA sequences, the origins of the genetic code, the construction of phylogenetic trees, and the early theory of sequence alignment

[1] No references before 1980 are provided in this chapter, with one sole exception.

has been provided elsewhere (Ouzounis and Valencia 2003). During the 1970s, the availability of data on protein sequences and structures fueled, for instance, the development of approaches for the analysis of mutation rates and solvent accessibility, respectively. Much effort was being made on the sequence alignment problem (Fitch and Smith 1983), hand-in-hand with the string comparison problem in computer science (Hall and Dowling 1980). The mutual influence of these two fields is a textbook case of multidisciplinary activity on the very same problem: Algorithms for the comparison of strings or macromolecular sequences were developed either by theoretically minded biologists or by computer scientists with a taste for application in a new field of research (Ouzounis and Valencia 2003). The latter years of that decade also saw the first data resources, primarily compilations of protein molecules, in anticipation of the first databases that emerged in the following years. Consequently, and following developments both in computer hardware and software, it had become possible to start storing, distributing, and analyzing data derived from biological experimentation that were amenable to computational analysis, the very definition of bioinformatics. These data were considered extremely valuable not to be shared by the entire community, and these were the seeds for the subsequent ethos of open-source and open-access principles that permeated this field. The agenda was set: there was a flurry of activity in the fields of sequence analysis and similarity searching (Gingeras and Roberts 1980; Wilbur and Lipman 1983), structure prediction (Richardson 1981; Kabsch and Sander 1984), molecular evolution (Doolittle 1981; Bajaj 1984), and molecular biology databases (Philipson 1988; Bernstein et al. 1977). A detailed citation list for bioinformatics during the 1980s has been published elsewhere (Ouzounis and Valencia 2003).

Until then, the primary objects of investigation were macromolecular sequences (DNA, RNA, or protein), typically analyzed as strings of symbols or related representations, and macromolecular structures, usually analyzed as sets of cartesian coordinates or by-products of these. More abstractly, it is reasonable to assume that the objects of study corresponded to single molecules or sets of molecules and their interconnections, in search of regularities at different levels. Such examples might be the identification of a promoter region across a handful of sequences from different species or a common folding pattern across a set of similarly folded protein structures. These limitations were mostly of a technical nature, constrained by the techniques of the day, and less of a conceptual nature. Computational biologists, twenty years ago, had to wait for another decade until high-throughput biology would help them realize their dream: the derivation of principles that govern the evolution and development of living organisms across multiple cells, tissues, organs, individuals, populations, and species.

3 CURRENT STATUS: THE SOFTWARE, DATA-DRIVEN ANALYSES OF ALL BIOLOGICAL INFORMATION TYPES

In the following decade, a number of key accomplishments include the establishment of rapid database searching using BLAST (Altschul et al. 1990), for example (one of the most highly cited biology publications of all time!), the matching of the entire protein sequence database (Gonnet et al. 1992), the detection of complex gene structures (Guigo et al. 1992), a more accurate homology modeling of protein structures (Levitt 1992), and sequence threading for proteins without homologs (Bowie et al. 1991). The explosion of computational techniques for the analysis of biological information in the 1990s dictates the inclusion of some of the above milestones in this section, about the current status of bioinformatics. The techniques for large-scale sequencing and other experimental measurements were being developed, and it was only a matter of time before bioinformatics took its proper place next to the new field of genomics (Benton 1996).

The first genome sequence that was profiled in its entirety, but never completed, was the genome of *Mycoplasma capricolum* (Bork et al. 1995), quickly followed by the landmark publication of the entire genome of *Haemopbilus influenzae* Rd (Pleischmann et al. 1995). Computational analysis would never be regarded as a nuisance for experimental biology again: the flood of sequence data necessitated fast, accurate, and accessible methods for analysis and interpretation (Andrade and Sander 1997). Very rapidly, and along the determination of more genome sequences, research developed toward the prediction of protein function from sequence (Bork and Koonin 1998 Wheelan and Boguski 1998 Andrade et al. 1999), the detection of metabolic pathways from reference genomes (Karp et al. 1996), and the comparative genomics of various species (Gogarten and Olendzenski 1999). The types of biological information expanded from single genes or proteins to entire gene families never observed earlier (Rawlings and Searls 1997), the interpretation of biochemical networks (Bono et al. 1998), the locations of genes along chromosomes, or annotation elements from free text (Ouzounis et al. 1996). Plenty of opportunities arose for the pharmaceutical industry, namely the identification of targets and candidates for the development of drugs, vaccines, diagnostic markers, and therapeutic proteins (Andrade and Sander 1997). Slowly but steadily, academic research topics became the subject of more applied research, for example in the field of nutritional research (Dellapenna 1999). Finally, two other developments that generated novel types of biological information, data with a familiar face but on a larger scale, were structural genomics (Orengo et al. 1999, Brenner 2001) and expression profiling (Ferea and Brown 1999). Applications of the above, for example modeling expression patterns, were quickly explored by medical researchers into aspects of human health and disease (Huang 1999).

Breakthroughs in whole-genome analysis were in the horizon, just before the end of the century. Methods for the analysis of proteins in cell or tissue samples (the new field of "proteomics") emerged (Chambers et al. 2000), the massive sequencing of entire genomes of free-living organisms (Broder and Venter 2000), the contextual detection of protein function networks (Eisenberg et al. 2000), e.g., based on gene fusion patterns (Enright 1999 et al.), and frameworks for the analysis of gene expression (Brasma and Vilo 2000). The field of bioinformatics was experiencing a tidal wave: before its establishment, there were already discussions for its future development and its place in biological research (Searls 2000) (Tsoka and Ouzounis 2000). Novel data types now included complex gene and genome structures of various cell types, including eukaryotes (Rawlings and Searls 1997), predicted functional annotations for a multitude of uncharacterized sequences (Bork and Koonin 1998), gene expression patterns of sets of genes under variable experimental conditions (Quackenbush 2001), genetic or protein interaction networks inferred from genome function and structure analyses, respectively (Pellegrini 2001), and finally the fusion of the field of genomics with genetics research for the identification of single-locus or multifactorial genetic diseases (Pang et al. 2000). Even more daring steps were taken, to include spatiotemporal information on expression patterns for entire organisms, for instance in embryology (Davidson and Baldock 2001).

With such a frenetic activity, it was then reasonable to proclaim that the impact on medicine will be monumental, with information processing holding a central role in the simulation of molecular processes in cells and the inference of drug effects in humans based on their genetic backgrounds (Sander 2000). More specifically, in clinical (cancer) research and future practice, applications of computational approaches include genomics-based prognosis (Kallioniemi 2001), early diagnosis and cancer biomarker discovery (Hondermarck et al. 2001), drug target discovery (Pavelie and Gall-Troselj 2001), treatment and therapeutic targeting (Schultze and Vonderheide 2001), drug development (Basik et al. 2003), clinical drug resistance (Damaraju et al. 2002), as well as clinical data management (Haque et al. 2002). Finally, specialized resources such as the Cancer Genome Anatomy Project (CGAP) were developed to support cancer research (Strausberg et al. 2001).

Following the above passage about applications of computational approaches in cancer research, it is worth providing a current status report on bioinformatics, albeit briefly, due to the massive literature that abounds. Inevitably, only research of the last few years will be cited in this section, so that an up-to-date view can be provided. Currently, the issue of sequence annotation and functional inference is still with us (Ouzounis and Karp 2002). We argued previously that only community-based annotation can solve the accuracy and update problems that permeate more "generic" sequence annotation resources (Tsoka and Ouzounis 2000). From individual genes and proteins to their ensuing functional association networks (Valencia and Pazos 2002), we have uncovered the new world of "systems" biology (Kirschner 2005). The discovery of entire classes of proteins with related structure and function is still in progress, e.g., the kinases of the human genome (Manning et al. 2002). This type of classification systems for protein structure and function (Ouzounis et al. 2003) will continue to play a significant role in the molecular dissection of human health and disease. Other aspects of molecular information include not only gene and protein families or classes, but also comparative genome analyses at close range (Hardison 2003), metabolic databases (Tsoka and Ouzounis 2003), integrated networks for gene/protein interactions (Kanehisa and Bork 2003), free-text information to support annotation efforts in genomics (Janssen et al. 2005), and more solid microarray data analyses (Allison et al. 2006). Resources that capture and distribute information for sets of proteins, e.g., UniProt (Apweiler et al. 2004), or genomes, e.g., Ensembl (Birney et al. 2004), will continue to support data analysis in modern biology. Other computational approaches that will play an increasingly important role include the structure determination of proteins on a genome-wide scale (Chandonia and Brenner 2006), the construction of designed proteins (Park et al. 2004), the detection of "orthologs" across species and larger phylogenetic distances (Koonin 2005), the inference of ancestral states for genome content and structure (Ouzounis 2005), the comparative genomics of gene expression and annotation patterns (Lopez-Bigas et al. 2006), and finally the delineation of genome dynamics, presently better understood for bacterial genomes (Ochman and Davalos 2006).

Despite the risks of not being an expert in the field of cancer research, it will nevertheless be useful to discuss some cancer-specific, recent bioinformatics work that connects to the topics elsewhere in this book. Such examples are not exhaustive and include the computational analysis of epigenetic effects in cancer biology (Yang and Lee 2004), such as DNA methylation patterns and histone modifications (Lee 2003), the contextual simulations of cell cycle in neoplasm development (Alberghina et al. 2004), the development of formalisms for such simulations (Christopher et al. 2004), the analysis of single-nucleotide polymorphisms (SNPs) (Clifford et al. 2004), cancer target discovery (Desany and Zhang 2004), cancer biomarker discovery (Rhodes and Chinnaiyan 2004), and even bioinformatics training and education (Umar 2004). Also, resources such as the Mouse Tumor Biology Database (Krupke 2005) will become increasingly important, alongside gene expression experiments (Rhodes and Chinnaiyan 2005, Segal et al. 2005), novel therapeutics research (Mount and Pandey 2005), targeted proteomics technology (Pasadas et al. 2005), and connection to other areas such as immunology (Strausberg 2005). Finally, the connection of cancer research to systems biology (Khalil and Hill 2005), or the realization that cancer research actually is a systems biology disease (Hornberg et al. 2006), is too recent to be assessed at this point, but they appear to be very promising avenues for future research. In all, bioinformatics is expected to continue its fascinating interplay with the field of genomics in cancer research, namely cancer bioinformatics and oncogenomics (Strausberg et al. 2004), respectively.

4 FUTURE TRENDS: THE HARDWARE, COMPUTING WITH LIVING MATTER, AND POTENTIAL APPLICATIONS

Evidently, any treatise on future trends must be wrong. It is also invariably the hardest section on any such presentation. However, we will attempt here, on the basis of the past history and more recent developments, to provide a more personal assessment on the future of bioinformatics research and its application to human health and disease. Like a weather forecast, some of the predictions will be rather obvious and some rather esoteric (and probably wrong). The purpose of this epilogue is not so much to attempt some crystal-gazing of an unpredictable future, as to pave the way for a consolidation of this field with other disciplines, now themselves experiencing an unparalleled explosion. In anticipation of this fusion, we can then foster more interdisciplinary actions and collaborations that were proven to be so successful in the development of bioinformatics, with mathematics, computer science, physics, chemistry, and biology all contributing toward the creation of a truly unique discipline at the crossroads of modern science.

It should be evident that bioinformatics developed in close association with other technological capabilities in biological sciences, primarily data generation and analysis. In the early days, the primary focus of the field was the analysis of molecular sequences and structures, followed by entire genomes and transcriptional profiles. As more data types are being revealed on a large scale, including supramolecular complexes, cellular compartments, tissue-specific variation, anatomical features and population diversity, the challenges for computational analyses on multiple scales and even more complex integration are mounting. Despite more data and faster computers, the primary driving force as in any intellectual activity will be a host of new ideas and approaches that will allow the further development of the field and its future establishment at the crown of biological sciences. This was the obvious part of a perspective for possible future trends.

The less obvious part might be the total and complete embedding of the field of bioinformatics in any aspect of biological research, beyond recognition. There are risks and opportunities here: the danger will be that the field might become unrecognizable in a few years, when all biomedical and biological research will be heavily relying on computational analysis of any biological information type. When computation permeates every nook and cranny of biology, the notion of a separate field might become obsolete.

The opportunities, however, will come this time from the engineering sciences in general, and potentially from microfabrication technology in particular. The next generation of bioinformatics analyses as applied in biological and biomedical research might occur outside the more traditional computing environments we have become accustomed to. With recent advances in the biomolecular engineering (Ryu and Nam 2000), the microengineering of cellular interactions (Folch and Toner 2000), the adaptation of semiconductor tools to spatially organize cellular environments (Bhadriraju and Chen 2002), the development of single-cell assay technologies (Lidstrom and Meldrum 2003), the construction of cell-based biosensors (Park and Shuler 2003), the eagerly anticipated minimal genome designs (Werner 2003), the advances in microtissue engineering (Kelm and Fussenegger 2004) and three dimensional tissue fabrication (Tsang and Bhatia 2004), the visualization of diffusion-enabled molecular movement in cells (Weiss and Nilsson 2004), and the various applications of nanotechnology in biological imaging (Fu et al. 2005), we are now facing a brave new world, where computation becomes a process that can occur in real time and beyond silicon chips. All this microscale technology might possibly provide the platform of bioinformatics of the future. Instead of separating measurement science (e.g., sequencing) from analysis (e.g., computing), it is conceivable that some of the measuring devices will have substantial computational capabilities in such a way that the bulk of biological data analysis might happen *in situ*. These microdevices would require complex communication, control and sensor capabilities, real-time programming capacity, sophisticated interfaces, and "swarming"[2] tactics. It does not take a leap of imagination to realize the huge potential of this type of applications in the field of biological (O'Brien et al. 2003, Montemagno 2004) and biomedical (Cavalcanri 2003 Bogunia-Kubik and Sugisaka 2002) research that can transform the notion of systems biology toward predictive and preventative medicine (Hood et al. 2004).

REFERENCES

Alberghina, L., Chiaradonna, F. and Vanoni, M. Systems biology and the molecular circuits of cancer. *Chembiochem* **5**, 1322–33 (2004).

Allison, D. B., Cui, X., Page, G. P. and Sabripour, M. Microarray data analysis: from disarray to consolidation and consensus. *Nat Rev Genet* **7**, 55–65 (2006).

Altschul, S. F., Gish, W., Miller, W., Myers, E. W. and Lipman, D. J. Basic local alignment search tool. *J Mol Biol* **215**, 403–10 (1990).

[2] Swarming is a seemingly amorphous, but deliberately structured, coordinated, strategic way to perform military strikes from all directions.

Andrade, M. A. and Sander, C. Bioinformatics: from genome data to biological knowledge. *Curr Opin Biotechnol* **8**, 675–83 (1997).

Andrade, M. A. et al. Automated genome sequence analysis and annotation. *Bioinformatics* **15**, 391–412 (1999).

Apweiler, R., Bairoch, A. and Wu, C. H. Protein sequence databases. *Curr Opin Chem Biol* **8**, 76–80 (2004).

Bajaj, M. and Blundell, T. Evolution and the tertiary structure of proteins. *Annu Rev Biophys Bioeng* **13**, 453–92 (1984).

Basik, M., Mousses, S. and Trent, J. Integration of genomic technologies for accelerated cancer drug development. *Biotechniques* **35**, 580–**2**, 584, 586 passim (2003).

Benton, D. Bioinformatics-principles and potential of a new multidisciplinary tool. *Trends Biotechnol* **14**, 261–72 (1996).

Bernstein, F. C. et al. The Protein Data Bank: a computer-based archival file for macromolecular structures. *J Mol Biol* **112**, 535–42 (1977).

Bhadriraju, K. and Chen, C. S. Engineering cellular microenvironments to improve cellbased drug testing. *Drug Discov Today* **7**, 612–20 (2002).

Birney, E. et al. An overview of Ensembl. *Genome Res* **14**, 925–8 (2004).

Bogunia-Kubik, K. and Sugisaka, M. From molecular biology to nanotechnology and nanomedicine. *Biosystems* **65**, 123–38 (2002).

Boguski, M. S. Bioinformatics. *Curr Opin Genet Dev* **4**, 383–8 (1994).

Bono, H., Ogata, H., Goto, S. and Kanehisa, M. Reconstruction of amino acid biosynthesis pathways from the complete genome sequence. *Genome Res* **8**, 203–10 (1998).

Bork, P. and Koonin, E. V. Predicting functions from protein sequences-where are the bottlenecks? *Nat Genet* **18**, 313–8 (1998).

Bork, P. et al. Exploring the Mycoplasma capricolum genome: a minimal cell reveals its physiology. *Mol Microbiol* **16**, 955–67 (1995).

Bowie, J. U., Luthy, R. and Eisenberg, D. A method to identify protein sequences that fold into a known three-dimensional structure. *Science* **253**, 164–70 (1991).

Brazma, A. and Vilo, J. Gene expression data analysis. *FEBS Lett* **480**, 17–24 (2000).

Brenner, S. E. A tour of structural genomics. *Nat Rev Genet* **2**, 801–9 (2001).

Broder, S. and Venter, J. C. Sequencing the entire genomes of free-living organisms: the foundation of pharmacology in the new millennium. *Annu Rev Pharmacol Toxicol* **40**, 97–132 (2000).

Cavalcanti, A. Assembly automation with evolutionary nanorobots and sensor-based control applied to nanomedicine. *IEEE Trans Nanotech* **2**, 82–87 (2003)

Chambers, G., Lawrie, L., Cash, P. and Murray, G. I. Proteomics: a new approach to the study of disease. *J Pathol* **192**, 280–8 (2000).

Chandonia, J. M. and Brenner, S. E. The impact of structural genomics: expectations and outcomes. *Science* **311**, 347–51 (2006).

Christopher, R. et al. Data-driven computer simulation of human cancer cell. *Ann N Y Acad Sci* **1020**, 132–53 (2004).

Clifford, R. J. et al. Bioinformatics tools for single nucleotide polymorphism discovery and analysis. *Ann N Y Acad Sci* **1020**, 101–9 (2004).

Damaraju, S., Sawyer, M. and Zanke, B. Genomic approaches to clinical drug resistance. *Cancer Treat Res* **112**, 347–72 (2002).

Davidson, D. and Baldock, R. Bioinformatics beyond sequence: mapping gene function in the embryo. *Nat Rev Genet* **2**, 409–17 (2001).

DellaPenna, D. Nutritional genomics: manipulating plant micronutrients to improve human health. *Science* **285**, 375–9 (1999).

Desany, B. and Zhang, Z. Bioinformatics and cancer target discovery. *Drug Discov Today* **9**, 795–802 (2004).

Doolittle, R. F. *Similar amino acid sequences: chance or common ancestry? Science* **214**, 149–59 (1981).

Eisenberg, D., Marcotte, E. M., Xenarios, I. and Yeates, T. O. Protein function in the post-genomic era. *Nature* **405**, 823–6 (2000).

Enright, A. J., Iliopoulos, I., Kyrpides, N. C. and Ouzounis, C. A. Protein interaction maps for complete genomes based on gene fusion events. *Nature* **402**, 86–90 (1999).

Ferea, T. L. and Brown, P. O. Observing the living genome. *Curr Opin Genet Dev* **9**, 715–22 (1999).

Fields, C. Informatics for ubiquitous sequencing. *Trends Biotechnol* **14**, 286–9 (1996).

Fitch, W. M. and Smith, T. F. Optimal sequence alignments. *Proc Natl Acad Sci U S A* **80**, 1382–86 (1983).

Fleischmann, R. D. et al. Whole-genome random sequencing and assembly of Haemophilus influenzae Rd. *Science* **269**, 496–512 (1995).

Folch, A. and Toner, M. Microengineering of cellular interactions. *Annu Rev Biomed Eng* **2**, 227–56 (2000).

Fu, A., Gu, W., Larabell, C. and Alivisatos, A. P. Semiconductor nanocrystals for biological imaging. *Curr Opin Neurobiol* **15**, 568–75 (2005).

Gingeras, T. R. and Roberts, R. J. Steps toward computer analysis of nucleotide sequences. *Science* **209**, 1322–8 (1980).

Gogarten, J. P. and Olendzenski, L. Orthologs, paralogs and genome comparisons. *Curr Opin Genet Dev* **9**, 630–6 (1999).

Gonnet, G. H., Cohen, M. A. and Benner, S. A. Exhaustive matching of the entire protein sequence database. *Science* **256**, 1443–5 (1992).

Guigo, R., Knudsen, S., Drake, N. and Smith, T. Prediction of gene structure. *J Mol Biol* **226**, 141–57 (1992).

Hall, P. A. V. and Dowling, G. R. Approximate string matching. *Comput Surv* **12**, 381–402 (1980).

Haque, S., Mital, D. and Srinivasan, S. Advances in biomedical informatics for the management of cancer. *Ann N Y Acad Sci* **980**, 287–97 (2002).

Hardison, R. C. Comparative genomics. PLoS Biol 1, E58 (2003).

Hondermarck, H. et al. Proteomics of breast cancer for marker discovery and signal pathway profiling. *Proteomics* **1**, 1216–32 (2001).

Hood, L., Heath, J. R., Phelps, M. E. and Lin, B. Systems biology and new technologies enable predictive and preventative medicine. *Science* **306**, 640–3 (2004).

Hornberg, J. J., Bruggeman, F. J., Westerhoff, H. V. and Lankelma, J. Cancer: a Systems Biology disease. *Biosystems* **83**, 81–90 (2006).

Huang, S. Gene expression profiling, genetic networks, and cellular states: an integrating concept for tumorigenesis and drug discovery. *J Mol Med* **77**, 469–80 (1999).

Janssen, P., Goldovsky, L., Kunin, V., Darzentas, N. and Ouzounis, C. A. Genome coverage, literally speaking. The challenge of annotating 200 genomes with 4 million publications. *EMBO Rep* **6**, 397–9 (2005).

Kabsch, W. and Sander, C. On the use of sequence homologies to predict protein structure: identical pentapeptides can have completely different conformations. *Proc Natl Acad Sci U S A* **81**, 1075–8 (1984).

Kallioniemi, O. P. Biochip technologies in cancer research. *Ann Med* **33**, 142–7 (2001).

Kanehisa, M. and Bork, P. Bioinformatics in the post-sequence era. Nat Genet 33 Suppl, 305–10 (2003).

Karp, P. D., Ouzounis, C. and Paley, S. HinCyc: a knowledge base of the complete genome and metabolic pathways of H. influenzae. *Proc Int Conf Intell Syst Mol Biol* **4**, 116–24 (1996).

Kelm, J. M. and Fussenegger, M. Microscale tissue engineering using gravity-enforced cell assembly. *Trends Biotechnol* **22**, 195–202 (2004).

Khalil, I. G. and Hill, C. Systems biology for cancer. *Curr Opin Oncol* **17**, 44–8 (2005).

Kirschner, M. W. The meaning of systems biology. *Cell* **121**, 503–4 (2005).

Koonin, E. V. Orthologs, paralogs, and evolutionary genomics. *Annu Rev Genet* **39**, 309–38 (2005).

Krupke, D. et al. The Mouse Tumor Biology Database: integrated access to mouse cancer biology data. *Exp Lung Res* **31**, 259–70 (2005).

Lee, M. P. Genome-wide analysis of epigenetics in cancer. *Ann N Y Acad Sci* **983**, 101–9 (2003).

Levitt, M. Accurate modeling of protein conformation by automatic segment matching. *J Mol Biol* **226**, 507–33 (1992).

Lidstrom, M. E. and Meldrum, D. R. Life-on-a-chip. *Nat Rev Microbiol* **1**, 158–64 (2003).

Lopez-Bigas, N., Blencowe, B. J. and Ouzounis, C. A. Highly consistent patterns for inherited human diseases at the molecular level. *Bioinformatics* **22**, 269–77 (2006).

Manning, G., Whyte, D. B., Martinez, R., Hunter, T. and Sudarsanam, S. The protein kinase complement of the human genome. *Science* **298**, 1912–34 (2002).

Montemagno, C. D. Integrative technology for the twenty-first century. *Ann N Y Acad Sci* **1013**, 38–49 (2004).

Mount, D. W. and Pandey, R. Using bioinformatics and genome analysis for new therapeutic interventions. *Mol Cancer Ther* **4**, 1636–43 (2005).

O'Brien, T. P. et al. Genome function and nuclear architecture: from gene expression to nanoscience. *Genome Res* **13**, 1029–41 (2003).

Ochman, H. and Davalos, L. M. The nature and dynamics of bacterial genomes. *Science* **311**, 1730–3 (2006).

Orengo, C. A., Todd, A. E. and Thornton, J. M. From protein structure to function. Curr Opin Struct Biol **9**, 374–82 (1999).

Ouzounis, C. Two or three myths about bioinformatics. *Bioinformatics* **16**, 187–9 (2000).

Ouzounis, C. Bioinformatics and the theoretical foundations of molecular biology.Bioinformatics **18**, 377–8 (2002).

Ouzounis, C., Casari, G., Sander, C., Tamames, J. and Valencia, A. Computational comparisons of model genomes. *Trends Biotechnol* **14**, 280–5 (1996).

Ouzounis, C. A. Ancestral state reconstructions for genomes. Curr Opin Genet Dev **15**, 595–600 (2005).

Ouzounis, C. A. and Karp, P. D. The past, present and future of genome-wide reannotation. Genome Biol 3, COMMENT2001 (2002).

Ouzounis, C. A. and Valencia, A. Early bioinformatics: the birth of a discipline - a personal view. *Bioinformatics* **19**, 2176–90 (2003).

Ouzounis, C. A., Coulson, R. M., Enright, A. J., Kunin, V. and Pereira-Leal, J. B. Classification schemes for protein structure and function. *Nat Rev Genet* **4**, 508–19 (2003).

Pang, C. P., Baum, L. and Lam, D. S. Hunting for disease genes in multi-functional diseases. Clin Chem Lab Med **38**, 819–25 (2000).

Park, S., Yang, X. and Saven, J. G. Advances in computational protein design. Curr Opin Struct Biol **14**, 487–94 (2004).

Park, T. H. and Shuler, M. L. Integration of cell culture and microfabrication technology. *Biotechnol Prog* **19**, 243–53 (2003).

Pavelic, K. and Gall-Troselj, K. Recent advances in molecular genetics of breast cancer. *J Mol Med* **79**, 566–73 (2001).

Pellegrini, M. Computational methods for protein function analysis. Curr Opin Chem Biol **5**, 46–50 (2001).

Philipson, L. The DNA data libraries. *Nature* **332**, 676 (1988).

Posadas, E. M., Simpkins, F., Liotta, L. A., MacDonald, C. and Kohn, E. C. Proteomic analysis for the early detection and rational treatment of cancer-realistic hope? Ann Oncol **16**, 16–22 (2005).

Quackenbush, J. Computational analysis of microarray data. *Nat Rev Genet* **2**, 418–27 (2001).

Rawlings, C. J. and Searls, D. B. Computational gene discovery and human disease. Curr Opin Genet Dev **7**, 416–23 (1997).

Rhodes, D. R. and Chinnaiyan, A. M. Bioinformatics strategies for translating genomewide expression analyses into clinically useful cancer markers. *Ann N Y Acad Sci* **1020**, 32–40 (2004).

Rhodes, D. R. and Chinnaiyan, A. M. Integrative analysis of the cancer transcriptome. Nat Genet 37 Suppl, S31–7 (2005).

Richardson, J. S. The anatomy and taxonomy of protein structure. *Adv Protein Chem* **34**, 167–339 (1981).

Ryu, D. D. and Nam, D. H. Recent progress in biomolecular engineering. *Biotechnol Prog* **16**, 2–16 (2000).

Sander, C. Genomic medicine and the future of health care. *Science* **287**, 1977–8 (2000).

Schultze, J. L. and Vonderheide, R. H. From cancer genomics to cancer immunotherapy: toward second-generation tumor antigens. *Trends Immunol* **22**, 516–23 (2001).

Searls, D. B. Bioinformatics tools for whole genomes. Annu Rev Genomics Hum Genet **1**, 251–79 (2000).

Segal, E., Friedman, N., Kaminski, N., Regev, A. and Koller, D. From signatures to models: understanding cancer using microarrays. Nat Genet 37 Suppl, S38–45 (2005).

Strausberg, R. L. Tumor microenvironments, the immune system and cancer survival. *Genome Biol* **6**, 211 (2005).

Strausberg, R. L., Greenhut, S. F., Grouse, L. H., Schaefer, C. F. and Buetow, K. H. In silico analysis of cancer through the Cancer Genome Anatomy Project. Trends Cell Biol 11, S66–71 (2001).

Strausberg, R. L., Simpson, A. J., Old, L. J. and Riggins, G. J. Oncogenomics and the development of new cancer therapies. *Nature* **429**, 469–74 (2004).

Tsang, V. L. and Bhatia, S. N. Three-dimensional tissue fabrication. Adv Drug Deliv Rev **56**, 1635–47 (2004).

Tsoka, S. and Ouzounis, C. A. Metabolic database systems for the analysis of genomewide function. *Biotechnol Bioeng* **84**, 750–5 (2003).

Tsoka, S. and Ouzounis, C. A. Recent developments and future directions in computational genomics. *FEBS Lett* **480**, 42–8 (2000).

Umar, A. Applications of bioinformatics in cancer detection: a lexicon of bioinformatics terms. *Ann N Y Acad Sci* **1020**, 263–76 (2004).

Valencia, A. and Pazos, F. Computational methods for the prediction of protein interactions. Curr Opin Struct Biol **12**, 368–73 (2002).

Weiss, M. and Nilsson, T. In a mirror dimly: tracing the movements of molecules in living cells. *Trends Cell Biol* **14**, 267–73 (2004).

Werner, E. In silico multicellular systems biology and minimal genomes. *Drug Discov Today* **8**, 1121–7 (2003).

Wheelan, S. J. and Boguski, M. S. Late-night thoughts on the sequence annotation problem. *Genome Res* **8**, 168–9 (1998).

Wilbur, W. J. and Lipman, D. J. Rapid similarity searches of nucleic acid and protein data banks. *Proc Natl Acad Sci U S A* **80**, 726–30 (1983).

Yang, H. H. and Lee, M. P. Application of bioinformatics in cancer epigenetics. *Ann N Y Acad Sci* **1020**, 67–76 (2004).

2 The Statistical Design and Interpretation of Microarray Experiments

Kevin K. Dobbin and Richard M. Simon

ABSTRACT

This chapter reviews major issues related to design and interpretation of microarray experiments. Important aspects of design covered include identification of experimental objectives, treatment of batch effects, selection of replication and pooling levels, determining sample size for class comparison and class prediction, and optimal allocation of samples to arrays and labels in dual label experiments. Aspects of interpretation focus on class prediction issues, including case selection, external vs. internal validation, and pitfalls in cross-validation.

Key Words: Microarrays, Experimental design, Sample size, Prediction, Validation

1 EXPERIMENTAL OBJECTIVES OF MICROARRAY STUDIES

The objective of a microarray experiment should be to test a specific hypothesis or set of hypotheses. The hypothesis will in general be related to overall gene expression patterns, and not to the expression of individual genes. For example, the hypothesis could be that some genes are differentially expressed in different phenotypes, stages, or prognostic groups, of a particular cancer. The purpose of the experiment is to prove or disprove the hypothesis and, if it is true, to identify a candidate list of differentially expressed genes for further study. Many microarray experiments have objectives that fall into one of three broad categories: class comparison, class discovery and class prediction. The word "class" appearing in each category indicates that the objectives are not merely concerned with drawing conclusions about differences in expression between individual RNA samples, but, more broadly, in identifying differences in gene expression due to the phenotype of the sample, an exposure to which the samples were subjected, or, in cell line experiments, to the conditions in which the cells were grown;

G.J. Gordon (ed.), *Cancer Drug Discovery and Development: Bioinformatics in Cancer and Cancer Therapy*, DOI: 10.1007/978-1-59745-576-3_2, © Humana Press, a part of Springer Science + Business Media, LLC 2009

hence the "class" of a sample refers, depending on the context, to the phenotype, exposure, or growth conditions.

In *class comparison* experiments the objective is to identify genes that are differentially expressed in different classes of samples. The class labels in this type of experiment are known ahead of time and are determined independent of the gene expression patterns. In experiments with a *class discovery* objective, the class labels are not known ahead of time but are determined from the gene expression data by cluster analysis of the samples. The goal in class discovery is to find subgroups of samples that share a similar gene expression profile. Cluster analysis is also sometimes applied to the genes, in which case the goal is to identify subgroups of genes that share a similar pattern of expression over the samples. Finally, in *class prediction* experiments the objective is to create a gene expression-based predictor which can be applied to future samples to predict class membership. The classes will often have some potential utility in making clinical decisions, such as a predictor of who will and who will not benefit from adjuvant chemotherapy (Paik et al., 2004).

2 QUALITY CONTROL AND CONFOUNDING

Gene expression measurement using microarrays is a complex process with many steps, and the associated protocols are detailed. Changes in some aspects of a protocol can influence the resulting gene expression measurements (see, e.g., Bammler et al., 2005). In order to avoid confusing measurement changes attributable to shifts in technical aspects of the assay with real changes in gene expression in the samples, an attempt should be made to assay all arrays in an experiment in as uniform a manner as possible, including common protocols, laboratory personnel, and analysis times to ensure the machine settings and other laboratory conditions are uniform. But it is often not possible to run a large number of arrays at the same time and in identical conditions, so microarrays are split up into different batches, with each batch run at the same time under the most uniform conditions possible. The batches could represent different laboratories, different personnel in the same laboratory, or the same personnel at different times in the same laboratory. Attempts should be made to run different batches in as uniform a manner as possible. Uniform protocols and measurement conditions can produce reproducible results even across different laboratories (Dobbin et al., 2005).

When multiple batches are involved in an experiment, it is important to avoid confounding the batch effects with the effects of interest. Confounding occurs when it cannot be determined to which of two or more causes an observed difference in gene expression is attributable. For example, if all tumors are arrayed on one day and all normal tissues on another day, then it will be impossible to know if differences between batches are due to cancer-specific differences or batch anomalies. For this reason, the design should include samples from each class in each batch, ideally an equal number from each class.

When data are grouped in batches, preliminary exploratory analyses of the microarrays can be used to assess whether there are systematic differences between any of the batches. Quality metrics associated with some software analysis tools can be used to compare batch quality. But quality metrics will not cover all possible technical problems. Systematic differences between batches can also be identified by looking at the distance between each pair of samples to see if the samples in the same batch tend to be closer together than the samples in different batches, indicating a "batch" effect. Typical distance metrics are one minus the correlation or the Euclidean distance between the gene expression vectors. Graphical tools are often useful for these types of preliminary analyses, and include multidimensional scaling (MDS) plots, principal components plots, and cluster analyses. If, in these analyses, samples separate out by batch, or if there are one or two batches that separate out from the rest, then this indicates potential quality control or consistency problems. An attempt should be made to identify these types of quality control issues before the formal data analysis is carried out. Otherwise the study can lead to erroneous conclusions. Even if quality control problems are identified in later downstream data analysis, going back to fix these problems at that point can compromise the statistical validity of the entire analysis.

3 REPLICATION LEVEL AND POOLING FOR CLASS COMPARISON AND CLASS DISCOVERY

Currently, most microarray experiments use either dual label or single label arrays. In a dual label system, two RNA samples are separately labeled with different fluorescent dyes and hybridized to the same array. Then the array is scanned twice, once for each color, resulting in two image files and two quantitative measurements for each gene or feature represented on the arrays. In single label systems, a single sample is hybridized to each array, labeled with a single dye, and the microarray is scanned once. There are differences in the data structure between dual label and single label systems that have important implications for some aspects of design and data analysis. But there are also design considerations that the two share. Here, we will treat aspects of design common to both systems, and then treat aspects of design unique to each, separately.

3.1 Replication Level

The level at which replicates are performed in an experiment will determine the scope of the inferences that can be drawn. Replication may be divided into two levels, technical replicates and biological replicates. Technical replicates occur when the same sample is measured multiple times. When an experiment involves only technical replicates, then valid conclusions can only be made about the differences in gene expression of the individual samples in the study. No valid conclusions about the larger populations from which the samples have come, or about the effect of an exposure or growth condition, can be drawn. Biological replicates involve independent replication on different samples, or, in the context of cell culture experiments, independent replication of the growing and harvesting of the cells. In an experiment that involves multiple biological replicates from each of several different classes, inferences apply not just to the samples under study but to the populations from which those samples were drawn, or the effect of the exposure to which they were subjected or the condition under which they were grown. Hence, inferences from experiments with biological replication are much stronger than those from studies with only technical replicates. For class comparison, class discovery, or class prediction experiments, replication should be done as much as possible at the level of biological replication. Although technical replicates can be

Table 1
Equal cost experiments in each row of table, comparing no technical replicates to one technical replicate per sample

Cost ratio of arrays to samples (c_a/c_o)	*Arrays per group (r=1)*	*Arrays per group (r=2)*	*Power to detect twofold change (r=1) (%)*	*Power to detect twofold change (r=2) (%)*
0.05	22	42	86	88
0.2	21	36	83	79
1	21	28	83	59
5	22	24	86	46

Here r is the number of technical replicates per samples. Set the significance level to 0.001. Biological variance $\delta^2 = 4/10$ and technical variance $\delta^2 = 1/10$

informative and helpful for quality assurance or process control, they generally come at the cost of precision and accuracy due to loss of arrays for biological replication.

For example, suppose the goal is class comparison. Let c_a represent the cost of measuring an aliquot of RNA and c_0 represent the cost of obtaining a sample of RNA from an independent biological replicate. Then if there are narrays in an experiment, and m distinct biological samples, the total cost of the experiment is $c = nc_a + mc_0$. We computed power for equal cost experiments (see Appendix) and the results appear in Table 1. As Table 1 shows, in general technical replicates result in less power to detect differentially expressed genes; only if samples are much more expensive than arrays is the power better when technical replicates are performed. For example, if the array assay cost is one twentieth the cost of obtaining an independent biological sample (first row of the table), then, performing a single technical replicate per sample will result in slightly improved power. But, otherwise, technical replicates result in loss of power, sometimes significant.

3.2 Pooling

Pooling samples is an option in class comparison experiments. One can pool RNA from several sources into the same test tube. The resulting RNA mixture will have a gene expression pattern that is a combination of all the contributing samples. It is generally assumed that an aliquot from this combination RNA mixture will have a gene expression profile that is roughly equal to the average profile of all the contributing samples, and that this relation will be roughly preserved on the log-transformed data (Kendziorski et al., 2003). Pooling potentially represents a way to assay the same number of samples with fewer microarrays, and thus reduce cost. For example, if one wanted to assay 15 samples from one class, one could pool three samples per array, requiring only five arrays instead of the usual 15. But it turns out that reducing the number of arrays not only reduces the cost, but also reduces the statistical power to detect differentially expressed genes (Shih et al., 2004). So there is a tradeoff.

Perhaps the most important point to make about pooling is that one should never pool all the samples from a class together into one pool. The problem with this strategy is that it does not allow for any biological replication. When one has a single mixed pool from a class, then all one can perform is technical replication and, as discussed above in the section on replication level, no valid conclusions about the effect of class (sample type, exposure, growth conditions) on gene expression can be drawn. The fact that the pool is a mixture of different RNA samples does not get around the fundamental problem that there is no way to estimate the biological variation.

Under the usual assumptions made about pooling, multiple independent[1] pools from each class can produce valid class comparisons. But when pooling is used to reduce the number of arrays required and at the same time there is a reduction in the number of degrees of freedom for statistical inference, this means increased noise in the comparisons, and reduced power to detect differential expression. In order to offset this increased noise, it is necessary to assay additional samples to recover the loss of power. There is a tradeoff between the number of microarrays required for an experiment and the number of samples required. An example of the tradeoff is shown in Fig.1, where it can be seen that the number of samples required increases faster than the number of arrays required decreases, particularly when the number of samples pooled per array is more than 2. The result is that pooling will generally make sense only if sample cost is less than

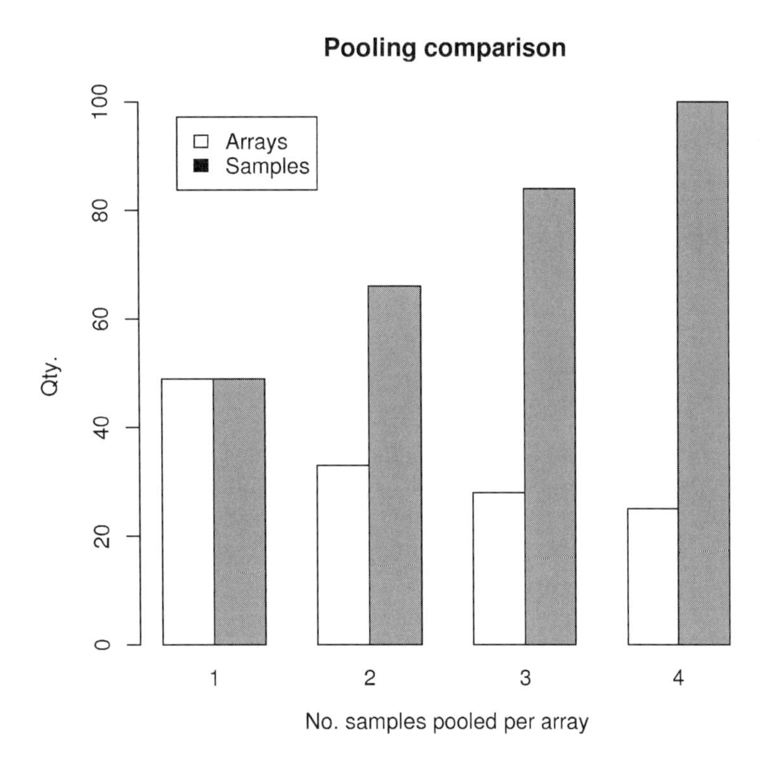

Fig. 1. Number of arrays and samples required for various pooling levels. An independent pool is constructed for each array, so that no sample is represented on more than one array. Settings were significance level $\alpha = 0.001$, power $1 - \beta = 0.95$, effect size $\sigma = 1$ (twofold change in expression), biological variance $\tau^2 = 4/10$ and technical variance $\sigma^2 = 1/10$

[1]Two pools are independent if they share no samples in common.

microarray cost (Kendziorski et al., 2003; Shih et al., 2004; Dobbin and Simon, 2005). This is one reason why pooling is rarely performed with valuable human specimens.

4 SAMPLE SIZE FOR SINGLE-LABEL CLASS COMPARISON EXPERIMENTS

We have presented sample size formulae for a variety of common microarray experimental situations (Dobbin and Simon, 2005). Here we present an example of a formula for a class comparison study. In class comparison, each gene is evaluated individually to determine whether it is differentially expressed. If α is the probability of a false-positive result for a particular gene (significance level), $1-\beta$ is the power to detect a difference of size δ in gene expression on the log base 2 scale (so that $\delta=1$ represents a twofold change in expression), and $\tau_g^2+\sigma_g^2$ (representing a sum of biological and technical variation) is the variation in expression for gene g across the different arrays (each with a different sample from the same class), then an overall sample size of

$$\min\left(n : n \geq 4\left[\frac{t_{n-2,\alpha/2}+t_{n-2,\beta}}{\delta}\right]^2(\tau_g^2+\sigma_g^2)\right)$$

is required. Here $t_{n-2,\alpha/2}$ is the $\alpha/2$ percentile of the $t$distribution with $n-2$ degrees of freedom. The min $(n:...)$ notation means that the smallest $n$satisfying the inequality should be used for the sample size; this can be calculated iteratively by starting with n=4 and increasing $n$until the inequality is satisfied; alternatively, software such as nQuery Advisor®can be used to do the calculation.

The formula can be used to control the number of false-positive findings for a specified power, or a false discovery rate (FDR) approximation formula used to provide approximate control of the FDR. In either case, parameter settings must be chosen to obtain a sample size estimate. The significance level α is the probability that a gene is determined to be significantly differentially expressed between the classes when, in fact, it is not. In many studies the samples are relatively homogeneous, so that it is likely that the vast majority of genes are not differentially expressed. Since α is the probability of mistakenly concluding that one of these thousands of genes is really differentially expressed, α should be small enough to control the number of false positives. For example, if arrays have 22,000 genes, then using α=0.001 results in an expected 22 false-positive genes.[2] The power, $1-\beta$, to detect a differentially expressed gene can be set to 0.90 or 0.95, so that one expects to detect 90% or 95% of the truly differentially expressed genes. Setting δ=1 sets the limit of detection of differential expression at twofold. The variance parameter $\tau^2+\sigma^2$represents the within-class variance of expression for a gene; this can be best estimated from a previous similar study. In human studies, we have found that a common median for these variances is 0.50. Some examples are given in Table 2. Alternative sample size methods to control the FDR based on simulation (Li et al., 2005) or parametric mixture models (Hu et al., 2005; Jung, 2005) have been presented.

[2]The Bonferroni method for determining α is considered too conservative for gene expression studies (Simon and Dobbin, 2003)

Table 2
Ten thousand genes, 50 of which are differentially expressed by δ=1, i.e., twofold

α	$1-\beta$	*n per group*	*FDR*	*E[#FD]*	*E[#TD]*
0.001	0.95	28	0.17	9.95	47.5
0.001	0.90	24	0.18	9.95	45
0.005	0.95	22	0.51	49.75	47.5
0.005	0.90	19	0.53	49.75	45
0.01	0.95	20	0.68	99.5	47.5
0.01	0.90	17	0.69	99.5	45

Uses within-class variance among samples: $\tau_g^2 + \sigma_g^2 = 0.50$

5 DESIGN AND SAMPLE SIZE FOR DUAL-LABEL CLASS COMPARISON EXPERIMENTS

Dual-label experiments have unique design issues because individual samples are paired together onto arrays. The reason samples are paired together and hybridized to the same array is to allow for the mathematical elimination of noise attributable to spot effects. Spot effects refer to variation due to the size and location of the spot, and the spread of the labeled probe mixture over the surface of the array. Hence, spot effects consist of a combination of a number of elements that add up to a significant source of variation. Their elimination via a two dye system results in a much less noisy experiment and greatly improved estimates of differential expression. Single-label arrays such as the Affymetrix GeneChips™ use different technology which reduces the technical noise elements that contribute to the spot effects to such an extent that there is no longer an advantage to mathematically eliminating these effects, so that only one dye is required.

The structure of the data in dual-label microarray experiments, with very precise comparisons between the two different samples on an array, and very noisy comparisons of samples on different arrays, is a common one in statistical literature and is called a block structure. The optimal design for class comparison studies in this type of structure is a balanced block design. When there are just two classes, a balanced block design pairs one sample from each class together on each array, and labels the samples so that half the samples from a class are labeled with one dye and the other half with the other dye. More than two classes lead to balanced incomplete block designs, and examples of these types of designs can be found in introductory textbooks on experimental design.[3] The balanced block design will provide greater power and efficiency than other designs that use the same number of arrays (Dobbin and Simon, 2002). For class discovery objectives, where class labels are not known ahead of time, a reference design will almost always be the best design choice (Dobbin and Simon, 2002). A reference design uses a

[3]One must be careful to adjust these designs for microarray experiments by ensuring that half the samples from each class are labeled with each dye.

single-labeled reference sample, which is split into multiple subsamples and hybridized to each array. The nonreference samples are all labeled with the other dye.[4] The reference sample serves as a standard for comparing the expression of any pair of nonreference samples. The reference design is well suited to class discovery because the performance of cluster analysis depends critically on accurately measuring distances between all these pairs of samples. For more complex situations, where there are several objectives or several potential ways of classifying samples, design considerations become more complex (see Dobbin and Simon, 2002; Dobbin et al., 2003 for further discussion).

Sample size formulae for dual-label class comparison experiments using reference designs, block designs, technical replicates and pooled designs appear in the work of Dobbin and Simon (2005).

6 SPECIAL ISSUES RELATED TO DESIGN AND INTERPRETATION OF CLASS PREDICTION STUDIES

As attempts are made to move gene expression signatures towards clinical application, there is a growing emphasis on building predictive models from microarray data. Here we discuss issues specifically related to the context of class prediction studies.

6.1 Case Selection

The cases should, as much as possible, be representative of the population for which the predictor is being developed. To this end, it is important to clearly define what clinical decisions will be impacted if an effective predictor is discovered. For example, if the predictor has the potential to affect the decision of whether or not to add adjuvant chemotherapy to surgical treatment for stage II colon cancer patients, but is unlikely to impact treatment decisions for stage III colon cancer patients, then the sample should consist of stage II colon cancer patients.

6.2 Developing a Predictor

Once an appropriate case set has been identified, there are many different methods for constructing a predictor, including support vector machines, weighted voting methods, partial least squares, compound covariate prediction, nearest neighbor, diagonal linear discriminant analysis, and empirical Bayes methods. Dudoit et al. (2002) compared several of these methods, where it was found that simpler methods such as diagonal linear discriminant analysis may be as good as more complex alternatives. Almost all methods of predictor development for gene expression data involve a dimension

[4] Note that reference designs do not require dye swapping of individual arrays, i.e.,that is, running the same two samples a second time with the sample labels reversed, except in special situations where the study has a secondary objective of comparing reference to non-reference samples (Dobbin et al., 2003).

reduction step. This dimension reduction can serve two purposes (1) it can reduce the random variation caused by the inclusion of large numbers of "noise genes" in the predictor and (2) if the dimension is reduced by gene selection, then the potential clinical utility of the predictor is greatly increased because the predictor can be constructed from a smaller subset of genes (Simon, 2005). Also, the second aspect mentioned above is usually biologically plausible because samples will overall be homogeneous. For example, Paik et al. (2005) used gene expression data to identify a predictor based on a subset of 21 genes, which they were able to subsequently verify in expedited fashion using multiplex real-time RT-PCR on retrospective paraffin-embedded specimens. Because of the practical advantages to the second aspect mentioned above, we will assume dimension reduction is achieved by gene selection.

6.3 Validating a Predictor

Development of a multigene predictor from high dimensional DNA microarray data requires a series of studies. A preliminary study that assesses the viability of developing a predictor with good performance characteristics is a start – there is, after all, no guarantee that a good predictor exists. At the other extreme would be a large, prospective confirmation study meant to establish a fully developed predictor's utility in assisting with clinical decision making. What constitutes an adequate validation for a study will depend in part on where on this spectrum the study resides. For smaller preliminary studies, *internal validation* may be adequate; for larger confirmatory studies, *external validation* is required (Simon, 2005). *Internal validation* occurs in preliminary studies when the data is split into a test set and a separate training set, and the predictor is developed on the training set and validated on the test set of samples. Internal validation, if performed properly and if there are no intrinsic biases[5] in the data, can provide nearly unbiased estimates of predictor performance. *External validation* occurs when the data used to estimate the predictor performance comes from a truly independent dataset not used to develop the predictor. External validation is crucial for verifying that there are no intrinsic biases in the dataset used to develop the predictor, and to establish a realistic assessment of the clinical utility of the predictor (Simon, 2005).

Internal validation can be performed using a single test set and a single training set, or in a sequential manner using multiple test set/training set pairs in cross-validation. Splitting the data into a single test set and training set allows for the most flexible approach to predictor construction, because it only requires that the predictor be constructed a single time, on the training set. On the other hand, cross-validation requires that the predictor be constructed "from scratch" on each of the training sets. This approach will only be feasible if there is an automatic algorithm that allows the predictor to be constructed without any human input other than perhaps setting up of the initial input parameters prior to data analysis. If this is the type of predictor being developed, then cross-validation will be preferable to the single test set and training set approach.

[5] Intrinsic biases can be introduced when technical aspects of the assay or processing vary systematically with class.

Hence the method by which the predictor will be developed should determine how the internal validation will be performed.

If one performs internal validation using crossvalidation methods, then the resulting estimates of the performance of the predictor can be severely biased if one preselects the set of informative genes to be used (Ambroise and McLachlan, 2002; Simon et al., 2003). The bias is due to data reuse because the left-out sample was used to select the informative genes, and then reused to evaluate the predictor. The potential severity of the resulting bias can be surprisingly large (see, e.g., Simon et al., 2003; Fig.1). For example, in Simon et al. (2003), microarray data were simulated in which there was no true difference between the classes, so that no predictor existed that could perform better than chance. Yet, when prediction accuracy was estimated using crossvalidation on preselected genes (i.e., incorrectly), the estimated accuracy of the predictor was 100% in over 90% of the simulations. When the genes were not preselected (i.e., correct cross-validation), then the estimated accuracy of the predictor centered around 50%, a coin toss, as is correct.

6.4 Sample Size

There is no widely accepted method for determining the sample size that is adequate for developing a gene expression-based classifier from microarray data. Mukherjee et al. (2005) presented a learning curve method for classifier development and applied it to a number of microarray datasets. They found that in general 10–20 samples may be adequate for morphological classification problems and 50 or more for treatment outcome classification. Dobbin and Simon (2007) developed a model-based approach which suggested that in many situations 20–30 samples per class may be adequate for developing a classifier. More work in this area seems needed as interest in the development of predictive markers from microarray data grows.

APPENDIX

If there are r technical replicates for each sample, then $c = n(c_a + c_o / r)$. Two experiments cost the same if $n_1(c_a + c_o / 1) = n_r(c_a + c_o / r), \frac{n_1}{n_r} = \frac{c_a + c_o / r}{c_a + c_o} = \frac{c_a / c_o + 1/r}{c_a / c_o + 1}$, where n_1 and n_r are the number of arrays when 1 and r technical replicates per sample are performed, respectively.

REFERENCES

Ambroise C and McLachlan GJ. Selection bias in gene extraction on the basis of microarray gene-expression data. Proc Natl Acad Sci USA 2002; 99:6562–6

Bammler T, Beyer RP, Bhattacharya S, et al. Standardizing global gene expression analysis between laboratories and across platforms. Nat Methods 2005;2:351–6

Dobbin K, Simon R. Comparison of microarray designs for class comparison and class discovery. Bioinformatics 2002;18:1438–45

Dobbin K, Simon R. Sample size determination in microarray experiments for class comparison and prognostic classification. Biostatistics 2005;6:27–38

Dobbin K, Simon R. Sample size planning for developing classifiers using high dimensional DNA microarray data. Biostatistics 2007;8:101–17

Dobbin K, Shih JH, Simon R. Statistical design of reverse dye microarrays. Bioinformatics 2003;19:803–10

Dobbin KK, Beer DG, Meyerson M, et al. Interlaboratory comparability study of cancer gene expression analysis using oligonucleotide microarrays. Clin Cancer Res 2005;11:565–72

Dudoit S, Fridlyand J, and Speed TP. Comparison of discrimination methods for the classification of tumors using gene expression data. J Am Stat Assoc 2002;97:77–87

Hu J, Zou F, Wright FA. Practical FDR-based sample size calculations in microarray experiments. Bioinformatics 2005;21:3264–72

Jung SH. Sample size for FDR-control in microarray data analysis. Bioinformatics 2005;21:3097–104

Kendziorski CM, Zhang Y, Lan H, Attie AD. The efficiency of pooling mRNA in microarray experiments. Biostatistics 2003;4:465–77

Li SS, Bigler J, Lampe JW, Potter JD, Feng Z. FDR-controlling testing procedures and sample size determination for microarrays. Stat Med 2005;24:2267–80

Mukherjee S, Tamayo P, Rogers S, et al. Estimating dataset size requirements for classifying DNA microarray data. J Comput Biol 2005;10:119–42

Paik S, Shak S, Tang G, et al. A multigene assay to predict recurrence of tamoxifen-treated, node-negative breast cancer. N Engl J Med 2004;351:2817–26

Shih JH, Michalowska AM, Dobbin K, Ye Y, Qiu TH, Green JE. Effects of pooling mRNA in microarray class comparisons. Bioinformatics 2004;20:3318–25

Simon R. Roadmap for developing and validating therapeutically relevant genomic classifiers. J Clin Oncol 2005;23:7332–41

Simon RM, Dobbin K. Experimental design of DNA microarray experiments. Biotechniques 2003;34:516–21

Simon R, Radmacher MD, Dobbin K, McShane LM. Pitfalls in the use of DNA microarray data for diagnostic and prognostic classification. J Natl Cancer Inst 2003;95:14–8

3 Whole-Genome Analysis of Cancer

Steven A. Enkemann, James M. McLoughlin, Eric H. Jensen, and Timothy J. Yeatman

ABSTRACT

With the completion of the human genome project, a multitude of techniques have been invented to evaluate large portions of the human genome simultaneously. Investigations are typically focused on the genome, transcriptome or proteome to identify unique characteristics that may explain the origin of human disease and potentially predict future outcomes. The goal of these investigative pursuits is to eventually individualize treatment for each patient based on their unique gene expression patterns.

Key Words: Genome, Transcriptome, Proteome, Functional genomics, Microarray, GeneChip, Sequencing, Cancer

1 INTRODUCTION

The term "gene" was first used in 1909 to describe a functional unit of inheritance (Johannsen, 1909) and by the early 1920s it was commonly used by German scientists (Carlson, 1966). At that time, a gene was defined by the characteristics that could be passed on from generation to generation. Thus, the true definition of a gene is an inherited factor that determines a trait, and the "genome" is defined as the sum total of these inherited factors for an organism. More than 30 years after the term "gene" came into use, Avery, McCarty, and McCloud first demonstrated that the inherited factor was DNA (Avery et al., 1944). Thus, a gene is more than a sequence of DNA. It is irrevocably tied to the measurable activity that it encodes. It is the functional consequences of an inherited piece of DNA.

Biologists now learn the central dogma of biology at an early age: DNA directs the transcription of RNA molecules, which in turn direct the translation of proteins (Fig. 1). These proteins produce activities that are ultimately observed as functional consequences for the cell or the organism. These phenotypes are the final measures that define genes. Therefore, the study of genes, which in recent times is referred to as "functional genomics," involves all aspects of the study of cells and organisms, from the DNA itself to the outward characteristics that it encodes. Functional genomics is the

G.J. Gordon (ed.), *Cancer Drug Discovery and Development: Bioinformatics in Cancer and Cancer Therapy*, DOI: 10.1007/978-1-59745-576-3_3,

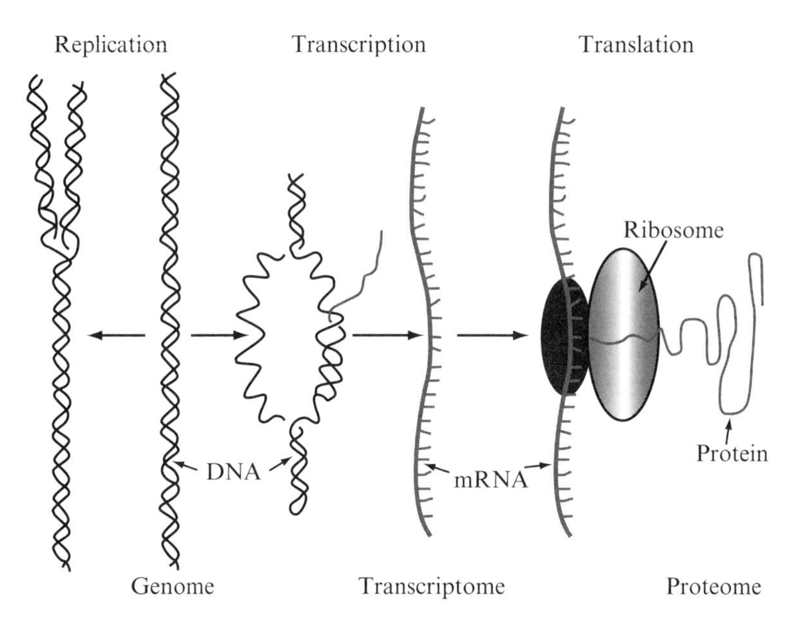

Fig. 1. The analysis of whole-genome expression. The analysis of gene expression can occur at any stage

study of genes, their encoded proteins, and their associated biological processes. The discipline of functional genomics seeks to unravel the secrets of how the genome functions over time as well as in different biological states. It seeks to investigate the functional importance of all genes. However, rather than investigate those genes one at a time, it usually involves techniques that investigate large groups of genes at one time.

The complete sequencing of the human genome opened many new avenues of scientific research and new mechanisms for investigation. It pioneered the concept of big science in the biological disciplines, where large sums of data were generated and advanced computational algorithms were required to process these results. It also opened the genome to investigation by multiple new techniques. Some of these new techniques required computers rather than traditional laboratory methods to evaluate the genome. For example, instead of starting with a phenotype and tracing it back to a sequence of DNA in order to discover the genes, it is now possible to start with the genome sequence and look for structures indicative of promoters, open reading frames, splice junctions, and homology to known sequences in order to discover new genes. It is no longer necessary to know a phenotype in order to define a new gene. In fact, most genes are now discovered based on their DNA sequence. Comparative genome analysis allows one to quickly find a homologous gene in a second organism once it has been identified in an evolutionarily-related organism. This technique can also be used to quickly identify differences between closely related organisms. It has been used to identify pathogenicity factors by comparing pathogenic and nonpathogenic bacterial strains (Dobrindt, 2005; Dobrindt et al., 2003; Schmidt et al., 2004; Wick et al., 2005).

Very similar concepts have been used for the study of cancer. For example, the comparison of tumor genomes relative to normal tissue has been done for years through karyotyping.

The 3 billion base pairs of the human genome are just beginning to reveal the true complexity of the human organism (Schmidt et al., 2004). The human genome contains approximately 25,000 to 35,000 genetic loci (Ewing and Green, 2000; Lander et al., 2001). These regions, from which transcripts arise, make up only 1.5% of the total genome sequence. The noncoding regions may one day prove to be important in the organization of the functional organism, but currently they are often referred to as "junk" DNA (Cawley et al., 2004; Graveley, 2001; Kapranov et al., 2002; Lander et al., 2001; Venter et al., 2001). The 35,000 genetic loci give rise to an estimated 120,000 transcripts due to variations in the start and stopping sites and alternative splicing of the transcripts (Camargo et al., 2001; Das et al., 2001; Liang et al., 2000). From these transcripts perhaps 1 million uniquely functional proteins could be produced when one considers posttranslational changes and these functional proteins can give rise to even more observable phenotypes (Anderson, 2005). Therefore the number of "genes" in a human can be numbered as low as 25,000, if one only counts genetic loci encoding possible proteins, or enumerated at greater than 1 million if one counts the resultant proteins or the phenotypes encoded by these genetic loci. As biologists migrate to a systems approach for studying organisms it is clear that high-throughput techniques and bioinformatics tools are required to generate, store, and process the information gathered from many thousands of genes at once.

2 THE GENOME

To the current generation of scientists, the genome is thought of as the complete DNA sequence of the germ line cells of an organism. Knowledge of this DNA sequence can provide valuable information about the functions and malfunctions of genes. In organisms such as humans the genome is subdivided into chromosomes, which can be visualized in dividing cells. Smaller units, such as chromosomal bands can also be observed on a microscopic level. These units have been used for years to study the influence of the genome on cellular function (Jackson, 1978; Martin and Hoehn, 1974; Pogosianz and Prigogina, 1972). With the completion of the human genome we can also evaluate the genome on the nucleotide level. Whole gene analysis is used to identify regions of the genome that might contain genes that influence certain disease processes. Single nucleotide changes in, or near, a gene are frequently the causes of inherited or spontaneous mutations that lead to the development of cancers (Claus, 1995; Den Otter et al., 1990; Weinberg, 1983). Therefore, combinations of whole genomic and focused approaches are useful for discovering genetic influences related to cancer.

It has been known for years that mutations in oncogenes and tumor suppressors can influence the development of cancer. Large-scale sequencing of the genome can be used to identify specific mutations in a single tumor or to identify the spectrum of mutations found in many cancers (Capella et al., 1991; Casey et al., 2005; Frank et al., 1999; Li et al., 1998). The promise of these investigations is that a single tumor might be identified by the types of mutations it contains and that this knowledge could lead to better choices for treatment. It is also known that large genomic rearrangements are associated with cancer. The largest of these changes are observable in the microscope but smaller changes can now be resolved with functional genomic techniques (Beheshti et al., 2003; Hoque et al., 2003; Mundle and Sokolova, 2004; Squire et al., 2003; Weiss et al., 2003).

3 THE TRANSCRIPTOME

The term *transcriptome* was coined to describe the large number of transcripts that can result from copying and splicing together portions of the genome (Velculescu et al., 1997). The *transcriptome* represents the multitude of RNA messages that encode the various proteins necessary to express the function of a gene. The term is often used to describe the sum total of all transcripts in the cell under specific biological conditions and is also used to describe all the possible transcripts under all possible conditions. Many more transcripts exist than the number of genetic loci that encode them.

The study of the transcriptome will define the large picture of the functional cell. Differences in the transcriptome lead to differences in the functionality of a cell, tissue, or organism. Elements of the transcriptome have been identified, one at a time, for many years, but can now be evaluated in large groups because of microarray technology and large-scale sequencing projects. Scientists can now get a big picture view of a cell or tissue. With bioinformatics techniques one can use the big picture or drill down to the differences of individual transcripts (Kapranov et al., 2002; Liu, 2005). With respect to cancer, knowledge of the transcriptome has been used to define subclasses of tumors which will aid in treatment decisions (Golub et al., 1999; Perou et al., 2000; Ramaswamy et al., 2001). Large groups of transcripts might be used for classification, but smaller groups, or even a single transcript, might provide insight into the prognosis of the patient. For example, knowledge of the expression level of the estrogen receptor (ER) aids in determining the clinical course of some breast tumors (Leal et al., 1995; Perin et al., 1996). Other tumors might have a similar biological marker. The benefit of this technology is that screening many transcripts improves the chance of identifying the one important transcriptional marker.

Another important aspect of the transcriptome is the phenomenon of alternative splicing. Over 40% of genes in the human genome are thought to undergo alternative splicing (Brett et al., 2000; Mironov et al., 1999; Modrek and Lee, 2003). Changing the combination of the exons joined together and the location of starting and stopping points for the transcript produces variation in protein expression. Alterations in the type of transcripts produced by splicing can contribute to cancer (Mercatante and Kole, 2000; Milani et al., 2006). Alternative splicing can sometimes explain the heterogeneity of tumor response to therapy among populations of the same tumor type (Mercatante and Kole, 2000). Gene expression arrays are being used to identify these variants and to potentially correlate them to clinical responses (Bracco and Kearsey, 2003; Veuger et al., 2002).

4 PROTEOME

To date, the majority of biological investigations have involved the study of proteins, protein:protein interactions, and posttranslational modifications. The totality of all proteins in an organism is now called the proteome (Kahn, 1995). In parallel to the genome and the transcriptome, the proteome is now investigated with large-scale and high-throughput techniques. Both antibody-based techniques and mass spectrometry have proven to be extremely powerful in localizing, quantitating, and structurally characterizing individual proteins or small groups of proteins in an experiment. However, the true potential in these methods lies in the ability to elucidate the structure and function of large numbers of proteins within one experiment. The large-scale analysis of the protein composition of a cell, tissue, or fluid is often referred to as proteomics (Petricoin and Liotta, 2004; Stults and Arnott, 2005; Wulfkuhle et al., 2003). High-throughput mechanisms for studying the proteome include evaluating many proteins in a single cell or tissue. Alternatively, one might study a single protein in a multitude of tissue samples by techniques such as tissue microarrays or protein microarrays.

The proteome changes in many ways due to transcriptional changes and posttranslational modifications. It is perhaps the most revealing arena for understanding how a gene functions and how a cell interacts with the local environment. In fact, some cellular responses occur completely within the proteome and do not involve changes in the genome or transcriptome. Protein modifications and protein:protein interactions mediate many of the cellular responses to environmental stimuli. Proteomic experiments aim to generate new hypotheses by studying the changes in expression levels or posttranslational modifications of proteins in response to a stimulus. In evaluating the proteome, one might look for changes in a single protein or for a panel of candidate biomarkers that signify a cellular response to treatment or a difference between groups of samples. For the analysis of cancer one might look for differences in the proteome due to the mutation of a single gene, following drug treatment, or between groups of patients separated by their clinical characteristics (e.g., histology or survival outcome) (Dephoure et al., 2005; Soreghan et al., 2003). Evaluation of the proteome may give molecular insight into mechanisms of drug resistance and might provide markers for the early detection of diseases (Alexander et al., 2004; Bhattacharyya et al., 2004; Hondermarck et al., 2001; Petricoin et al., 2005; Zhang et al., 2004).

5 METHODS FOR INVESTIGATING THE-OMES

5.1 Sequencing Techniques

5.1.1 Expressed Sequence Tags

One of the secondary benefits of the human genome project was that high-throughput sequencing became available for hypothesis driven science. This has led to a number of sequence-based techniques that have expanded our understanding of biology in general and problems like cancer, specifically. One of the first uses of high-throughput sequencing was to shift from the genome to the transcriptome. Many laboratories combined the concept of generating complementary DNA (cDNA) libraries and sequencing to create libraries of expressed sequence tags (ESTs) (Kawamoto et al., 2000; Okazaki et al., 2002; Stapleton et al., 2002). ESTs are single sequencing reads of approximately 300–700 nucleotides taken from cDNA clones. They were first described in 1991 as a potential tool to interrogate the human genome for potential variation (Adams et al., 1991). It was not long before this endeavor shifted from simply cataloging sequences found in various cell types to comparisons between cell types (Carulli et al., 1998; Kawamoto et al., 2000; Lindlof, 2003). The sequence information from EST sequencing projects is useful for identifying previously undefined genes and many of the designations used to catalog the unknown sequences have been maintained in the current gene nomenclature. Prefixes, such as FLJ (full-length long Japan), MGC (Mammalian Gene Collection), and KIAA, which in Japanese stands for the Kazusa DNA Research Institute reference character, are commonly seen in published gene lists. These designations allow one to bioinformatically track information related to these clones until an official gene name is agreed upon. Despite such problems as poor sequence quality, chimerism, and vector or intronic contamination, the EST projects have contributed immensely to the discovery and cataloging of human gene sequences. The cancer phenotype has been studied through the comparison of EST sequence pools (Krizman et al., 1999; Reis et al., 2005). For example, Asmann et al. (2002) used available EST databases to electronically profile the differences between normal and malignant prostate tissues and to develop potential molecular markers for the detection of cancer.

5.1.2 Serial Analysis of Gene Expression

Another sequencing technique that examines the transcriptome is serial analysis of gene expression (SAGE) (Velculescu et al., 1995). The SAGE technique involves enzymatically removing short sequences from cDNA copies of the transcript pool and concatenating them together in a large string that is then sequenced. The short sequences are markers for specific transcripts. The frequency of the short tags within the sequencing data is a measure of the abundance of that transcript within the original transcriptome. This technology has been used to compare expressions across experimental conditions (Sengoelge et al., 2005; Zucchi et al., 2004). SAGE can be used to assess the expression of previously unknown genes, as it does not depend on prior sequence knowledge. However, the limitations of SAGE may outweigh its advantages. The technology relies on the complexity of cloning and requires high-throughput sequencing; so it can be expensive. Sometimes, there is also difficulty in identifying specific transcripts due to the short tag size; often 14–21 bp. Some SAGE applications also rely on a sequenced genome or a significant cDNA database for interpretation (Liu, 2005). Matsuzaki et al. (2005) used the SAGE libraries and EST databases to identify unique genes differentially expressed in melanoma cells and melanocytes. They were also able to identify subsets of patients that had initiated an immune response to uniquely over expressed proteins. This response may eventually explain the varied clinical response seen in patients.

5.1.3 Massively Parallel Signature Sequencing

Massively parallel signature sequencing (MPSS) functions similar to SAGE but uses a novel restriction-ligation, bar code identification, and bead-based detection system to identify 17–20 bp tags of every transcript in an RNA sample (Brenner et al., 2000). Similar to SAGE, MPSS is complex and sophisticated and is also limited by the number of ambiguously assigned tags. However, MPSS does provide greater sequence coverage than SAGE. MPSS can sequence over a million clones in one read making it most useful for low frequency transcripts below the detection limits of oligonucleotide microarrays. MPSS, like SAGE, does not need knowledge of putative transcripts to be functional and therefore can be used to discover unknown genes (Shah et al., 2006). Chen et al. (2005b) used MPSS to identify transcript variants in the CT45 gene in testicular cancer. They uncovered specific variants that were highly expressed in germ cell tumors compared to normal testicular tissue.

Massively parallel sequencing techniques are continuing to improve in their speed and accuracy in sequencing the genome. Although it currently costs between $10 and $25 million to fully sequence the human genome, initiatives are in place to promote new techniques that would significantly decrease the cost. One promising new technology appears to have high throughput and high accuracy. Margulies et al. (2005) have reported a highly parallel sequencing system that can sequence up to 25 million nucleotides in 4 h. The speed, accuracy, and price of this system may one day lead to genome sequencing as a common laboratory event.

5.1.4 Single Nucleotide Polymorphisms

Sequencing of diverse human genomes has already led to the development of a public database of more than 1 million well-documented common variations in the human genome. This HapMap project is continuing to search for genetic variation in the form of single nucleotide polymorphisms (SNPs) that accounts for human diversity (Thorisson and Stein, 2003). Counting rare SNPs, more than 2 million differences have been documented and with more sequenced genomes the total number is estimated to reach >10 million (Altshuler et al., 2005; Botstein and Risch, 2003; Gu et al., 1998; McVean et al., 2005; Sherry et al., 2001). There are a variety of techniques that can make use of this information with respect to cancer. Individual SNPs can be used as signposts in the genome for linking specific genotypes to diseases. The SNP data can also be used to detect the large genomic deletions and insertions that occur in some cancers (Hoque et al., 2003). Variation found within the coding regions of some genes can affect the activity of the proteins they encode and polymorphisms found in the regulatory regions can effect the level of gene expression (Marsh, 2005). There are already instances where the amount of protein produced or the specific polymorphism expressed affects the patient's response to chemotherapy (Landi et al., 2003). The field of pharmacogenomics specifically deals with the relationship between an individual's response to chemicals and the polymorphisms in his genome (Bomgaars and McLeod, 2005; Marsh, 2005; Turesky, 2004). Sequencing is traditionally used to identify individual point mutations and SNPs, but once specific variants are known to exist they can be detected using microarray technology.

5.2 *Microarray Techniques*

Perhaps the most promising new technique for functional genomics is microarray technology. This technology relies on the complementary joining of two nucleic acid strands to form a duplex. Because of the sequence specificity of this bonding the technology can be used to identify specific sequences among a pool of billions of different sequences. By multiplexing the hybridization, thousands of sequences can be identified at the same time from the sample pool. Microarray technology can be used to screen entire transcriptomes or sequence individual nucleotides within the genome. Although usually based on prior sequence knowledge, the technique can also be used for gene discovery. Typically it is used to probe the transcriptome under directed experimentation. A thousand to several thousand genes can be analyzed with a relatively small amount of RNA so the technique is widely applicable and relatively inexpensive. However, the microarray platforms are limited to those genes chosen for the array so there is no guarantee that the gene of interest is represented on the array. A typical high-density array may also contain many unknown genes further limiting analysis. The signals from microarrays are considered to be less reliable and less accurate than a quantitative RT-PCR analysis. But since 12,000–60,000 genes are appraised in one experiment a little accuracy is sacrificed for the information-rich output of microarrays.

5.2.1 Complementary DNA Gene Expression Microarrays

Gene expression arrays are designed to capture information about the transcription of as many genes as possible with each experiment. In cDNA arrays the probes used to capture this information are generated from transcripts pulled from tissues. Messenger RNA from cells is converted to cDNA fragments using reverse transcription and cloned into plasmids. The clones are then sequenced and cataloged so different clones will be used for each spot. The probes are generated by amplifying the clones using PCR and spotting them onto a two-dimensional surface to form the array using either ceramic, stainless steel, or titanium spotting pins or ink jets (Bertucci et al., 1999; Chen et al., 1998; Schena et al., 1995). The important feature is that the probes are placed in a precise arrangement so the identity of each probe is known. Experimental mRNA samples are then evaluated by generating cDNA labeled with either a fluorescent or radioactive tag and hybridizing the labeled sample to the array. Transcripts in the experimental sample will hybridize to the appropriate spot and the resultant signal is detected using an appropriate scanner. The scanned image of the entire array must then be processed to generate the appropriate signal intensity for each gene represented on the array. The majority of the work associated with a microarray experiment deals with the processing and evaluation of the scanned array data (Armstrong and van de Wiel, 2004; Leung and Cavalieri, 2003; Quackenbush, 2001).

More than 10,000 papers have been published using microarray technology. The first microarray paper used cDNA arrays (Schena et al., 1995) and these arrays have been used liberally to evaluate cancer (Bucca et al., 2004; Khan et al., 1999). Every aspect of tumorigenesis has been investigated and possibly every tumor site has been considered (Al Moustafa et al., 2002; Alevizos et al., 2001; Baris et al., 2005; Durkin et al., 2004; Elek et al., 2000; Larramendy et al., 2002; Lee et al., 2004; Onda et al., 2004; Rihn et al., 2000; Selaru et al., 2002; Sorlie et al., 2001; Wolf et al., 2000; Yang et al., 2004). Some laboratories have used microarrays to evaluate chromosomal rearrangements that occur with the emergence of the disease (Heiskanen et al., 2001; Jiang et al., 2004; Pollack et al., 1999; Wrobel et al., 2005; Yi et al., 2005). Others have examined the transcriptional profile of inherited tumors (Hedenfalk et al., 2001, 2002). Some have looked for markers indicative of the emergence or progression of the disease while other laboratories have focused on utilizing the transcriptome to subclassify both distinctly different and histologically similar tumors (Bittner et al., 2000; Bull et al., 2001; Cao et al., 2004; Eschrich et al., 2005; Halvorsen et al., 2005; Hegde et al., 2001; Hu et al., 2005; Khanna et al., 2001; Mousses et al., 2001; Ono et al., 2000). Many have tried to understand or predict the response of a tumor to chemotherapeutic agents or other treatment methods (Akervall et al., 2004; Chang et al., 2002; Kihara et al., 2001; Kudoh et al., 2000; Matsuyama et al., 2006; Mousses et al., 2001; Nakeff et al., 2002; Peehl et al., 2004; Samimi et al., 2005; Smith, 2002; Takata et al., 2005). The goal of most of these investigations is to improve the treatment of patients by getting an understanding of cancer on a molecular level.

5.2.2 Oligonucleotide Microarrays

A second basic approach in assessing the whole-genome profile is by using oligonucleotide-based arrays (Lockhart et al., 1996). For this style of microarray small DNA sequences ranging from 20 to 70 nucleotides in length are synthesized directly on a support medium or are robotically spotted as for cDNA arrays. Again, the sequence and location of each oligonucleotide is known for the array and the experimental use of these arrays is similar to that of the cDNA arrays.

Although similar, there are several important differences between cDNA arrays and some of the oligonucleotide-based arrays. Typically, cDNA arrays have longer probes, which can produce more stable binding and more intense signaling compared to the shorter sequence oligonucleotide-based arrays (Stillman and Tonkinson, 2001). The tradeoff is that cDNA probes cannot easily evaluate transcript variants (Castle et al., 2003; Kampa et al., 2004). A gene might be identified but not the specific transcript. Most cDNA array methods require the use of two contrasting labeled RNA samples to provide experimental information. This is often accomplished by using a reference RNA pool to provide the contrasting signal for comparisons (Gadgil et al., 2005; Novoradovskaya et al., 2004). In contrast, the Affymetrix GeneChip is unique compared to most arrays in that a single-labeled cRNA sample is hybridized to a chip rather than having both a reference and test sample (Fig. 2) (Lockhart et al., 1996). Most

Fig. 2. A portion of the Affymetrix GeneChip U133+ Array. A phycoerythrin fluorescent dye is used to stain the arrays, but information is captured in gray scale. Data is interpreted from the gray scale measurements. The above gray scale image is a typical computerized picture generated by the computer. It is not necessary to generate color or a picture to interpret the data

cDNA arrays are produced from clones held in bacterial stocks. This means that a desired clone is readily available for further biological work. However, clone management is time consuming and cumbersome and cDNA arrays produce a higher incidence of crosshybridization, PCR-generated contamination, and misidentified spots. Oligonucleotide arrays can only be produced once sufficient sequence information in known about the genome or transcriptome of an organism and so the organisms available for this technology are limited. On the other hand, oligonucleotides can be synthesized for any portion of a known genome, so they can be used for purposes other than transcriptome analysis.

Any sequence information can be used to produce an oligonucleotide-based array. There are now tiling arrays that contain probes that completely cover an entire chromosome or genome. These arrays are useful for probing the transcriptome to identify all potential transcripts from the genome or for high-density mapping of chromosomal aberrations (Kapranov et al., 2005; Roversi et al., 2005). There are also exon arrays, which can be used to identify the splice variants existing in the transcriptome of a particular cell (Nagao et al., 2005). Individual genes can even be sequenced with microarrays to look for de novo mutations or genetic polymorphisms (Huang et al., 2005; Wikman et al., 2000).

Oligonucleotide-based microarrays have been used to study cancer in many of the same ways as cDNA-based arrays. Nearly every organ site has been analyzed and every aspect of tumor development and response to therapy has been evaluated with this technology (Agrawal et al., 2003; Bhattacharjee et al., 2001; Bonome et al., 2005; Dressman et al., 2006; Frederiksen et al., 2003; Freije et al., 2004; Ginos et al., 2004; Gyorffy et al., 2006; Huang et al., 2006). Because, of the overlap between studies performed with cDNA arrays and oligonucleotide arrays many laboratories have studied results from both technologies with the hope of producing a more robust gene expression signature of various cancer types (Bloom et al., 2004; Carter et al., 2005; Kuo et al., 2002; Warnat et al., 2005). However, some features of the oligonucleotide arrays make them favorable for certain applications. For example, Kasamatsu et al. (2005) were able to evaluate both the genome and transcriptome in the same tumors with high-density oligonucleotide arrays and found a correlation between some of the differences in gene expression and genomic anomalies in adenoid cystic carcinomas. The commercial arrays are more consistent from batch to batch and generally have more features per array. They can also be used by any laboratory; with the implications that, results could be replicated in a different laboratory. For these reasons investigators are turning to commercial arrays to combine results or to directly compare results from multiple institutions (Bammler et al., 2005; Dobbin et al., 2005; Irizarry et al., 2005). The larger commercial arrays are also more frequently used by groups attempting to build gene expression-based classification schemes for tumor type, stage, or survival chances (Dressman et al., 2006; Frederiksen et al., 2003; Freije et al., 2004; Ginos et al., 2004; Gyorffy et al., 2006; Huang et al., 2006). The array data can provide a deeper understanding of cancer, and it is also useful for a deeper understanding of the technology itself. Several cancer data sets have been used well by the bioinformatics community to test and compare analytic approaches (Bhattacharjee et al., 2001; Golub et al., 1999). Because of their commercial nature, the oligonucleotide-based arrays are more likely to provide a clinical use for microarrays in cancer treatment.

5.2.3 ChIP-on-Chip Technology

To understand how the transcriptome is regulated one needs to identify DNA-binding sites of transcription factors. The technique known as ChIP-on-chip combines chromatin immunoprecipitation with microarray chips. Chromatin immunoprecipitation (ChIP) is performed by in situ crosslinking of a specific transcription factor to its DNA-binding site. The DNA is sheared and the DNA fragments, with the specific transcription factor bound to them, are then immunoprecipitated by a transcription factor-specific antibody. The DNA segments are then amplified with PCR, labeled, and hybridized to tiled arrays. With this technology the DNA-binding sites for specific transcription factors can be assessed over the whole genome (Horak and Snyder, 2002). Jin et al. used ChIP-on-chip to verify in silico predictions of target genes regulated by the ER alpha (Jin et al., 2004) and Jen and Cheung (2005) were able to identify new targets of the p53 gene in response to ionizing radiation. The technique can also be extended to evaluate other components of the genome or associated chromatin. The patterns of DNA methylation related to disease status can now be evaluated in a whole-genome approach (Wilson et al., 2006).

5.2.4 SNP Arrays

SNPs are the normal genetic variations that differentiate individuals from each other. These single base differences may be the cause of some tumors and the reason many tumors do not respond to therapy. Mutations in oncogenes and tumor suppressors are part of this genetic variation, but genetic variation in other genes might also be important in the tumorigenic process (Turesky, 2004). For example, Hein (2002) demonstrated that a polymorphism in the NAT2 gene can have a variable influence in the modification of aromatic amine metabolism. If NAT2 is rapidly acetylated, the resulting phenotype is a patient at a higher risk for colon cancer, whereas slower acetylation results in a higher risk to develop bladder cancer. The study of these polymorphisms usually involves sequencing or PCR-based techniques when considered one at a time. But once an array of candidates is identified the analysis can be performed with microarray for thousands of SNPs at a time. One array already in production evaluates polymorphisms in the Cytochrome P450 genes that might influence how individuals detoxify chemotherapeutic agents (Jain, 2005).

SNPs can also serve as signposts in the genome. High-density arrays of SNPs can be used to look for large-scale deletions and amplifications in the genome (Matsuzaki et al., 2004a,b). Loss of heterozygosity occurs when a region of a chromosome is deleted. This commonly occurs in cancer and SNP arrays can be used to quickly identify this phenomenon (Dumur et al., 2003; Huang et al., 2004). For example, Zhao et al. (2005) evaluated 70 patients with small cell and nonsmall cell lung cancer. They were able to identify examples of loss of heterozygosity and instances of genomic amplification that might explain the increased growth potential of the specific tumors.

5.3 Protein Analysis Techniques

5.3.1 Tissue Arrays

In the investigative push to understand human cancers, attempts were made to assess multiple disease processes simultaneously. The original concept of simultaneous evaluation of multiple tissues is credited to Hector Battifora who published in 1986 his technique for a multitumor tissue block known at the time as the "sausage block" (Battifora, 1986). Eventually, a large number of small tissue segments would be implanted into a single paraffin block to create an array of as many as one thousand tissues. Up to 200 consecutive 5-μm sections could then be taken from these blocks allowing for multiple staining and microscopic evaluations of the tumors. Any of the microscopic techniques is then available for hundreds of tumors with the same amount of time, effort, and reagents normally required for one tissue sample. Kononen et al. first described the tissue microarray in 1998 and multiple studies have followed confirming the convenience and effectiveness of this new approach (Kononen et al., 1998; Moch et al., 1999; Mucci et al., 2000; Richter et al., 2000; Schraml et al., 1999). Techniques such as H&E staining, immunohistochemistry, fluorescent in situ hybridization (FISH), in situ PCR, and others can be performed. Therefore, tissue microarrays can be used to evaluate the genome, transcriptome, or proteome. However, the usefulness of the results is dependant on the care used to create the array. It is vital that each core sample represents the original tumor well or bias can result. Arrays produced with larger cores are more likely to capture the area of interest while smaller diameter discs allow for more selective microdissection of the important tumor components of the tissues in question as well as for minimizing the amount of tissue removed from the original specimen (Skacel et al., 2002). Any subsample of the original tumor may misrepresent the true heterogeneity of the tumor and arrays may not contain surrounding tissues which may contribute to the tumorigenic process (Liotta and Kohn, 2001). Despite this, multiple studies have shown a similar sensitivity for tissue microarrays relative to the original tumor blocks in detecting genetic aberrations (Camp et al., 2000; Sallinen et al., 2000; Schraml et al., 1999). The benefit of tissue microarrays lies in the ability to perform multiple experiments simultaneously on multiple tumors and on serial sections.

Tissue microarrays may soon become the first place one tests a new hypothesis related to the pathological evaluation of tumors as large numbers of tumors can be screened in a short time (Skacel et al., 2002). This technique will allow one to quickly test and refine new pathological measures of cancer subtypes. For example, Chin-Chen et al. used tissue arrays to aid in the immunohistochemical pathologic differentiation of hepatocellular carcinoma, renal cell carcinoma, and adrenocorticoid carcinoma (Pan et al., 2005). These three tumor sites are often difficult to differentiate histologically. The tissue array platform offered a quick way to evaluate the overall sensitivity of their technique. Tissue arrays also allow different institutions to screen the same samples with a new technique and evaluate the effect of the observer bias in reporting staining results (Mengel et al., 2002; von Wasielewski et al., 2002).

5.3.2 PROTEOMICS

Proteomics is loosely defined as any study of the proteome, but the term is usually used to imply the study of a large population of proteins on the tissue, cellular, or subcellular level (Petricoin and Liotta, 2004; Somiari et al., 2005; Stults and Arnott, 2005; Wulfkuhle et al., 2003). A complex mixture of proteins, such as serum, plasma, tissue, or a cellular extract may be resolved into smaller complexes of protein by liquid chromatography (e.g., affinity, ion exchange, or reverse phase separations) or gel-based techniques (e.g., isoelectric focusing and SDS-PAGE, commonly known as 2D electrophoresis). Each chromatographic fraction or gel spot can then be interrogated with liquid chromatography coupled to tandem mass spectrometry peptide sequencing (LC-MS/MS) and database searching to determine its protein constituents (Fig. 3). Similar fractions from multiple samples can also be compared semiquantitatively with mass spectrometry profiling or quantitatively with differential fluorescence techniques or isotopic labeling. The analysis by mass spectrometry has been applied to complex mixtures such as whole plasma or tissue (Caprioli, 2005; Steel et al., 2003) or to more purified complexes such as individual spots on a two-dimensional gel (Greengauz-Roberts et al., 2005; Wulfkuhle et al., 2003).

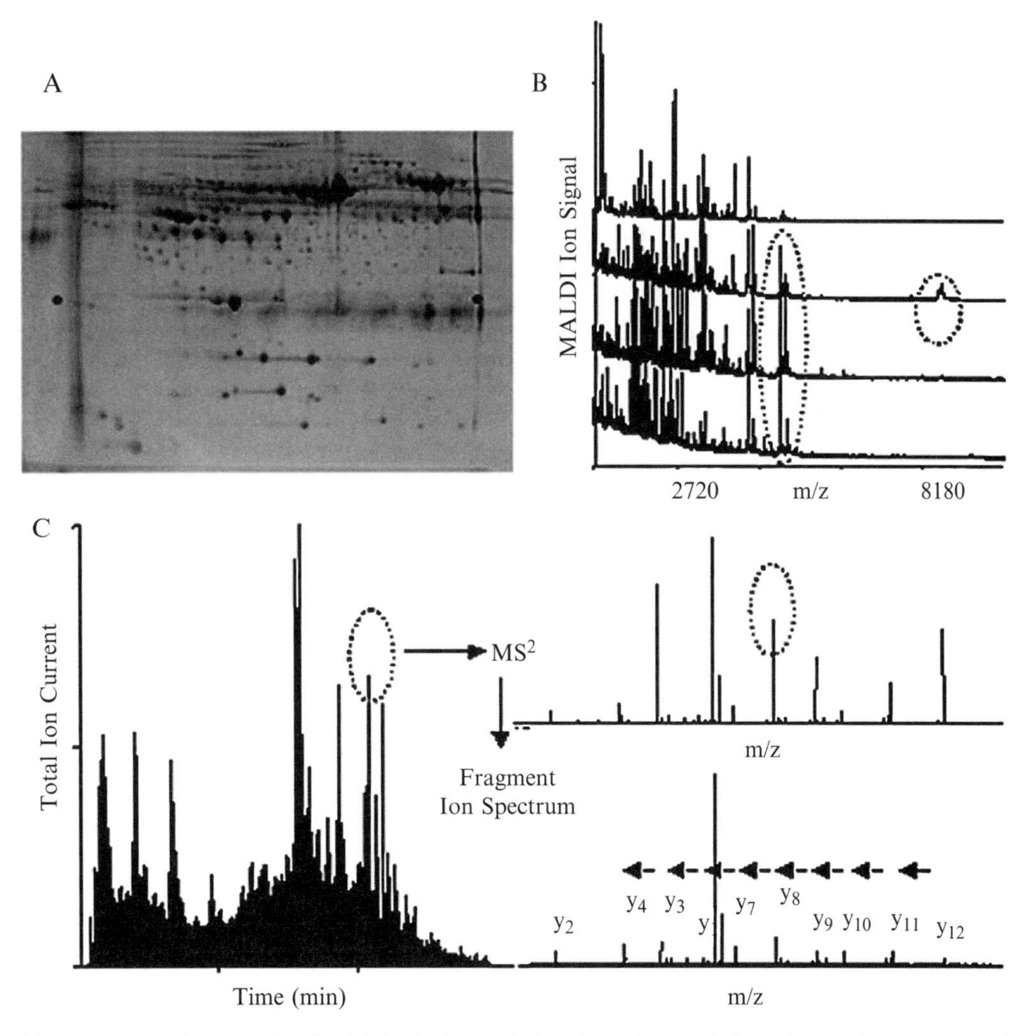

Fig. 3. Proteomic analysis of a biological sample involves the resolution of complex mixtures of proteins into individual polypeptides and fragments of polypeptides. (**a**) Picture of a biological sample following two-dimensional gel electrophoresis. Individual spots can be evaluated for intensity and composition. (**b**) Individual mass spectrometry traces can be compared to identify differences between samples such as those circled. (**c**) Individual peaks of an ion spectrum can be further fractionated to identify the peptide fragment, and ultimately the protein, responsible for the differences seen in a sample

One highly promising arena for proteomic analysis is in the search for clinical biomarkers (Alaiya et al., 2005; Bhattacharyya et al., 2004; Chen et al., 2005a; Hondermarck et al., 2001; Srinivas et al., 2001; Steel et al., 2003; Zhang et al., 2004). Bhattacharyya et al. (2004) were able to detect pancreatic cancer with high accuracy using high-throughput proteomics on the serum from patients. On the other hand, Chen et al. (2005a) used a similar approach to identify many proteins associated with pancreatic cancer that might one day be detected by less expensive techniques. Zhang et al. (2004) used proteomics in a multi-institutional study of women with ovarian cancer, benign pelvic masses, or no pathology. They uncovered three distinct protein markers unique to ovarian cancer that could conceivably be used for early detection tumor markers. The combination of these three proteins with CA-125 was much more sensitive in detecting ovarian cancer than using CA-125 alone.

Proteomics approaches have also been used to subtype tumors for a histological diagnosis whether directly from tumor samples or in attempts at early detection (Borczuk et al., 2004; Seike et al., 2005; Steel et al., 2003). Direct tissue analysis with mass spectrometry is feasible at the single cell level (Danna and Nolan, 2006) and has been used on microdissected samples for purer tumor analysis (Greengauz-Roberts et al., 2005; Jain, 2002). The technique has even been used to probe multiple locations across microscopic tissue slices (Chaurand et al., 2004). This allows one to look at both the tumor and the surrounding tissue for changes related to tumor growth or to look for microscopic tumors in an otherwise normal looking tissue sample. These kinds of studies will probably provide valuable information about the emergence of tumors from microscopic disease sites that cannot be obtained by any other method.

6 CLINICAL APPLICATION

During the evolution of a cancer cell, the diseased genome undergoes a series of genetic changes that drastically alters the cellular metabolism. Many genes lose their function or take on new roles to promote cellular growth, invasion, metastasis and angiogenesis. Each tumor follows a different path towards tumorigenesis, so the genetic variation between seemingly similar tumors can be significant. These differences may be the underlying reason for variability in the way tumors progress and the way they respond to attempts at eradication. A full understanding of any tumor will require a complete understanding of the functional genomics of that tumor. There is no question that the tools used to comprehend the functional genomics of human cells will one day be routine techniques for the clinical diagnosis and treatment of cancer (Table 1) (Mount and Pandey, 2005; Yeatman, 2003).

Currently, most cancers are identified based on their pathological appearance under the microscope. This microscopic examination also includes a crude analysis of the genome for some cancer types. Karyotyping is an important component of the diagnosis of many hematological malignancies (Bayani and Squire, 2002; Jotterand and Parlier, 1996). Yet other forms of cancer also are known to possess chromosomal anomalies (Gronwald et al., 2005; Kimura et al., 2004; Meijer et al., 1998; Micci et al., 2004). It is likely that they too benefit from genomic evaluation. Current tools allow for genomic screening at a higher resolution (Jones et al., 2005; Nakao et al., 2004). This

Table 1
Techniques used to analyze gene expression

Genome	Karyotyping (including spectral karyotyping) High-throughput sequencing Fluorescent in situ hybridization Microarrays: Comparative genome hybridization Single nucleotide polymorphism arrays
Transcriptome	Serial analysis of gene expression Massively parallel signature sequencing Expressed sequence tag library analysis cDNA arrays Oligonucleotide arrays ChIP-on-chip assays
Proteome	Mass spectrometry (MS) 2D gel electrophoresis plus MS Column/batch chromatography Tissue arrays Protein arrays

resolution could even progress to the level of individual nucleotides, if one knew where to look (Jain, 2005; Landi et al., 2003; Wikman et al., 2000).

The transcriptome also reveals useful information about cancer and may one day become a clinical tool. Many studies have been published providing evidence for the clinical relevance of microarray data. Because microarrays evaluate thousands of genes, the focus is on the complex gene expression signature of a tumor and how it might relate to clinical outcomes. Several observations are emerging from the transcriptional analysis of cancer. It is easy to distinguish between normal tissue and the cancers that arise within these tissues. Recognized histological subtypes of cancer are also visible within the gene expression profiles of those tumors. Histologically similar tumor types may have distinct subtypes based on the gene expression profiles and these subtypes may one day influence the treatment of the tumors.

Many groups have been successful at identifying different types of cancer based on the gene expression profile of the tumors (Bucca et al., 2004; Cao et al., 2004; Elek et al., 2000; Halvorsen et al., 2005; Hu et al., 2005; Khan et al., 1999; Lee et al., 2004; Smith, 2002; Sorlie et al., 2001; Wrobel et al., 2005; Zhang and Ji, 2005). Some studies have tackled more difficult classification problems and can reliably delineate recognized subgroups and find others not yet recognized by histology (Bhattacharjee et al., 2001; Bloom et al., 2004; Golub et al., 1999; Ramaswamy et al., 2001). Specifically, some groups have shown that gene expression profiles can be used to identify tumors of unknown origin. Ramaswamy et al. (2001) had a 78% accuracy in predicting the identity of tumors of unknown origin; while, Bloom et al. (2004) showed an 84% accuracy using both cDNA and oligonucleotide microarrays results. The poorly differentiated tumors were the least likely to classify correctly, suggesting that they had dedifferentiated significantly from the tissue of origin. Perou et al. evaluated 65 surgical specimens from 42 breast cancer patients (Perou et al., 2000). They noted significant gene expression differences between patients suggesting that several subtypes might exist.

Tumor subtypes might explain the differences seen in the therapeutic responses of patients. Several authors have identified gene signatures that might predict a positive or negative response to therapy (Cheok et al., 2003; Kakiuchi et al., 2004; McLean et al., 2004; Staunton et al., 2001). McLean et al. (2004) used oligonucleotide microarrays to generate a 31 gene classifiers that differentiated between CML patients who had a complete response to imatinib and those with minimal or no response. They then could predict response with 93.4% sensitivity and 87.7% positive predictive value. Likewise, Kakiuchi et al. (2004) evaluated advanced nonsmall cell lung cancer and found 51 genes that predicted a response to gefitinib (Iressa). Cheok et al. (2003) evaluated over 9,000 genes expressed in AML and followed the changes in expression before and after therapy with methotrexate and mercaptopurine, given alone or in combination. They discovered that the gene response varied based on treatment, including whether one drug was given or both. The most successful predictors of response seem to track with the biological basis of the treatment. The presence or absence of the target for the drug is detectable in the gene expression signature and is most predictive of the outcome. As more designer drugs come into the clinic the need for transcriptome analysis and classification will become more critical (Rhodes and Chinnaiyan, 2005).

Many laboratories are also using microarray data to identify genes that may influence tumor progression, metastatic potential, and survival outcomes (Agrawal et al., 2003; Henshall et al., 2003; Ramaswamy et al., 2003; Sanchez-Carbayo et al., 2003; van 'T Veer et al., 2002; van de Vijver et al., 2002; Vasselli et al., 2003). van de Vijver et al. (2002) used a 70 gene classifier to evaluate 295 patients with stage I or stage II breast cancer. The classifier identified 180 patients with a poor-prognosis signature and 115 with a good prognosis signature. The 10-year survival was 54% in the poor-prognosis group and 94% in the good prognosis group. The 10-year probability of remaining distant-disease free was 50% in the poor-prognosis group and 85% in the good prognosis group. Thus, there is hope for the prediction of clinical responses based on gene expression patterns in tumors. In a different tumor type, Dave et al. (2004) examined specimens from patients with follicular lymphoma. Using a gene classifier, they could group patients into four survival groups with different mean survival times. The classification genes appear to be expressed by nonmalignant immune cells infiltrating the tumor once again demonstrating that there may be a biological basis for the differential response.

7 SUMMARY

Over the past century, the approach to human disease has changed dramatically. What was known to early geneticists of the twentieth century was that living beings inherited and expressed unique phenotypes. The disease known as cancer was understood to be a unique phenotype of uncontrolled cellular proliferation. Once scientists discovered that DNA was the origin of phenotypic expressions, research shifted towards understanding the role of gene expression and mutations in the development of cancer.

With the completion of the human genome project, the focus has again shifted. There has been an expansion of the search for factors related to the development and viability of cancer cells. A higher resolution view of the genome is now considered and a larger view of the transcriptome or proteome of the cancer cell is evaluated for clinical decisions. In the near future it is likely that the scientific approach to cancer will involve an examination of a tumor's entire genome, transcriptome, and proteome to suggest treatment and predict outcomes.

Individualized medicine is the next realm for cancer diagnosis and treatment. Millions of bits of data will be captured for each individual case of cancer. This data will come from high throughput and/or high output techniques such as microarray technology and proteomics. To capture, store, process, and analyze this data computers are required. The most important new player in the fight against cancer is the computer scientist. Bioinformatics specialists have joined clinical investigations to process and analyze data from the emerging high-throughput technologies. It is their expertise that will generate the software necessary to synthesize the massive amount of data into a concrete treatment plan for each patient.

The use of high-throughput technologies in the study of cancer has just begun. The most promising conclusion from the first wave of results is that the technology is likely to work. Standardization, such as in the methods of tissue collection or RNA processing, is improving the reproducibility of experimental findings. There is also promise in the attempts to predict response on the basis of gene expression profiling. When a biological basis exists for the treatment mechanism, this biological signal is visible in the gene expression profile. The histological subtypes identified by pathologists are also discernable by gene expression profiling. Thus biological differences are resolved by an informatics analysis of a tumor. But informatics is also revealing that the issues affecting cancer are enormous. It is already known that cancer is a wide spectrum of diseases. Functional genomics is going to expand the subtypes of cancer even further. This diversity might currently explain the variation seen in response to treatment and the risk of recurrence. The variation might also mean that individual tumors get individual treatment in the future. Although most claims still require more follow-up work to verify genetic signatures, it is clear that the scientific community has embraced the promise of high-throughput sequencing, microarray technology, and other large-scale techniques for investigating many questions related to cancer.

REFERENCES

Adams M.D., Kelley J.M., Gocayne J.D., Dubnick M., Polymeropoulos M.H., Xiao H., Merril C.R., Wu A., Olde B., Moreno R.F., et al. 1991. Complementary DNA sequencing: expressed sequence tags and human genome project. Science 252: 1651–1656

Agrawal D., Chen T., Irby R., Quackenbush J., Chambers A.F., Szabo M., Cantor A., Coppola D. and Yeatman T.J. 2003. Osteopontin identified as colon cancer tumor progression marker. CR Biol 326: 1041–1043

Akervall J., Guo X., Qian C.N., Schoumans J., Leeser B., Kort E., Cole A., Resau J., Bradford C., Carey T., Wennerberg J., Anderson H., Tennvall J. and Teh B.T. 2004. Genetic and expression profiles of squamous cell carcinoma of the head and neck correlate with cisplatin sensitivity and resistance in cell lines and patients. Clin Cancer Res 10: 8204–8213

Al Moustafa A.E., Alaoui-Jamali M.A., Batist G., Hernandez-Perez M., Serruya C., Alpert L., Black M.J., Sladek R. and Foulkes W.D. 2002. Identification of genes associated with head and neck carcinogenesis by cDNA microarray comparison between matched primary normal epithelial and squamous carcinoma cells. Oncogene 21: 2634–2640

Alaiya A., Al-Mohanna M. and Linder S. 2005. Clinical cancer proteomics: promises and pitfalls. J Proteome Res 4: 1213–1222

Alevizos I., Mahadevappa M., Zhang X., Ohyama H., Kohno Y., Posner M., Gallagher G.T., Varvares M., Cohen D., Kim D., Kent R., Donoff R.B., Todd R., Yung C.M., Warrington J.A. and Wong D.T. 2001. Oral cancer in vivo gene expression profiling assisted by laser capture microdissection and microarray analysis. Oncogene 20: 6196–6204

Alexander H., Stegner A.L., Wagner-Mann C., Du Bois G.C., Alexander S. and Sauter E.R. 2004. Proteomic analysis to identify breast cancer biomarkers in nipple aspirate fluid. Clin Cancer Res 10: 7500–7510

Altshuler D., Brooks L.D., Chakravarti A., Collins F.S., Daly M.J. and Donnelly P. 2005. A haplotype map of the human genome. Nature 437: 1299–1320

Anderson L. 2005. Candidate-based proteomics in the search for biomarkers of cardiovascular disease. J Physiol 563: 23–60

Armstrong N.J. and van de Wiel M.A. 2004. Microarray data analysis: from hypotheses to conclusions using gene expression data. Cell Oncol 26: 279–290

Asmann Y.W., Kosari F., Wang K., Cheville J.C. and Vasmatzis G. 2002. Identification of differentially expressed genes in normal and malignant prostate by electronic profiling of expressed sequence tags. Cancer Res 62: 3308–3314

Avery O.T., MacLeod C.M. and McCarty M. 1944. Studies on the chemical nature of the substance inducing transformation of Pneumococcal types. J Exp Med 79: 137–158

Bammler T., Beyer R.P., Bhattacharya S., Boorman G.A., Boyles A., Bradford B.U., Bumgarner R.E., Bushel P.R., Chaturvedi K., Choi D., Cunningham M.L., Deng S., Dressman H.K., Fannin R.D., Farin F.M., Freedman J.H., Fry R.C., Harper A., Humble M.C., Hurban P., Kavanagh T.J., Kaufmann W.K., Kerr K.F., Jing L., Lapidus J.A., Lasarev M.R., Li J., Li Y.J., Lobenhofer E.K., Lu X., Malek R.L., Milton S., Nagalla S.R., O'Malley J P., Palmer V.S., Pattee P., Paules R.S., Perou C.M., Phillips K., Qin L.X., Qiu Y., Quigley S.D., Rodland M., Rusyn I., Samson L.D., Schwartz D.A., Shi Y., Shin J.L., Sieber S.O., Slifer S., Speer M.C., Spencer P.S., Sproles D.I., Swenberg J.A., Suk W.A., Sullivan R.C., Tian R., Tennant R.W., Todd S.A., Tucker C.J., Van Houten B., Weis B.K., Xuan S. and Zarbl H. 2005. Standardizing global gene expression analysis between laboratories and across platforms. Nat Methods 2: 351–356

Baris O., Mirebeau-Prunier D., Savagner F., Rodien P., Ballester B., Loriod B., Granjeaud S., Guyetant S., Franc B., Houlgatte R., Reynier P. and Malthiery Y. 2005. Gene profiling reveals specific oncogenic mechanisms and signaling pathways in oncocytic and papillary thyroid carcinoma. Oncogene 24: 4155–4161

Battifora H. 1986. The multitumor (sausage) tissue block: novel method for immunohistochemical antibody testing. Lab Invest 55: 244–248

Bayani J.M. and Squire J.A. 2002. Applications of SKY in cancer cytogenetics. Cancer Invest 20: 373–386

Beheshti B., Braude I., Marrano P., Thorner P., Zielenska M. and Squire J.A. 2003. Chromosomal localization of DNA amplifications in neuroblastoma tumors using cDNA microarray comparative genomic hybridization. Neoplasia 5: 53–62

Bertucci F., Bernard K., Loriod B., Chang Y.C., Granjeaud S., Birnbaum D., Nguyen C., Peck K. and Jordan B.R. 1999. Sensitivity issues in DNA array-based expression measurements and performance of nylon microarrays for small samples. Hum Mol Genet 8: 1715–1722

Bhattacharjee A., Richards W.G., Staunton J., Li C., Monti S., Vasa P., Ladd C., Beheshti J., Bueno R., Gillette M., Loda M., Weber G., Mark E.J., Lander E.S., Wong W., Johnson B.E., Golub T.R., Sugarbaker D.J. and Meyerson M. 2001. Classification of human lung carcinomas by mRNA expression profiling reveals distinct adenocarcinoma subclasses. Proc Natl Acad Sci USA 98: 13790–13795

Bhattacharyya S., Siegel E.R., Petersen G.M., Chari S.T., Suva L.J. and Haun R.S. 2004. Diagnosis of pancreatic cancer using serum proteomic profiling. Neoplasia 6: 674–686

Bittner M., Meltzer P., Chen Y., Jiang Y., Seftor E., Hendrix M., Radmacher M., Simon R., Yakhini Z., Ben-Dor A., Sampas N., Dougherty E., Wang E., Marincola F., Gooden C., Lueders J., Glatfelter A.,

Pollock P., Carpten J., Gillanders E., Leja D., Dietrich K., Beaudry C., Berens M., Alberts D. and Sondak V. 2000. Molecular classification of cutaneous malignant melanoma by gene expression profiling. Nature 406: 536–540

Bloom G., Yang I.V., Boulware D., Kwong K.Y., Coppola D., Eschrich S., Quackenbush J. and Yeatman T.J. 2004. Multi-platform, multi-site, microarray-based human tumor classification. Am J Pathol 164: 9–16

Bomgaars L. and McLeod H.L. 2005. Pharmacogenetics and pediatric cancer. Cancer J 11: 314–323

Bonome T., Lee J.Y., Park D.C., Radonovich M., Pise-Masison C., Brady J., Gardner G.J., Hao K., Wong W.H., Barrett J.C., Lu K.H., Sood A.K., Gershenson D.M., Mok S.C. and Birrer M.J. 2005. Expression profiling of serous low malignant potential, low-grade, and high-grade tumors of the ovary. Cancer Res 65: 10602–10612

Borczuk A.C., Shah L., Pearson G.D., Walter K.L., Wang L., Austin J.H., Friedman R.A. and Powell C.A. 2004. Molecular signatures in biopsy specimens of lung cancer. Am J Respir Crit Care Med 170: 167–174

Botstein D. and Risch N. 2003. Discovering genotypes underlying human phenotypes: past successes for mendelian disease, future approaches for complex disease. Nat Genet 33(Suppl): 228–237

Bracco L. and Kearsey J. 2003. The relevance of alternative RNA splicing to pharmacogenomics. Trends Biotechnol 21: 346–353

Brenner S., Johnson M., Bridgham J., Golda G., Lloyd D.H., Johnson D., Luo S., McCurdy S., Foy M., Ewan M., Roth R., George D., Eletr S., Albrecht G., Vermaas E., Williams S.R., Moon K., Burcham T., Pallas M., DuBridge R.B., Kirchner J., Fearon K., Mao J. and Corcoran K. 2000. Gene expression analysis by massively parallel signature sequencing (MPSS) on microbead arrays. Nat Biotechnol 18: 630–634

Brett D., Hanke J., Lehmann G., Haase S., Delbruck S., Krueger S., Reich J. and Bork P. 2000. EST comparison indicates 38% of human mRNAs contain possible alternative splice forms. FEBS Lett 474: 83–86

Bucca G., Carruba G., Saetta A., Muti P., Castagnetta L. and Smith C.P. 2004. Gene expression profiling of human cancers. Ann NY Acad Sci 1028: 28–37

Bull J.H., Ellison G., Patel A., Muir G., Walker M., Underwood M., Khan F. and Paskins L. 2001. Identification of potential diagnostic markers of prostate cancer and prostatic intraepithelial neoplasia using cDNA microarray. Br J Cancer 84: 1512–1519

Camargo A.A., Samaia H.P., Dias-Neto E., Simao D.F., Migotto I.A., Briones M.R., Costa F.F., Nagai M.A., Verjovski-Almeida S., Zago M.A., Andrade L.E., Carrer H., El-Dorry H.F., Espreafico E.M., Habr-Gama A., Giannella-Neto D., Goldman G.H., Gruber A., Hackel C., Kimura E.T., Maciel R.M., Marie S.K., Martins E.A., Nobrega M.P., Paco-Larson M.L., Pardini M.I., Pereira G.G., Pesquero J.B., Rodrigues V., Rogatto S.R., da Silva I.D., Sogayar M.C., Sonati M.F., Tajara E.H., Valentini S.R., Alberto F.L., Amaral M.E., Aneas I., Arnaldi L.A., de Assis A.M., Bengtson M.H., Bergamo N.A., Bombonato V., de Camargo M.E., Canevari R.A., Carraro D.M., Cerutti J.M., Correa M.L., Correa R.F., Costa M.C., Curcio C., Hokama P.O., Ferreira A.J., Furuzawa G.K., Gushiken T., Ho P.L., Kimura E., Krieger J.E., Leite L.C., Majumder P., Marins M., Marques E.R., Melo A.S., Melo M.B., Mestriner C.A., Miracca E.C., Miranda D.C., Nascimento A.L., Nobrega F.G., Ojopi E.P., Pandolfi J.R., Pessoa L.G., Prevedel A.C., Rahal P., Rainho C.A., Reis E.M., Ribeiro M.L., da Ros N., de Sa R.G., Sales M.M., Sant'anna S.C., dos Santos M.L., da Silva A.M., da Silva N.P., Silva W.A., Jr., da Silveira R.A., Sousa J.F., Stecconi D., Tsukumo F., Valente V., Soares F., Moreira E.S., Nunes D.N., Correa R.G., Zalcberg H., Carvalho A.F., Reis L.F., Brentani R.R., Simpson A.J. and de Souza S.J. 2001. The contribution of 700,000 ORF sequence tags to the definition of the human transcriptome. Proc Natl Acad Sci USA 98: 12103–12108

Camp R.L., Charette L.A. and Rimm D.L. 2000. Validation of tissue microarray technology in breast carcinoma. Lab Invest 80: 1943–1949

Cao Q.J., Belbin T., Socci N., Balan R., Prystowsky M.B., Childs G. and Jones J.G. 2004. Distinctive gene expression profiles by cDNA microarrays in endometrioid and serous carcinomas of the endometrium. Int J Gynecol Pathol 23: 321–329

Capella G., Cronauer-Mitra S., Pienado M.A. and Perucho M. 1991. Frequency and spectrum of mutations at codons 12 and 13 of the c-K-ras gene in human tumors. Environ Health Perspect 93: 125–131

Caprioli R.M. 2005. Deciphering protein molecular signatures in cancer tissues to aid in diagnosis, prognosis, and therapy. Cancer Res 65: 10642–10645

Carlson E.A. 1966. The Gene: A Critical History. Saunders, Philadelphia, xi, 301 p

Carter S.L., Eklund A.C., Mecham B.H., Kohane I.S. and Szallasi Z. 2005. Redefinition of Affymetrix probe sets by sequence overlap with cDNA microarray probes reduces cross-platform inconsistencies in cancer-associated gene expression measurements. BMC Bioinformatics 6: 107

Carulli J.P., Artinger M., Swain P.M., Root C.D., Chee L., Tulig C., Guerin J., Osborne M., Stein G., Lian J. and Lomedico P.T. 1998. High throughput analysis of differential gene expression. J Cell Biochem Suppl 30–31: 286–296

Casey G., Lindor N.M., Papadopoulos N., Thibodeau S.N., Moskow J., Steelman S., Buzin C.H., Sommer S.S., Collins C.E., Butz M., Aronson M., Gallinger S., Barker M.A., Young J.P., Jass J.R., Hopper J.L., Diep A., Bapat B., Salem M., Seminara D. and Haile R. 2005. Conversion analysis for mutation detection in MLH1 and MSH2 in patients with colorectal cancer. JAMA 293: 799–809

Castle J., Garrett-Engele P., Armour C.D., Duenwald S.J., Loerch P.M., Meyer M.R., Schadt E.E., Stoughton R., Parrish M.L., Shoemaker D.D. and Johnson J.M. 2003. Optimization of oligonucleotide arrays and RNA amplification protocols for analysis of transcript structure and alternative splicing. Genome Biol 4: R66

Cawley S., Bekiranov S., Ng H.H., Kapranov P., Sekinger E.A., Kampa D., Piccolboni A., Sementchenko V., Cheng J., Williams A.J., Wheeler R., Wong B., Drenkow J., Yamanaka M., Patel S., Brubaker S., Tammana H., Helt G., Struhl K. and Gingeras T.R. 2004. Unbiased mapping of transcription factor binding sites along human chromosomes 21 and 22 points to widespread regulation of noncoding RNAs. Cell 116: 499–509

Chang B.D., Swift M.E., Shen M., Fang J., Broude E.V. and Roninson I.B. 2002. Molecular determinants of terminal growth arrest induced in tumor cells by a chemotherapeutic agent. Proc Natl Acad Sci USA 99: 389–394

Chaurand P., Sanders M.E., Jensen R.A. and Caprioli R.M. 2004. Proteomics in diagnostic pathology: profiling and imaging proteins directly in tissue sections. Am J Pathol 165: 1057–1068

Chen J.J., Wu R., Yang P.C., Huang J.Y., Sher Y.P., Han M.H., Kao W.C., Lee P.J., Chiu T.F., Chang F., Chu Y.W., Wu C.W. and Peck K. 1998. Profiling expression patterns and isolating differentially expressed genes by cDNA microarray system with colorimetry detection. Genomics 51: 313–324

Chen R., Yi E.C., Donohoe S., Pan S., Eng J., Cooke K., Crispin D.A., Lane Z., Goodlett D.R., Bronner M.P., Aebersold R. and Brentnall T.A. 2005a. Pancreatic cancer proteome: the proteins that underlie invasion, metastasis, and immunologic escape. Gastroenterology 129: 1187–1197

Chen Y.T., Scanlan M.J., Venditti C.A., Chua R., Theiler G., Stevenson B.J., Iseli C., Gure A.O., Vasicek T., Strausberg R.L., Jongeneel C.V., Old L.J. and Simpson A.J. 2005b. Identification of cancer/testis-antigen genes by massively parallel signature sequencing. Proc Natl Acad Sci USA 102: 7940–7945

Cheok M.H., Yang W., Pui C.H., Downing J.R., Cheng C., Naeve C.W., Relling M.V. and Evans W.E. 2003. Treatment-specific changes in gene expression discriminate in vivo drug response in human leukemia cells. Nat Genet 34: 85–90

Claus E.B. 1995. The genetic epidemiology of cancer. Cancer Surv 25: 13–26

Danna E.A. and Nolan G.P. 2006. Transcending the biomarker mindset: deciphering disease mechanisms at the single cell level. Curr Opin Chem Biol 10: 20–27

Das M., Burge C.B., Park E., Colinas J. and Pelletier J. 2001. Assessment of the total number of human transcription units. Genomics 77: 71–78

Dave S.S., Wright G., Tan B., Rosenwald A., Gascoyne R.D., Chan W.C., Fisher R.I., Braziel R.M., Rimsza L.M., Grogan T.M., Miller T.P., LeBlanc M., Greiner T.C., Weisenburger D.D., Lynch J.C., Vose J., Armitage J.O., Smeland E.B., Kvaloy S., Holte H., Delabie J., Connors J.M., Lansdorp P.M., Ouyang Q., Lister T.A., Davies A.J., Norton A.J., Muller-Hermelink H.K., Ott G., Campo E., Montserrat E., Wilson W.H., Jaffe E.S., Simon R., Yang L., Powell J., Zhao H., Goldschmidt N., Chiorazzi M. and Staudt L.M. 2004. Prediction of survival in follicular lymphoma based on molecular features of tumor-infiltrating immune cells. N Engl J Med 351: 2159–2169

Den Otter W., Koten J.W., Van der Vegt B.J., Beemer F.A., Boxma O.J., Derkinderen D.J., De Graaf P.W., Huber J., Lips C.J., Roholl P.J., et al. 1990. Oncogenesis by mutations in anti-oncogenes: a view. Anticancer Res 10: 475–487

Dephoure N., Howson R.W., Blethrow J.D., Shokat K.M. and O’Shea E.K. 2005. Combining chemical genetics and proteomics to identify protein kinase substrates. Proc Natl Acad Sci USA 102: 17940–17945

Dobbin K.K., Beer D.G., Meyerson M., Yeatman T.J., Gerald W.L., Jacobson J.W., Conley B., Buetow K.H., Heiskanen M., Simon R.M., Minna J.D., Girard L., Misek D.E., Taylor J.M., Hanash S., Naoki K., Hayes D.N., Ladd-Acosta C., Enkemann S.A., Viale A. and Giordano T.J. 2005. Interlaboratory comparability study of cancer gene expression analysis using oligonucleotide microarrays. Clin Cancer Res 11: 565–572

Dobrindt U. 2005. (Patho-)Genomics of *Escherichia coli*. Int J Med Microbiol 295: 357–371

Dobrindt U., Agerer F., Michaelis K., Janka A., Buchrieser C., Samuelson M., Svanborg C., Gottschalk G., Karch H. and Hacker J. 2003. Analysis of genome plasticity in pathogenic and commensal *Escherichia coli* isolates by use of DNA arrays. J Bacteriol 185: 1831–1840

Dressman H.K., Hans C., Bild A., Olson J.A., Rosen E., Marcom P.K., Liotcheva V.B., Jones E.L., Vujaskovic Z., Marks J., Dewhirst M.W., West M., Nevins J.R. and Blackwell K. 2006. Gene expression profiles of multiple breast cancer phenotypes and response to neoadjuvant chemotherapy. Clin Cancer Res 12: 819–826

Dumur C.I., Dechsukhum C., Ware J.L., Cofield S.S., Best A.M., Wilkinson D.S., Garrett C.T. and Ferreira-Gonzalez A. 2003. Genome-wide detection of LOH in prostate cancer using human SNP microarray technology. Genomics 81: 260–269

Durkin A.J., Bloomston M., Yeatman T.J., Gilbert-Barness E., Cojita D., Rosemurgy A.S. and Zervos E.E. 2004. Differential expression of the Tie-2 receptor and its ligands in human pancreatic tumors. J Am Coll Surg 199: 724–731

Elek J., Park K.H. and Narayanan R. 2000. Microarray-based expression profiling in prostate tumors. In Vivo 14: 173–182

Eschrich S., Yang I., Bloom G., Kwong K.Y., Boulware D., Cantor A., Coppola D., Kruhoffer M., Aaltonen L., Orntoft T.F., Quackenbush J. and Yeatman T.J. 2005. Molecular staging for survival prediction of colorectal cancer patients. J Clin Oncol 23: 3526–3535

Ewing B. and Green P. 2000. Analysis of expressed sequence tags indicates 35,000 human genes. Nat Genet 25: 232–234

Frank T.S., Deffenbaugh A.M., Hulick M. and Gumpper K. 1999. Hereditary susceptibility to breast cancer: significance of age of onset in family history and contribution of BRCA1 and BRCA2. Dis Markers 15: 89–92

Frederiksen C.M., Knudsen S., Laurberg S. and Orntoft T.F. 2003. Classification of Dukes’ B and C colorectal cancers using expression arrays. J Cancer Res Clin Oncol 129: 263–271

Freije W.A., Castro-Vargas F.E., Fang Z., Horvath S., Cloughesy T., Liau L.M., Mischel P.S. and Nelson S.F. 2004. Gene expression profiling of gliomas strongly predicts survival. Cancer Res 64: 6503–6510

Gadgil M., Lian W., Gadgil C., Kapur V. and Hu W.S. 2005. An analysis of the use of genomic DNA as a universal reference in two channel DNA microarrays. BMC Genomics 6: 66

Ginos M.A., Page G.P., Michalowicz B.S., Patel K.J., Volker S.E., Pambuccian S.E., Ondrey F.G., Adams G.L. and Gaffney P.M. 2004. Identification of a gene expression signature associated with recurrent disease in squamous cell carcinoma of the head and neck. Cancer Res 64: 55–63

Golub T.R., Slonim D.K., Tamayo P., Huard C., Gaasenbeek M., Mesirov J.P., Coller H., Loh M.L., Downing J.R., Caligiuri M.A., Bloomfield C.D. and Lander E.S. 1999. Molecular classification of cancer: class discovery and class prediction by gene expression monitoring. Science 286: 531–537

Graveley B.R. 2001. Alternative splicing: increasing diversity in the proteomic world. Trends Genet 17: 100–107

Greengauz-Roberts O., Stoppler H., Nomura S., Yamaguchi H., Goldenring J.R., Podolsky R.H., Lee J.R. and Dynan W.S. 2005. Saturation labeling with cysteine-reactive cyanine fluorescent dyes provides increased sensitivity for protein expression profiling of laser-microdissected clinical specimens. Proteomics 5: 1746–1757

Gronwald J., Jauch A., Cybulski C., Schoell B., Bohm-Steuer B., Lener M., Grabowska E., Gorski B., Jakubowska A., Domagala W., Chosia M., Scott R.J. and Lubinski J. 2005. Comparison of genomic abnormalities between BRCAX and sporadic breast cancers studied by comparative genomic hybridization. Int J Cancer 114: 230–236

Gu Z., Hillier L. and Kwok P.Y. 1998. Single nucleotide polymorphism hunting in cyberspace. Hum Mutat 12: 221–225

Gyorffy B., Surowiak P., Kiesslich O., Denkert C., Schafer R., Dietel M. and Lage H. 2006. Gene expression profiling of 30 cancer cell lines predicts resistance towards 11 anticancer drugs at clinically achieved concentrations. Int J Cancer 118: 1699–1712

Halvorsen O.J., Oyan A.M., Bo T.H., Olsen S., Rostad K., Haukaas S.A., Bakke A.M., Marzolf B., Dimitrov K., Stordrange L., Lin B., Jonassen I., Hood L., Akslen L.A. and Kalland K.H. 2005. Gene expression profiles in prostate cancer: association with patient subgroups and tumour differentiation. Int J Oncol 26: 329–336

Hedenfalk I., Duggan D., Chen Y., Radmacher M., Bittner M., Simon R., Meltzer P., Gusterson B., Esteller M., Kallioniemi O.P., Wilfond B., Borg A. and Trent J. 2001. Gene-expression profiles in hereditary breast cancer. N Engl J Med 344: 539–548

Hedenfalk I.A., Ringner M., Trent J.M. and Borg A. 2002. Gene expression in inherited breast cancer. Adv Cancer Res 84: 1–34

Hegde P., Qi R., Gaspard R., Abernathy K., Dharap S., Earle-Hughes J., Gay C., Nwokekeh N.U., Chen T., Saeed A.I., Sharov V., Lee N.H., Yeatman T.J. and Quackenbush J. 2001. Identification of tumor markers in models of human colorectal cancer using a 19,200-element complementary DNA microarray. Cancer Res 61: 7792–7797

Hein D.W. 2002. Molecular genetics and function of NAT1 and NAT2: role in aromatic amine metabolism and carcinogenesis. Mutat Res 506–507: 65–77

Heiskanen M., Kononen J., Barlund M., Torhorst J., Sauter G., Kallioniemi A. and Kallioniemi O. 2001. CGH, cDNA and tissue microarray analyses implicate FGFR2 amplification in a small subset of breast tumors. Anal Cell Pathol 22: 229–234

Henshall S.M., Afar D.E., Hiller J., Horvath L.G., Quinn D.I., Rasiah K.K., Gish K., Willhite D., Kench J.G., Gardiner-Garden M., Stricker P.D., Scher H.I., Grygiel J.J., Agus D.B., Mack D.H. and Sutherland R.L. 2003. Survival analysis of genome-wide gene expression profiles of prostate cancers identifies new prognostic targets of disease relapse. Cancer Res 63: 4196–4203

Hondermarck H., Vercoutter-Edouart A.S., Revillion F., Lemoine J., el-Yazidi-Belkoura I., Nurcombe V. and Peyrat J.P. 2001. Proteomics of breast cancer for marker discovery and signal pathway profiling. Proteomics 1: 1216–1232

Hoque M.O., Lee C.C., Cairns P., Schoenberg M. and Sidransky D. 2003. Genome-wide genetic characterization of bladder cancer: a comparison of high-density single-nucleotide polymorphism arrays and PCR-based microsatellite analysis. Cancer Res 63: 2216–2222

Horak C.E. and Snyder M. 2002. ChIP-chip: a genomic approach for identifying transcription factor binding sites. Methods Enzymol 350: 469–483

Hu J., Bianchi F., Ferguson M., Cesario A., Margaritora S., Granone P., Goldstraw P., Tetlow M., Ratcliffe C., Nicholson A.G., Harris A., Gatter K. and Pezzella F. 2005. Gene expression signature for angiogenic and nonangiogenic non-small-cell lung cancer. Oncogene 24: 1212–1219

Huang J., Wei W., Zhang J., Liu G., Bignell G.R., Stratton M.R., Futreal P.A., Wooster R., Jones K.W. and Shapero M.H. 2004. Whole genome DNA copy number changes identified by high density oligonucleotide arrays. Hum Genomics 1: 287–299

Huang H.Y., Illei P.B., Zhao Z., Mazumdar M., Huvos A.G., Healey J.H., Wexler L.H., Gorlick R., Meyers P. and Ladanyi M. 2005. Ewing sarcomas with p53 mutation or p16/p14ARF homozygous deletion: a highly lethal subset associated with poor chemoresponse. J Clin Oncol 23: 548–558

Huang C.C., Cutcliffe C., Coffin C., Sorensen P.H., Beckwith J.B. and Perlman E.J. 2006. Classification of malignant pediatric renal tumors by gene expression. Pediatr Blood Cancer 46: 728–738

Irizarry R.A., Warren D., Spencer F., Kim I.F., Biswal S., Frank B.C., Gabrielson E., Garcia J.G., Geoghegan J., Germino G., Griffin C., Hilmer S.C., Hoffman E., Jedlicka A.E., Kawasaki E., Martinez-Murillo F., Morsberger L., Lee H., Petersen D., Quackenbush J., Scott A., Wilson M., Yang Y., Ye S.Q. and Yu W. 2005. Multiple-laboratory comparison of microarray platforms. Nat Methods 2: 345–350

Jackson L.G. 1978. Chromosomes and cancer: current aspects. Semin Oncol 5: 3–10

Jain K.K. 2002. Recent advances in oncoproteomics. Curr Opin Mol Ther 4: 203–209

Jain K.K. 2005. Applications of AmpliChip CYP450. Mol Diagn 9: 119–127

Jen K.Y. and Cheung V.G. 2005. Identification of novel p53 target genes in ionizing radiation response. Cancer Res 65: 7666–7673

Jiang F., Yin Z., Caraway N.P., Li R. and Katz R.L. 2004. Genomic profiles in stage I primary non small cell lung cancer using comparative genomic hybridization analysis of cDNA microarrays. Neoplasia 6: 623–635

Jin V.X., Leu Y.W., Liyanarachchi S., Sun H., Fan M., Nephew K.P., Huang T.H. and Davuluri R.V. 2004. Identifying estrogen receptor alpha target genes using integrated computational genomics and chromatin immunoprecipitation microarray. Nucleic Acids Res 32: 6627–6635

Johannsen W. 1909. Elemente der Exakten Erblichkeitslehre. Fischer, Jena

Jones A.M., Douglas E.J., Halford S.E., Fiegler H., Gorman P.A., Roylance R.R., Carter N.P. and Tomlinson I.P. 2005. Array-CGH analysis of microsatellite-stable, near-diploid bowel cancers and comparison with other types of colorectal carcinoma. Oncogene 24: 118–129

Jotterand M. and Parlier V. 1996. Diagnostic and prognostic significance of cytogenetics in adult primary myelodysplastic syndromes. Leuk Lymphoma 23: 253–266

Kahn P. 1995. From genome to proteome: looking at a cell’s proteins. Science 270: 369–370

Kakiuchi S., Daigo Y., Ishikawa N., Furukawa C., Tsunoda T., Yano S., Nakagawa K., Tsuruo T., Kohno N., Fukuoka M., Sone S. and Nakamura Y. 2004. Prediction of sensitivity of advanced non-small cell lung cancers to gefitinib (Iressa, ZD1839). Hum Mol Genet 13: 3029–3043

Kampa D., Cheng J., Kapranov P., Yamanaka M., Brubaker S., Cawley S., Drenkow J., Piccolboni A., Bekiranov S., Helt G., Tammana H. and Gingeras T.R. 2004. Novel RNAs identified from an in-depth analysis of the transcriptome of human chromosomes 21 and 22. Genome Res 14: 331–342

Kapranov P., Cawley S.E., Drenkow J., Bekiranov S., Strausberg R.L., Fodor S.P. and Gingeras T.R. 2002. Large-scale transcriptional activity in chromosomes 21 and 22. Science 296: 916–919

Kapranov P., Drenkow J., Cheng J., Long J., Helt G., Dike S. and Gingeras T.R. 2005. Examples of the complex architecture of the human transcriptome revealed by RACE and high-density tiling arrays. Genome Res 15: 987–997

Kasamatsu A., Endo Y., Uzawa K., Nakashima D., Koike H., Hashitani S., Numata T., Urade M. and Tanzawa H. 2005. Identification of candidate genes associated with salivary adenoid cystic carcinomas using combined comparative genomic hybridization and oligonucleotide microarray analyses. Int J Biochem Cell Biol 37: 1869–1880

Kawamoto S.,Yoshii J.,Mizuno K.,Ito K.,Miyamoto Y.,OhnishiT.,MatobaR.,HoriN.,Matsumoto Y.,Okumura T.,Nakao Y.,Yoshii H.,Arimoto J.,Ohashi H.,Nakanishi H., Ohno I., Hashimoto J., Shimizu K., Maeda K., Kuriyama H., Nishida K., Shimizu-Matsumoto A., Adachi W., Ito R., Kawasaki S. and Chae K.S. 2000. BodyMap: a collection of 3′ ESTs for analysis of human gene expression information. Genome Res 10: 1817–1827

Khan J., Saal L.H., Bittner M.L., Chen Y., Trent J.M. and Meltzer P.S. 1999. Expression profiling in cancer using cDNA microarrays. Electrophoresis 20: 223–229

Khanna C., Khan J., Nguyen P., Prehn J., Caylor J., Yeung C., Trepel J., Meltzer P. and Helman L. 2001. Metastasis-associated differences in gene expression in a murine model of osteosarcoma. Cancer Res 61: 3750–3759

Kihara C., Tsunoda T., Tanaka T., Yamana H., Furukawa Y., Ono K., Kitahara O., Zembutsu H., Yanagawa R., Hirata K., Takagi T. and Nakamura Y. 2001. Prediction of sensitivity of esophageal tumors to adjuvant chemotherapy by cDNA microarray analysis of gene-expression profiles. Cancer Res 61: 6474–6479

Kimura Y., Noguchi T., Kawahara K., Kashima K., Daa T. and Yokoyama S. 2004. Genetic alterations in 102 primary gastric cancers by comparative genomic hybridization: gain of 20q and loss of 18q are associated with tumor progression. Mod Pathol 17: 1328–1337

Kononen J., Bubendorf L., Kallioniemi A., Barlund M., Schraml P., Leighton S., Torhorst J., Mihatsch M.J., Sauter G. and Kallioniemi O.P. 1998. Tissue microarrays for high-throughput molecular profiling of tumor specimens. Nat Med 4: 844–847

Krizman D.B., Wagner L., Lash A., Strausberg R.L. and Emmert-Buck M.R. 1999. The Cancer Genome Anatomy Project: EST sequencing and the genetics of cancer progression. Neoplasia 1: 101–106

Kudoh K., Ramanna M., Ravatn R., Elkahloun A.G., Bittner M.L., Meltzer P.S., Trent J.M., Dalton W.S. and Chin K.V. 2000. Monitoring the expression profiles of doxorubicin-induced and doxorubicin-resistant cancer cells by cDNA microarray. Cancer Res 60: 4161–4166

Kuo W.P., Jenssen T.K., Butte A.J., Ohno-Machado L. and Kohane I.S. 2002. Analysis of matched mRNA measurements from two different microarray technologies. Bioinformatics 18: 405–412

Lander E.S., Linton L.M., Birren B., Nusbaum C., Zody M.C., Baldwin J., Devon K., Dewar K., Doyle M., FitzHugh W., Funke R., Gage D., Harris K., Heaford A., Howland J., Kann L., Lehoczky J., LeVine R., McEwan P., McKernan K., Meldrim J., Mesirov J.P., Miranda C., Morris W., Naylor J., Raymond C., Rosetti M., Santos R., Sheridan A., Sougnez C., Stange-Thomann N., Stojanovic N., Subramanian A., Wyman D., Rogers J., Sulston J., Ainscough R., Beck S., Bentley D., Burton J., Clee C., Carter N., Coulson A., Deadman R., Deloukas P., Dunham A., Dunham I., Durbin R., French L., Grafham D., Gregory S., Hubbard T., Humphray S., Hunt A., Jones M., Lloyd C., McMurray A., Matthews L., Mercer S., Milne S., Mullikin J.C., Mungall A., Plumb R., Ross M., Shownkeen R., Sims S., Waterston R.H., Wilson R.K., Hillier L.W., McPherson J.D., Marra M.A., Mardis E.R., Fulton L.A., Chinwalla A.T., Pepin K.H., Gish W.R., Chissoe S.L., Wendl M.C., Delehaunty K.D., Miner T.L., Delehaunty A., Kramer J.B., Cook L.L., Fulton R.S., Johnson D.L., Minx P.J., Clifton S.W., Hawkins T., Branscomb E., Predki P., Richardson P., Wenning S., Slezak T., Doggett N., Cheng J.F., Olsen A., Lucas S., Elkin C., Uberbacher E., Frazier M., Gibbs R.A., Muzny D.M., Scherer S.E., Bouck J.B., Sodergren E.J., Worley K.C., Rives C.M., Gorrell J.H., Metzker M.L., Naylor S.L., Kucherlapati R.S., Nelson D.L., Weinstock G.M., Sakaki Y., Fujiyama A., Hattori M., Yada T., Toyoda A., Itoh T., Kawagoe C., Watanabe H., Totoki Y., Taylor T., Weissenbach J., Heilig R., Saurin W., Artiguenave F., Brottier P., Bruls T., Pelletier E., Robert C., Wincker P., Smith D.R., Doucette-Stamm L., Rubenfield M., Weinstock K., Lee H.M., Dubois J., Rosenthal A., Platzer M., Nyakatura G., Taudien S., Rump A., Yang H., Yu J., Wang J., Huang G., Gu J., Hood L., Rowen L., Madan A., Qin S., Davis R.W., Federspiel N.A., Abola A.P., Proctor M.J., Myers R.M., Schmutz J., Dickson M., Grimwood J., Cox D.R., Olson M.V., Kaul R., Raymond C., Shimizu N., Kawasaki K., Minoshima S., Evans G.A., Athanasiou M., Schultz R., Roe B.A., Chen F., Pan H., Ramser J., Lehrach H., Reinhardt R., McCombie W.R., de la Bastide M., Dedhia N., Blocker H., Hornischer K., Nordsiek G., Agarwala R., Aravind L., Bailey J.A., Bateman A., Batzoglou S., Birney E., Bork P., Brown D.G., Burge C.B., Cerutti L., Chen H.C., Church D., Clamp M., Copley R.R., Doerks T., Eddy S.R., Eichler E.E., Furey T.S., Galagan J., Gilbert J.G., Harmon C., Hayashizaki Y., Haussler D., Hermjakob H., Hokamp K., Jang W., Johnson L.S., Jones T.A., Kasif S., Kaspryzk A., Kennedy S., Kent W.J., Kitts P., Koonin E.V., Korf I., Kulp D., Lancet D., Lowe T.M., McLysaght A., Mikkelsen T., Moran J.V., Mulder N., Pollara V.J., Ponting C.P., Schuler G., Schultz J., Slater G., Smit A.F., Stupka E., Szustakowski J., Thierry-Mieg D., Thierry-Mieg J., Wagner L., Wallis J., Wheeler R., Williams A., Wolf Y.I., Wolfe K.H., Yang S.P., Yeh R.F., Collins F., Guyer M.S., Peterson J., Felsenfeld A., Wetterstrand K.A., Patrinos A., Morgan M.J., de Jong P., Catanese J.J., Osoegawa K., Shizuya H., Choi S. and Chen Y.J. 2001. Initial sequencing and analysis of the human genome. Nature 409: 860–921

Landi S., Gemignani F., Gioia-Patricola L., Chabrier A. and Canzian F. 2003. Evaluation of a microarray for genotyping polymorphisms related to xenobiotic metabolism and DNA repair. Biotechniques 35: 816–820, 822, 824–817

Larramendy M.L., Niini T., Elonen E., Nagy B., Ollila J., Vihinen M. and Knuutila S. 2002. Overexpression of translocation-associated fusion genes of FGFRI, MYC, NPMI, and DEK, but absence of the translocations in acute myeloid leukemia. A microarray analysis. Haematologica 87: 569–577

Leal C.B., Schmitt F.C., Bento M.J., Maia N.C. and Lopes C.S. 1995. Ductal carcinoma in situ of the breast. Histologic categorization and its relationship to ploidy and immunohistochemical expression of hormone receptors, p53, and c-erbB-2 protein. Cancer 75: 2123–2131

Lee Y.F., John M., Falconer A., Edwards S., Clark J., Flohr P., Roe T., Wang R., Shipley J., Grimer R.J., Mangham D.C., Thomas J.M., Fisher C., Judson I. and Cooper C.S. 2004. A gene expression signature associated with metastatic outcome in human leiomyosarcomas. Cancer Res 64: 7201–7204

Leung Y.F. and Cavalieri D. 2003. Fundamentals of cDNA microarray data analysis. Trends Genet 19: 649–659

Li Y.J., Hoang-Xuan K., Zhou X.P., Sanson M., Mokhtari K., Faillot T., Cornu P., Poisson M., Thomas G. and Hamelin R. 1998. Analysis of the p21 gene in gliomas. J Neurooncol 40: 107–111

Liang F., Holt I., Pertea G., Karamycheva S., Salzberg S.L. and Quackenbush J. 2000. Gene index analysis of the human genome estimates approximately 120,000 genes. Nat Genet 25: 239–240

Lindlof A. 2003. Gene identification through large-scale EST sequence processing. Appl Bioinformatics 2: 123–129

Liotta L.A. and Kohn E.C. 2001. The microenvironment of the tumour-host interface. Nature 411: 375–379

Liu E.T. 2005. Genomic technologies and the interrogation of the transcriptome. Mech Ageing Dev 126: 153–159

Lockhart D.J., Dong H., Byrne M.C., Follettie M.T., Gallo M.V., Chee M.S., Mittmann M., Wang C., Kobayashi M., Horton H. and Brown E.L. 1996. Expression monitoring by hybridization to high-density oligonucleotide arrays. Nat Biotechnol 14: 1675–1680

Margulies M., Egholm M., Altman W.E., Attiya S., Bader J.S., Bemben L.A., Berka J., Braverman M.S., Chen Y.J., Chen Z., Dewell S.B., Du L., Fierro J.M., Gomes X.V., Godwin B.C., He W., Helgesen S., Ho C.H., Irzyk G.P., Jando S.C., Alenquer M.L., Jarvie T.P., Jirage K.B., Kim J.B., Knight J.R., Lanza J.R., Leamon J.H., Lefkowitz S.M., Lei M., Li J., Lohman K.L., Lu H., Makhijani V.B., McDade K.E., McKenna M.P., Myers E.W., Nickerson E., Nobile J.R., Plant R., Puc B.P., Ronan M.T., Roth G.T., Sarkis G.J., Simons J.F., Simpson J.W., Srinivasan M., Tartaro K.R., Tomasz A., Vogt K.A., Volkmer G.A., Wang S.H., Wang Y., Weiner M.P., Yu P., Begley R.F. and Rothberg J.M. 2005. Genome sequencing in microfabricated high-density picolitre reactors. Nature 437: 376–380

Marsh S. 2005. Thymidylate synthase pharmacogenetics. Invest New Drugs 23: 533–537

Martin G.M. and Hoehn H. 1974. Genetics and human disease. Hum Pathol 5: 387–405

Matsuyama R., Togo S., Shimizu D., Momiyama N., Ishikawa T., Ichikawa Y., Endo I., Kunisaki C., Suzuki H., Hayasizaki Y. and Shimada H. 2006. Predicting 5-fluorouracil chemosensitivity of liver metastases from colorectal cancer using primary tumor specimens: three-gene expression model predicts clinical response. Int J Cancer 119: 406–413

Matsuzaki H., Dong S., Loi H., Di X., Liu G., Hubbell E., Law J., Berntsen T., Chadha M., Hui H., Yang G., Kennedy G.C., Webster T.A., Cawley S., Walsh P.S., Jones K.W., Fodor S.P. and Mei R. 2004a. Genotyping over 100,000 SNPs on a pair of oligonucleotide arrays. Nat Methods 1: 109–111

Matsuzaki H., Loi H., Dong S., Tsai Y.Y., Fang J., Law J., Di X., Liu W.M., Yang G., Liu G., Huang J., Kennedy G.C., Ryder T.B., Marcus G.A., Walsh P.S., Shriver M.D., Puck J.M., Jones K.W. and Mei R. 2004b. Parallel genotyping of over 10,000 SNPs using a one-primer assay on a high-density oligonucleotide array. Genome Res 14: 414–425

Matsuzaki Y., Hashimoto S., Fujita T., Suzuki T., Sakurai T., Matsushima K. and Kawakami Y. 2005. Systematic identification of human melanoma antigens using serial analysis of gene expression (SAGE). J Immunother 28: 10–19

McLean L.A., Gathmann I., Capdeville R., Polymeropoulos M.H. and Dressman M. 2004. Pharmacogenomic analysis of cytogenetic response in chronic myeloid leukemia patients treated with imatinib. Clin Cancer Res 10: 155–165

McVean G., Spencer C.C. and Chaix R. 2005. Perspectives on Human Genetic Variation from the HapMap Project. PLoS Genet 1: e54

Meijer G.A., Hermsen M.A., Baak J.P., van Diest P.J., Meuwissen S.G., Belien J.A., Hoovers J.M., Joenje H., Snijders P.J. and Walboomers J.M. 1998. Progression from colorectal adenoma to carcinoma is associated with non-random chromosomal gains as detected by comparative genomic hybridisation. J Clin Pathol 51: 901–909

Mengel M., von Wasielewski R., Wiese B., Rudiger T., Muller-Hermelink H.K. and Kreipe H. 2002. Inter-laboratory and inter-observer reproducibility of immunohistochemical assessment of the Ki-67 labelling index in a large multi-centre trial. J Pathol 198: 292–299

Mercatante D. and Kole R. 2000. Modification of alternative splicing pathways as a potential approach to chemotherapy. Pharmacol Ther 85: 237–243

Micci F., Teixeira M.R., Haugom L., Kristensen G., Abeler V.M. and Heim S. 2004. Genomic aberrations in carcinomas of the uterine corpus. Genes Chromosomes Cancer 40: 229–246

Milani L., Fredriksson M. and Syvanen A.C. 2006. Detection of alternatively spliced transcripts in leukemia cell lines by minisequencing on microarrays. Clin Chem 52: 202–211

Mironov A.A., Fickett J.W. and Gelfand M.S. 1999. Frequent alternative splicing of human genes. Genome Res 9: 1288–1293

Moch H., Schraml P., Bubendorf L., Mirlacher M., Kononen J., Gasser T., Mihatsch M.J., Kallioniemi O.P. and Sauter G. 1999. High-throughput tissue microarray analysis to evaluate genes uncovered by cDNA microarray screening in renal cell carcinoma. Am J Pathol 154: 981–986

Modrek B. and Lee C.J. 2003. Alternative splicing in the human, mouse and rat genomes is associated with an increased frequency of exon creation and/or loss. Nat Genet 34: 177–180

Mount D.W. and Pandey R. 2005. Using bioinformatics and genome analysis for new therapeutic interventions. Mol Cancer Ther 4: 1636–1643

Mousses S., Wagner U., Chen Y., Kim J.W., Bubendorf L., Bittner M., Pretlow T., Elkahloun A.G., Trepel J.B. and Kallioniemi O.P. 2001. Failure of hormone therapy in prostate cancer involves systematic restoration of androgen responsive genes and activation of rapamycin sensitive signaling. Oncogene 20: 6718–6723

Mucci N.R., Akdas G., Manely S. and Rubin M.A. 2000. Neuroendocrine expression in metastatic prostate cancer: evaluation of high throughput tissue microarrays to detect heterogeneous protein expression. Hum Pathol 31: 406–414

Mundle S.D. and Sokolova I. 2004. Clinical implications of advanced molecular cytogenetics in cancer. Expert Rev Mol Diagn 4: 71–81

Nagao K., Togawa N., Fujii K., Uchikawa H., Kohno Y., Yamada M. and Miyashita T. 2005. Detecting tissue-specific alternative splicing and disease-associated aberrant splicing of the PTCH gene with exon junction microarrays. Hum Mol Genet 14: 3379–3388

Nakao K., Mehta K.R., Fridlyand J., Moore D.H., Jain A.N., Lafuente A., Wiencke J.W., Terdiman J.P. and Waldman F.M. 2004. High-resolution analysis of DNA copy number alterations in colorectal cancer by array-based comparative genomic hybridization. Carcinogenesis 25: 1345–1357

Nakeff A., Sahay N., Pisano M. and Subramanian B. 2002. Painting with a molecular brush: genomic/proteomic interfacing to define the drug action profile of novel solid-tumor selective anticancer agents. Cytometry 47: 72–79

Novoradovskaya N., Whitfield M.L., Basehore L.S., Novoradovsky A., Pesich R., Usary J., Karaca M., Wong W.K., Aprelikova O., Fero M., Perou C.M., Botstein D. and Braman J. 2004. Universal Reference RNA as a standard for microarray experiments. BMC Genomics 5: 20

Okazaki Y., Furuno M., Kasukawa T., Adachi J., Bono H., Kondo S., Nikaido I., Osato N., Saito R., Suzuki H., Yamanaka I., Kiyosawa H., Yagi K., Tomaru Y., Hasegawa Y., Nogami A., Schonbach C., Gojobori T., Baldarelli R., Hill D.P., Bult C., Hume D.A., Quackenbush J., Schriml L.M., Kanapin A., Matsuda H., Batalov S., Beisel K.W., Blake J.A., Bradt D., Brusic V., Chothia C., Corbani L.E., Cousins S., Dalla E., Dragani T.A., Fletcher C.F., Forrest A., Frazer K.S., Gaasterland T., Gariboldi M., Gissi C., Godzik A., Gough J., Grimmond S., Gustincich S., Hirokawa N., Jackson I.J., Jarvis E.D., Kanai A., Kawaji H., Kawasawa Y., Kedzierski R.M., King B.L., Konagaya A., Kurochkin I.V., Lee Y., Lenhard B., Lyons P.A., Maglott D.R., Maltais L., Marchionni L., McKenzie L., Miki H., Nagashima T., Numata K., Okido T., Pavan W.J., Pertea G., Pesole G., Petrovsky N., Pillai R., Pontius J.U., Qi D., Ramachandran S., Ravasi T., Reed J.C., Reed D.J., Reid J., Ring B.Z., Ringwald M., Sandelin A., Schneider C., Semple C.A., Setou M., Shimada K., Sultana R., Takenaka Y., Taylor M.S., Teasdale R.D., Tomita M., Verardo R., Wagner L., Wahlestedt C., Wang Y., Watanabe Y., Wells C., Wilming L.G., Wynshaw-Boris A., Yanagisawa M., Yang I., Yang L., Yuan Z., Zavolan M., Zhu Y., Zimmer A., Carninci P., Hayatsu N., Hirozane-Kishikawa T., Konno H., Nakamura M., Sakazume N., Sato K., Shiraki T., Waki K., Kawai J., Aizawa K., Arakawa T., Fukuda S., Hara A., Hashizume W., Imotani K., Ishii Y., Itoh M., Kagawa I., Miyazaki A., Sakai K., Sasaki D., Shibata K., Shinagawa A., Yasunishi A., Yoshino M., Waterston R., Lander E.S., Rogers J., Birney E. and Hayashizaki Y. 2002. Analysis of the mouse transcriptome based on functional annotation of 60,770 full-length cDNAs. Nature 420: 563–573

Onda M., Emi M., Yoshida A., Miyamoto S., Akaishi J., Asaka S., Mizutani K., Shimizu K., Nagahama M., Ito K., Tanaka T. and Tsunoda T. 2004. Comprehensive gene expression profiling of anaplastic thyroid cancers with cDNA microarray of 25 344 genes. Endocr Relat Cancer 11: 843–854

Ono K., Tanaka T., Tsunoda T., Kitahara O., Kihara C., Okamoto A., Ochiai K., Takagi T. and Nakamura Y. 2000. Identification by cDNA microarray of genes involved in ovarian carcinogenesis. Cancer Res 60: 5007–5011

Pan C.C., Chen P.C., Tsay S.H. and Ho D.M. 2005. Differential immunoprofiles of hepatocellular carcinoma, renal cell carcinoma, and adrenocortical carcinoma: a systemic immunohistochemical survey using tissue array technique. Appl Immunohistochem Mol Morphol 13: 347–352

Peehl D.M., Shinghal R., Nonn L., Seto E., Krishnan A.V., Brooks J.D. and Feldman D. 2004. Molecular activity of 1,25-dihydroxyvitamin D3 in primary cultures of human prostatic epithelial cells revealed by cDNA microarray analysis. J Steroid Biochem Mol Biol 92: 131–141

Perin T., Canzonieri V., Massarut S., Bidoli E., Rossi C., Roncadin M. and Carbone A. 1996. Immunohistochemical evaluation of multiple biological markers in ductal carcinoma in situ of the breast. Eur J Cancer 32A: 1148–1155

Perou C.M., Sorlie T., Eisen M.B., van de Rijn M., Jeffrey S.S., Rees C.A., Pollack J.R., Ross D.T., Johnsen H., Akslen L.A., Fluge O., Pergamenschikov A., Williams C., Zhu S.X., Lonning P.E., Borresen-Dale A.L., Brown P.O. and Botstein D. 2000. Molecular portraits of human breast tumours. Nature 406: 747–752

Petricoin E.F. and Liotta L.A. 2004. Proteomic approaches in cancer risk and response assessment. Trends Mol Med 10: 59–64

Petricoin E.F., III, Bichsel V.E., Calvert V.S., Espina V., Winters M., Young L., Belluco C., Trock B.J., Lippman M., Fishman D.A., Sgroi D.C., Munson P.J., Esserman L.J. and Liotta L.A. 2005. Mapping molecular networks using proteomics: a vision for patient-tailored combination therapy. J Clin Oncol 23: 3614–3621

Pogosianz H.E. and Prigogina E.L. 1972. Chromosome abnormalities and carcinogenesis. Neoplasma 19: 319–325

Pollack J.R., Perou C.M., Alizadeh A.A., Eisen M.B., Pergamenschikov A., Williams C.F., Jeffrey S.S., Botstein D. and Brown P.O. 1999. Genome-wide analysis of DNA copy-number changes using cDNA microarrays. Nat Genet 23: 41–46

Quackenbush J. 2001. Computational analysis of microarray data. Nat Rev Genet 2: 418–427

Ramaswamy S., Tamayo P., Rifkin R., Mukherjee S., Yeang C.H., Angelo M., Ladd C., Reich M., Latulippe E., Mesirov J.P., Poggio T., Gerald W., Loda M., Lander E.S. and Golub T.R. 2001. Multiclass cancer diagnosis using tumor gene expression signatures. Proc Natl Acad Sci USA 98: 15149–15154

Ramaswamy S., Ross K.N., Lander E.S. and Golub T.R. 2003. A molecular signature of metastasis in primary solid tumors. Nat Genet 33: 49–54

Reis E.M., Ojopi E.P., Alberto F.L., Rahal P., Tsukumo F., Mancini U.M., Guimaraes G.S., Thompson G.M., Camacho C., Miracca E., Carvalho A.L., Machado A.A., Paquola A.C., Cerutti J.M., da Silva A.M., Pereira G.G., Valentini S.R., Nagai M.A., Kowalski L.P., Verjovski-Almeida S., Tajara E.H., Dias-Neto E., Bengtson M.H., Canevari R.A., Carazzolle M.F., Colin C., Costa F.F., Costa M.C., Estecio M.R., Esteves L.I., Federico M.H., Guimaraes P.E., Hackel C., Kimura E.T., Leoni S.G., Maciel R.M., Maistro S., Mangone F.R., Massirer K.B., Matsuo S.E., Nobrega F.G., Nobrega M.P., Nunes D.N., Nunes F., Pandolfi J.R., Pardini M.I., Pasini F.S., Peres T., Rainho C.A., dos Reis P.P., Rodrigus-Lisoni F.C., Rogatto S.R., dos Santos A., dos Santos P.C., Sogayar M.C. and Zanelli C.F. 2005. Large-scale transcriptome analyses reveal new genetic marker candidates of head, neck, and thyroid cancer. Cancer Res 65: 1693–1699

Rhodes D.R. and Chinnaiyan A.M. 2005. Integrative analysis of the cancer transcriptome. Nat Genet 37(Suppl): S31–S37

Richter J., Wagner U., Kononen J., Fijan A., Bruderer J., Schmid U., Ackermann D., Maurer R., Alund G., Knonagel H., Rist M., Wilber K., Anabitarte M., Hering F., Hardmeier T., Schonenberger A., Flury R., Jager P., Fehr J.L., Schraml P., Moch H., Mihatsch M.J., Gasser T., Kallioniemi O.P. and Sauter G. 2000. High-throughput tissue microarray analysis of cyclin E gene amplification and overexpression in urinary bladder cancer. Am J Pathol 157: 787–794

Rihn B.H., Mohr S., McDowell S.A., Binet S., Loubinoux J., Galateau F., Keith G. and Leikauf G.D. 2000. Differential gene expression in mesothelioma. FEBS Lett 480: 95–100

Roversi G., Pfundt R., Moroni R.F., Magnani I., van Reijmersdal S., Pollo B., Straatman H., Larizza L. and Schoenmakers E.F. 2005. Identification of novel genomic markers related to progression to glioblastoma through genomic profiling of 25 primary glioma cell lines. Oncogene 25: 1571–1583

Sallinen S.L., Sallinen P.K., Haapasalo H.K., Helin H.J., Helen P.T., Schraml P., Kallioniemi O.P. and Kononen J. 2000. Identification of differentially expressed genes in human gliomas by DNA microarray and tissue chip techniques. Cancer Res 60: 6617–6622

Samimi G., Manorek G., Castel R., Breaux J.K., Cheng T.C., Berry C.C., Los G. and Howell S.B. 2005. cDNA microarray-based identification of genes and pathways associated with oxaliplatin resistance. Cancer Chemother Pharmacol 55: 1–11

Sanchez-Carbayo M., Socci N.D., Lozano J.J., Li W., Charytonowicz E., Belbin T.J., Prystowsky M.B., Ortiz A.R., Childs G. and Cordon-Cardo C. 2003. Gene discovery in bladder cancer progression using cDNA microarrays. Am J Pathol 163: 505–516

Schena M., Shalon D., Davis R.W. and Brown P.O. 1995. Quantitative monitoring of gene expression patterns with a complementary DNA microarray. Science 270: 467–470

Schmidt H., Hensel M., Dobrindt U., Agerer F., Michaelis K., Janka A., Buchrieser C., Samuelson M., Svanborg C., Gottschalk G., Karch H., Hacker J., Dobrindt U., Wick L.M., Qi W., Lacher D.W. and Whittam T.S. 2004. Pathogenicity islands in bacterial pathogenesis. Clin Microbiol Rev 17: 14–56

Schraml P., Kononen J., Bubendorf L., Moch H., Bissig H., Nocito A., Mihatsch M.J., Kallioniemi O.P. and Sauter G. 1999. Tissue microarrays for gene amplification surveys in many different tumor types. Clin Cancer Res 5: 1966–1975

Seike M., Kondo T., Fujii K., Okano T., Yamada T., Matsuno Y., Gemma A., Kudoh S. and Hirohashi S. 2005. Proteomic signatures for histological types of lung cancer. Proteomics 5: 2939–2948

Selaru F.M., Zou T., Xu Y., Shustova V., Yin J., Mori Y., Sato F., Wang S., Olaru A., Shibata D., Greenwald B.D., Krasna M.J., Abraham J.M. and Meltzer S.J. 2002. Global gene expression profiling in Barrett's esophagus and esophageal cancer: a comparative analysis using cDNA microarrays. Oncogene 21: 475–478

Sengoelge G., Luo W., Fine D., Perschl A.M., Fierlbeck W., Haririan A., Sorensson J., Rehman T.U., Hauser P., Trevick J.S., Kulak S.C., Wegner B. and Ballermann B.J. 2005. A SAGE-based comparison between glomerular and aortic endothelial cells. Am J Physiol Renal Physiol 288: F1290–F1300

Shah T., de Villiers E., Nene V., Hass B., Taracha E., Gardner M.J., Sansom C., Pelle R. and Bishop R. 2006. Using the transcriptome to annotate the genome revisited: application of massively parallel signature sequencing (MPSS). Gene 366: 104–108

Sherry S.T., Ward M.H., Kholodov M., Baker J., Phan L., Smigielski E.M. and Sirotkin K. 2001. dbSNP: the NCBI database of genetic variation. Nucleic Acids Res 29: 308–311

Skacel M., Skilton B., Pettay J.D. and Tubbs R.R. 2002. Tissue microarrays: a powerful tool for high-throughput analysis of clinical specimens: a review of the method with validation data. Appl Immunohistochem Mol Morphol 10: 1–6

Smith D.I. 2002. Transcriptional profiling develops molecular signatures for ovarian tumors. Cytometry 47: 60–62

Somiari R.I., Somiari S., Russell S. and Shriver C.D. 2005. Proteomics of breast carcinoma. J Chromatogr B Analyt Technol Biomed Life Sci 815: 215–225

Soreghan B.A., Yang F., Thomas S.N., Hsu J. and Yang A.J. 2003. High-throughput proteomic-based identification of oxidatively induced protein carbonylation in mouse brain. Pharm Res 20: 1713–1720

Sorlie T., Perou C.M., Tibshirani R., Aas T., Geisler S., Johnsen H., Hastie T., Eisen M.B., van de Rijn M., Jeffrey S.S., Thorsen T., Quist H., Matese J.C., Brown P.O., Botstein D., Eystein Lonning P. and Borresen-Dale A.L. 2001. Gene expression patterns of breast carcinomas distinguish tumor subclasses with clinical implications. Proc Natl Acad Sci USA 98: 10869–10874

Squire J.A., Pei J., Marrano P., Beheshti B., Bayani J., Lim G., Moldovan L. and Zielenska M. 2003. High-resolution mapping of amplifications and deletions in pediatric osteosarcoma by use of CGH analysis of cDNA microarrays. Genes Chromosomes Cancer 38: 215–225

Srinivas P.R., Kramer B.S. and Srivastava S. 2001. Trends in biomarker research for cancer detection. Lancet Oncol 2: 698–704

Stapleton M., Liao G., Brokstein P., Hong L., Carninci P., Shiraki T., Hayashizaki Y., Champe M., Pacleb J., Wan K., Yu C., Carlson J., George R., Celniker S. and Rubin G.M. 2002. The Drosophila gene collection: identification of putative full-length cDNAs for 70% of *D. melanogaster* genes. Genome Res 12: 1294–1300

Staunton J.E., Slonim D.K., Coller H.A., Tamayo P., Angelo M.J., Park J., Scherf U., Lee J.K., Reinhold W.O., Weinstein J.N., Mesirov J.P., Lander E.S. and Golub T.R. 2001. Chemosensitivity prediction by transcriptional profiling. Proc Natl Acad Sci USA 98: 10787–10792

Steel L.F., Shumpert D., Trotter M., Seeholzer S.H., Evans A.A., London W.T., Dwek R. and Block T.M. 2003. A strategy for the comparative analysis of serum proteomes for the discovery of biomarkers for hepatocellular carcinoma. Proteomics 3: 601–609

Stillman B.A. and Tonkinson J.L. 2001. Expression microarray hybridization kinetics depend on length of the immobilized DNA but are independent of immobilization substrate. Anal Biochem 295: 149–157

Stults J.T. and Arnott D. 2005. Proteomics. Methods Enzymol 402: 245–289

Takata R., Katagiri T., Kanehira M., Tsunoda T., Shuin T., Miki T., Namiki M., Kohri K., Matsushita Y., Fujioka T. and Nakamura Y. 2005. Predicting response to methotrexate, vinblastine, doxorubicin, and cisplatin neoadjuvant chemotherapy for bladder cancers through genome-wide gene expression profiling. Clin Cancer Res 11: 2625–2636

Thorisson G.A. and Stein L.D. 2003. The SNP Consortium website: past, present and future. Nucleic Acids Res 31: 124–127

Turesky R.J. 2004. The role of genetic polymorphisms in metabolism of carcinogenic heterocyclic aromatic amines. Curr Drug Metab 5: 169–180

Vasselli J.R., Shih J.H., Iyengar S.R., Maranchie J., Riss J., Worrell R., Torres-Cabala C., Tabios R., Mariotti A., Stearman R., Merino M., Walther M.M., Simon R., Klausner R.D. and Linehan W.M. 2003. Predicting survival in patients with metastatic kidney cancer by gene-expression profiling in the primary tumor. Proc Natl Acad Sci USA 100: 6958–6963

van't Veer L.J., Dai H., van de Vijver M.J., He Y.D., Hart A.A., Mao M., Peterse H.L., van der Kooy K., Marton M.J., Witteveen A.T., Schreiber G.J., Kerkhoven R.M., Roberts C., Linsley P.S., Bernards R. and Friend S.H. 2002. Gene expression profiling predicts clinical outcome of breast cancer. Nature 415: 530–536

Velculescu V.E., Zhang L., Vogelstein B. and Kinzler K.W. 1995. Serial analysis of gene expression. Science 270: 484–487

Velculescu V.E., Zhang L., Zhou W., Vogelstein J., Basrai M.A., Bassett D.E., Jr., Hieter P., Vogelstein B. and Kinzler K.W. 1997. Characterization of the yeast transcriptome. Cell 88: 243–251

Venter J.C., Adams M.D., Myers E.W., Li P.W., Mural R.J., Sutton G.G., Smith H.O., Yandell M., Evans C.A., Holt R.A., Gocayne J.D., Amanatides P., Ballew R.M., Huson D.H., Wortman J.R., Zhang Q., Kodira C.D., Zheng X.H., Chen L., Skupski M., Subramanian G., Thomas P.D., Zhang J., Gabor Miklos G.L., Nelson C., Broder S., Clark A.G., Nadeau J., McKusick V.A., Zinder N., Levine A.J., Roberts R.J., Simon M., Slayman C., Hunkapiller M., Bolanos R., Delcher A., Dew I., Fasulo D., Flanigan M., Florea L., Halpern A., Hannenhalli S., Kravitz S., Levy S., Mobarry C., Reinert K., Remington K., Abu-Threideh J., Beasley E., Biddick K., Bonazzi V., Brandon R., Cargill M., Chandramouliswaran I., Charlab R., Chaturvedi K., Deng Z., Di Francesco V., Dunn P., Eilbeck K., Evangelista C., Gabrielian A.E., Gan W., Ge W., Gong F., Gu Z., Guan P., Heiman T.J., Higgins M.E., Ji R.R., Ke Z., Ketchum K.A., Lai Z., Lei Y., Li Z., Li J., Liang Y., Lin X., Lu F., Merkulov G.V., Milshina N., Moore H.M., Naik A.K., Narayan V.A., Neelam B., Nusskern D., Rusch D.B., Salzberg S., Shao W., Shue B., Sun J., Wang Z., Wang A., Wang X., Wang J., Wei M., Wides R., Xiao C., Yan C., Yao A., Ye J., Zhan M., Zhang W., Zhang H., Zhao Q., Zheng L., Zhong F., Zhong W., Zhu S., Zhao S., Gilbert D., Baumhueter S., Spier G., Carter C., Cravchik A., Woodage T., Ali F., An H., Awe A., Baldwin D., Baden H., Barnstead M., Barrow I., Beeson K., Busam D., Carver A., Center A., Cheng M.L., Curry L., Danaher S., Davenport L., Desilets R., Dietz S., Dodson K., Doup L., Ferriera S., Garg N., Gluecksmann A., Hart B., Haynes J., Haynes C., Heiner C., Hladun S., Hostin D., Houck J., Howland T., Ibegwam C., Johnson J., Kalush F., Kline L., Koduru S., Love A., Mann F., May D., McCawley S., McIntosh T., McMullen I., Moy M., Moy L., Murphy B., Nelson K., Pfannkoch C., Pratts E., Puri V., Qureshi H., Reardon M., Rodriguez R., Rogers Y.H., Romblad D., Ruhfel B., Scott R., Sitter C., Smallwood M., Stewart E., Strong R., Suh E., Thomas R., Tint N.N., Tse S., Vech C., Wang G., Wetter J., Williams S., Williams M., Windsor S., Winn-Deen E., Wolfe K., Zaveri J., Zaveri K., Abril J.F., Guigo R., Campbell M.J., Sjolander K.V., Karlak B., Kejariwal A., Mi H., Lazareva B., Hatton T., Narechania A., Diemer K., Muruganujan A., Guo N., Sato S., Bafna V., Istrail S., Lippert R., Schwartz R., Walenz B., Yooseph S., Allen D., Basu A., Baxendale J., Blick L., Caminha M., Carnes-Stine J., Caulk P., Chiang Y.H., Coyne M., Dahlke C., Mays A., Dombroski M., Donnelly M., Ely D., Esparham S., Fosler C., Gire H., Glanowski S., Glasser K., Glodek A., Gorokhov M., Graham K., Gropman B., Harris M., Heil J., Henderson S., Hoover J., Jennings D., Jordan C., Jordan J., Kasha J., Kagan L., Kraft C., Levitsky A., Lewis M., Liu X., Lopez J., Ma D., Majoros W., McDaniel J., Murphy S., Newman M., Nguyen T., Nguyen N., Nodell M., Pan S., Peck J., Peterson M., Rowe W., Sanders R., Scott J., Simpson M., Smith T., Sprague A., Stockwell T., Turner R., Venter E., Wang M., Wen M., Wu D., Wu M., Xia A., Zandieh A. and Zhu X. 2001. The sequence of the human genome. Science 291: 1304–1351

Veuger M.J., Heemskerk M.H., Honders M.W., Willemze R. and Barge R.M. 2002. Functional role of alternatively spliced deoxycytidine kinase in sensitivity to cytarabine of acute myeloid leukemic cells. Blood 99: 1373–1380 van de Vijver M.J., He Y.D., van ’t Veer L.J., Dai H., Hart A.A., Voskuil D.W., Schreiber G.J., Peterse J.L., Roberts C., Marton M.J., Parrish M., Atsma D., Witteveen A., Glas A., Delahaye L., van der Velde T., Bartelink H., Rodenhuis S., Rutgers E.T., Friend S.H. and Bernards R. 2002. A gene-expression signature as a predictor of survival in breast cancer. N Engl J Med 347: 1999–2009

Warnat P., Eils R. and Brors B. 2005. Cross-platform analysis of cancer microarray data improves gene expression based classification of phenotypes. BMC Bioinformatics 6: 265

von Wasielewski R., Mengel M., Wiese B., Rudiger T., Muller-Hermelink H.K. and Kreipe H. 2002. Tissue array technology for testing interlaboratory and interobserver reproducibility of immunohistochemical estrogen receptor analysis in a large multicenter trial. Am J Clin Pathol 118: 675–682

Weinberg R.A. 1983. Alteration of the genomes of tumor cells. Cancer 51: 1971–1975

Weiss M.M., Snijders A.M., Kuipers E.J., Ylstra B., Pinkel D., Meuwissen S.G., van Diest P.J., Albertson D.G. and Meijer G.A. 2003. Determination of amplicon boundaries at 20q13.2 in tissue samples of human gastric adenocarcinomas by high-resolution microarray comparative genomic hybridization. J Pathol 200: 320–326

Wick L.M., Qi W., Lacher D.W. and Whittam T.S. 2005. Evolution of genomic content in the stepwise emergence of *Escherichia coli* O157:H7. J Bacteriol 187: 1783–1791

Wikman F.P., Lu M.L., Thykjaer T., Olesen S.H., Andersen L.D., Cordon-Cardo C. and Orntoft T.F. 2000. Evaluation of the performance of a p53 sequencing microarray chip using 140 previously sequenced bladder tumor samples. Clin Chem 46: 1555–1561

Wilson I.M., Davies J.J., Weber M., Brown C.J., Alvarez C.E., MacAulay C., Schubeler D. and Lam W.L. 2006. Epigenomics: mapping the methylome. Cell Cycle 5: 155–158

Wolf M., El-Rifai W., Tarkkanen M., Kononen J., Serra M., Eriksen E.F., Elomaa I., Kallioniemi A., Kallioniemi O.P. and Knuutila S. 2000. Novel findings in gene expression detected in human osteosarcoma by cDNA microarray. Cancer Genet Cytogenet 123: 128–132

Wrobel G., Roerig P., Kokocinski F., Neben K., Hahn M., Reifenberger G. and Lichter P. 2005. Microarray-based gene expression profiling of benign, atypical and anaplastic meningiomas identifies novel genes associated with meningioma progression. Int J Cancer 114: 249–256

Wulfkuhle J.D., Paweletz C.P., Steeg P.S., Petricoin E.F., III and Liotta L. 2003. Proteomic approaches to the diagnosis, treatment, and monitoring of cancer. Adv Exp Med Biol 532: 59–68

Yang L.Y., Wang W., Peng J.X., Yang J.Q. and Huang G.W. 2004. Differentially expressed genes between solitary large hepatocellular carcinoma and nodular hepatocellular carcinoma. World J Gastroenterol 10: 3569–3573

Yeatman T.J. 2003. The future of clinical cancer management: one tumor, one chip. Am Surg 69: 41–44

Yi Y., Mirosevich J., Shyr Y., Matusik R. and George A.L., Jr. 2005. Coupled analysis of gene expression and chromosomal location. Genomics 85: 401–412

Zhang L.H. and Ji J.F. 2005. Molecular profiling of hepatocellular carcinomas by cDNA microarray. World J Gastroenterol 11: 463–468

Zhang Z., Bast R.C., Jr., Yu Y., Li J., Sokoll L.J., Rai A.J., Rosenzweig J.M., Cameron B., Wang Y.Y., Meng X.Y., Berchuck A., Van Haaften-Day C., Hacker N.F., de Bruijn H.W., van der Zee A.G., Jacobs I.J., Fung E.T. and Chan D.W. 2004. Three biomarkers identified from serum proteomic analysis for the detection of early stage ovarian cancer. Cancer Res 64: 5882–5890

Zhao X., Weir B.A., LaFramboise T., Lin M., Beroukhim R., Garraway L., Beheshti J., Lee J.C., Naoki K., Richards W.G., Sugarbaker D., Chen F., Rubin M.A., Janne P.A., Girard L., Minna J., Christiani D., Li C., Sellers W.R. and Meyerson M. 2005. Homozygous deletions and chromosome amplifications in human lung carcinomas revealed by single nucleotide polymorphism array analysis. Cancer Res 65: 5561–5570

Zucchi I., Mento E., Kuznetsov V.A., Scotti M., Valsecchi V., Simionati B., Vicinanza E., Valle G., Pilotti S., Reinbold R., Vezzoni P., Albertini A. and Dulbecco R. 2004. Gene expression profiles of epithelial cells microscopically isolated from a breast-invasive ductal carcinoma and a nodal metastasis. Proc Natl Acad Sci USA 101: 18147–18152

4 Bioinformatics Approaches to the Analysis of the Transcriptome of Animal Models of Cancer

Mark J. Hoenerhoff, Aleksandra M. Michalowski, Ting-Hu Qiu, and Jeffery E. Green

ABSTRACT

The development of genetically engineered mouse (GEM) models of human disease have played an integral role in understanding the mechanisms of action of many classes of genes involved in cancer development and progression. Their development has been critical in exploring the complexity of interactions of biological processes occurring in the entire organism, particularly when combined with recent global genomic approaches and bioinformatics. It has become apparent that breast cancer is a heterogeneous disease and multiple GEM models must be incorporated to represent the various forms of the human disease. Undoubtedly, these models and methods will be invaluable in the establishment of biomarkers and novel therapeutic approaches for patients with various subtypes of breast cancer.

Key Words: Mammary cancer, Genetically engineered mouse models, Gene expression profiling, Bioinformatics

1 RELEVANCE OF ANIMAL MODELS

The development of in vivo models of human cancer has been critical to understanding the mechanisms of human disease. Although studies using tissue culture systems in vitro are extremely useful for the dissection of molecular pathways at the cellular level, in vitro systems do not integrate the complexity of interactions occurring within the entire organism, as occurs in vivo in mouse models. Local and distant interactions between multiple cell types can be evaluated in the intact organism (Hennighausen, 2000; Kavanaugh and Green, 2003; Barkan, 2004). Genetically-engineered mouse (GEM) models have provided tremendous insights into the mechanisms of action of many classes of genes involved in cancer development and progression, including oncogenes, tumor suppressor genes, cell cycle regulatory genes, and growth factors (Kavanaugh et al., 2002; Barkan, 2004).

G.J. Gordon (ed.), *Cancer Drug Discovery and Development: Bioinformatics in Cancer and Cancer Therapy*, DOI: 10.1007/978-1-59745-576-3_4, © Humana Press, a part of Springer Science + Business Media, LLC 2009

The use of animal models in medical research can help overcome several difficulties inherent in human studies. The tremendous variation in genetic backgrounds within the human population often limits the interpretation of human studies or requires that very large numbers of patient samples be evaluated. Since GEM models are developed in well-defined genetic backgrounds, differences in genetic heterogeneity can be greatly reduced. Alternatively, crossing mice with a particular phenotype into alternative background strains where the phenotype is changed, is a powerful genetic approach for identifying genes that modify biological behaviors (Balmain and Nagase, 1998; Hunter et al., 2003; Kavanaugh and Green, 2003). In addition, many genetically engineered mice develop tumors after a predictable time period, and therefore the stage-specific alterations in oncogenic pathways, as well as responses to therapeutic agents, can be assessed and potentially translated into corresponding stages of human cancer progression. Such stage-specific studies are generally impossible to perform in human patients, since tumors are generally diagnosed at later stages and multiple sampling is not clinically justified (Kavanaugh and Green, 2003; Barkan, 2004). In this way, GEM models have been important in defining molecular signaling pathways associated with cancer promotion and progression, making GEM models important for use in assessing stage-specific responses in preclinical testing and prevention studies.

The introduction of specific genetic alterations in mice has led to scores of mouse tumor models. Many genes associated with breast cancer in humans can lead to mammary tumor formation when dysregulated in mice (Cardiff and Wellings, 1999; Cardiff, 2001, 2003; Barkan, 2004), and importantly, some of these tumors share morphological characteristics that are similar to lesions that occur in humans (Cardiff and Wellings, 1999; Cardiff, 2001, 2003). Although the use of transgenic mouse models is a powerful tool to study similar processes that occur in human cancer, there are still many limitations to these models. Truly representative disease models should reflect both the genotypic and phenotypic changes present in the human disease. Unfortunately, this is not often the case. Animal models of human cancer need to be evaluated based upon their similarity with the correlate human disease, and criteria for validation of these models need to be based on several comparative components (Fig. 1).

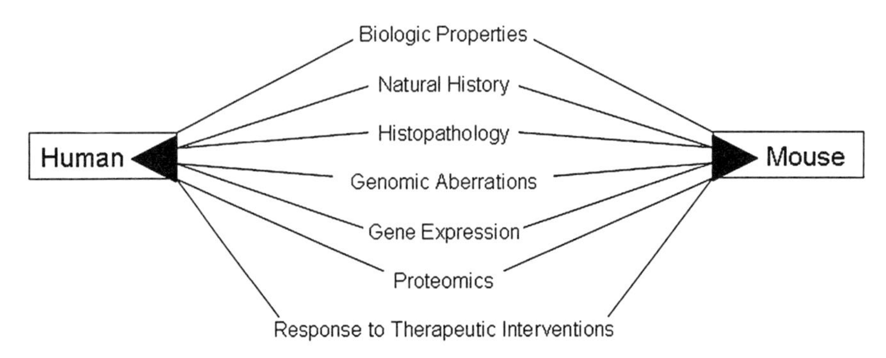

Fig. 1. Criteria for model validation. Relevant models of cancer should recapitulate important aspects of the human disease

The advent of new genomic technologies, such as global gene expression profiling and comparative genomic hybridization, provides an important means to compare the similarities and differences in gene expression and cancer evolution between mouse models and the human diseases that they are designed to represent. However, the extent to which these GEM models recapitulate the molecular pathways involved in human cancers on a global scale is only beginning to be assessed (Ried et al., 1995; Lee et al., 2005; Sweet-Cordero et al., 2005).

2 DESIGNING MICE IN OUR OWN MOLECULAR IMAGE

A variety of GEM models have been produced to recapitulate specific mutations that occur in the progression of human cancer. Importantly, GEM models have been engineered to mimic the human disease through (1) overexpression or activation of genes associated with human cancer development (oncogenes), (2) elimination of target (suppressor) genes through gene knock-out strategies, (3) generation of dominant negative proteins to disrupt the function of genes (Hutchinson and Muller, 2000; Kavanaugh and Green, 2003), and (4) combinations of the above. Many transgenic mouse models have been developed by targeting the misexpression of proto-oncogenes, growth factors, survival and regulatory pathways, and transcription factors (Table 1), although other mechanistic targets have also been used. In addition to transgenic animals that carry single genetic alterations, GEM models have been generated that carry compound genetic alterations to examine the interactions between multiple oncogenic pathways. The use of knock-out and knock-in strategies and conditional tissue-specific gene targeting has led to the creation of models that are more representative of the human disease and has further defined the biological functions of hundreds of genes (Hutchinson and Muller, 2000; Green and Hudson, 2005).

Gene knock-out strategies are used to mutate or eliminate a gene in the germline or in targeted somatic cells in order to evaluate the functional role of that gene, which often serves as a tumor suppressor in cancer promotion or progression (Hutchinson and Muller, 2000). Simple targeted knock-out models result in germline mutations that often result in embryonic lethality (Hutchinson and Muller, 2000; Green and Hudson, 2005). To overcome the problem of embryonic lethality and study gene function in a particular

Table 1
Examples of molecular pathways successfully targeted to generate transgenic mouse models of mammary cancer

Proto-oncogenes	*c-myc, c-neu, c-ha-ras*
Growth factors	FGF, TGFα, HGF
Growth factor receptors	ErbB2 (her2neu), RET-1, ERα
Signaling kinases/phosphatases	Ras, PTEN, Akt
Cell cycle mediators/DNA repair	p53, Cyclin D1, BRCA
Differentiation factors	Notch, Wnt, P-Cadherin

tissue, tissue-specific and temporally regulated conditionally induced mutations can be employed using *cre*—loxP technology, allowing genetic alterations to be introduced in specific tissues at particular developmental stages in the animal (Sauer, 1998; Hutchinson and Muller, 2000; Grisendi and Pandolfi, 2004).

In the *cre*—loxP system, the germline can be altered to include a pair of specific nucleotide sequences (loxP) that flank an engineered internal sequence; the loxP sites recombine in the presence of *cre* recombinase, thus excising the intervening internal sequence (Sauer, 1998; Hutchinson and Muller, 2000). To turn certain genes or mutations on or off in a tissue-specific and temporally controlled manner, variations of this system have been developed (Sauer, 1998; Gunther et al., 2002). In addition, constitutive tissue-specific gene-targeted models allow targeted deletion or constitutive overexpression of a gene in a particular tissue of interest, more reflective of the human disease.

Other conditional and targeted alterations in gene expression are designed to induce silence or overexpress a target gene in the presence of tetracycline or doxycycline (Gossen et al., 1995; Gunther et al., 2002). This method provides another powerful way of creating mouse models with genetic alterations that can be regulated at specific developmental time points or in the context of additional genetic alterations. The recent advent of advanced transgenic techniques using bacterial artificial chromosome (BAC) recombineering allows for the rapid introduction of targeted mutations into BAC clones (bacterial vectors that contain up to 200 kb of genomic DNA) (Copeland et al., 2001; Abe et al., 2004). New generations of mouse models using this technology will offer additional advantages in modeling human cancer in mice, particularly related to mimicking endogenous regulation of a genetic locus.

3 MODEL VALIDATION

Advanced genomic technologies have made possible the creation of mouse models that represent human diseases better. The relevance of an animal model for studying human disease needs to be based upon biologic validation, histopathologic validation, and molecular validation (Fig. 1).

3.1 Biologic Validation

Biologic validation of an animal model is based on how well the pathophysiology of the disease process in the mouse mimics that of the human condition. The natural history of the disease, including time course of progression, hormone responsiveness, metastatic rate, sites of metastasis, and so on, provides an important gauge to compare the biologic behavior of an animal model with that of the human disease. For instance, overexpression of ErbB2 has been reported in 25–30% of human breast cancers, and is associated with a high rate of metastasis. Similarly, in mouse models overexpressing her2neu in the mammary epithelium, solid mammary tumors develop that metastasize to the lungs (Jolicoeur et al., 1998; Jager et al., 2005). As another example, deletion of exon 11 of BRCA1 in the mouse results in a clinical progression similar to the human disease, where mammary tumors form after a long latency (Xu et al., 1999; Weaver et al., 2002). However, often there is relatively little information reported about the natural course of disease progression in GEM models (Cardiff et al., 2000).

Another means of assessing biologic validation is through the use of biomarkers (Cardiff, 2001). Similarities in immunohistochemical staining of comparable biomarkers in the mouse mammary gland and the respective human gland should be used to provide additional information on the appropriateness of the model for use in comparative research (Cardiff et al., 2000; Cardiff, 2001). Markers also potentially provide important insights into the cellular origins of tumors.

However, there are several important biological differences between the species that must be considered when using GEM models. For example, while more than half of human breast cancer is estrogen receptor (ER) positive, few GEM models develop ER positive tumors (Nandi et al., 1995; Cardiff et al., 2000; Cardiff, 2001). Recently, however, a model has been generated in which the p53 gene has been deleted, where tumors that develop are both ER positive and ER negative, recapitulating a critical aspect of human breast cancer biology (Medina et al., 2001; Lin et al., 2004). Additionally, promoters used to target transgenic expression in mice are not likely to induce levels of expression that occur in human cancers. For example, the WAP promoter and MMTV-LTR are hormonally regulated promoters, which may confound studies of oncogenesis in the hormone-responsive mammary gland (Hutchinson and Muller, 2000; Cardiff, 2001). Pregnancy enhances tumorigenesis in mice carrying transgenes using these promoters, whereas early pregnancy has a protective effect against breast cancer in women (Cardiff, 2001; Medina et al., 2001).

Metastatic breast cancer in mouse models is significantly different than what is observed in human patients. Although tumors from some GEM models metastasize to the lung, breast cancer in humans most often metastasizes to regional lymph nodes, and also commonly to the liver, lung, and bone (Cardiff and Wellings, 1999; Cardiff et al., 2000; Cardiff, 2001, 2003; Green et al., 2004). In the mouse, spread to regional lymph nodes is infrequent, and no mammary cancer models have yet been developed which demonstrate spontaneous metastatic mammary tumor spread to the bone (Cardiff and Wellings, 1999; Cardiff et al., 2000; Cardiff, 2001, 2003; Green et al., 2004).

3.2 Histopathologic Validation

The histopathologic characterization of certain types of cancer in GEM models has been well characterized and defined. For example, in GEM models of breast cancer, the current classification scheme employs a descriptive nomenclature incorporating a series of morphologic descriptors modified by biological potential, topographical distribution, histologic pattern, cytologic grade, inducers (etiology), and clinical contexts, when known (Cardiff et al., 2000). Similar histopathologic classification schemes have been applied for characterization of GEM tumors at various other organ sites as well (Weiss et al., 2002; Nikitin et al., 2004; Shappell et al., 2004). However, similar to the gene expression analysis and results of hierarchical clustering of mouse models in the study by Desai et al. (2002b), many transgenic mouse mammary tumors can been classified based on "signature tumor patterns." Signature tumor patterns are seen in association with transgenes for erbB, myc, ras, and ret-1, and can be differentiated from one another based on the particular histologic pattern and cytologic appearance (Cardiff et al., 2000). Similarly, some signature tumor phenotypes have been seen in human breast cancer, such as comedo ductal carcinoma in situ (DCIS), which possesses unique cellular atypia and central necrosis (Barnes et al., 1993; Cardiff et al., 2000), and also morphologic features that distinguish between BRCA1 and BRCA2 tumors (Lakhani et al., 1998; Cardiff et al., 2000). Importantly, the unique morphology of these tumors indicates that the biological phenotype can be correlated with a particular gene expression pattern (Cardiff and Wellings, 1999; Cardiff et al., 2000). Information on classification of mouse mammary tumors and signature tumors in transgenic mouse models of breast cancer is available at http://mammary.nih.gov/atlas/histology/jax-workshop/index.html.

3.3 Molecular Validation

Genomic technologies applied to mouse models of human cancer provide important information regarding the cancer transcriptome. This furthers our understanding of cancer pathogenesis, oncogene-specific expression signatures, potential molecular targets for therapies, and reveals valuable information in helping to select appropriate mouse models for specific experimental purposes (Kavanaugh et al., 2002; Kavanaugh and Green, 2003; Green et al., 2004; Ye et al., 2004). Gene expression profiling is a powerful tool that has advanced the classification and understanding of the molecular basis of cancer, and has identified new diagnostic categories not previously available through standard histopathologic methods of grading tumors (Perou et al., 1999; Luzzi et al., 2001; Liu, 2003; Seth et al., 2003; Sorlie et al., 2003; Cleator and Ashworth, 2004). Morphologically similar tumors can be molecularly subtyped, and correlations between the natural progression of the disease and the molecular subtype or response to therapy can be made (Perou et al., 1999; Sorlie et al., 2003). Microarray technology utilizing expression profiling will more completely define mouse models based on their molecular pathways, providing a more complete picture for comparison to the human disease (Fargiano et al., 2003; Ye et al., 2004). By comparing gene expression between transgenic models through array data analysis, researchers can evaluate each model in terms of its genetic relationship to the human disease and determine which model is most appropriate for study based on their molecular profile (Kavanaugh et al., 2002; Kavanaugh and Green, 2003).

Initial studies from our laboratory revealed that although tumors resulting from transgenic manipulation share many molecular characteristics (including induction of cell cycle regulators, glycolytic pathways, metabolic regulators, zinc finger proteins and protein tyrosine phophatases), distinct patterns of gene expression can be identified based upon the initiating oncogene. This provides a "signature" gene expression pattern that discriminates each model based upon the initiating oncogenic event (Desai et al., 2002b; Kavanaugh et al., 2002; Kavanaugh and Green, 2003). For example, common and oncogene-specific events associated with mammary carcinogenesis have been identified in c-myc, c-neu, c-ha-ras, PyMT, and SV40 transgenic mouse models (Desai et al., 2002b), and more recently in p53 and BRCA1 knock-out models (A.M. Michalowski et al., unpublished data). Models can be separated into several distinct groups based on oncogene-specific differential gene expression. All MMTV-myc tumors clustered into one group, while all C3(1)-Tag and WAP-Tag tumors clustered into a second, and the MMTV-neu, MMTV-ras, and MMTV-PyMT clustered into a third (Desai et al., 2002b).

Comparative gene expression profiling of oncogene-derived tumors will broaden the understanding of oncogene-specific pathways, a study difficult to perform in the human disease owing to the potential multiplicity of genetic changes associated with human breast cancer (Desai et al., 2002b). This data is available at http://gedp.nci.nih.gov/dc/index.jsp. For human patients, some studies have used similar strategies comparing multiple gene expression patterns rather than a single gene expression pattern to add power in defining characteristics of an individual disease, and predicting clinical outcomes, in order to resolve biological heterogeneity (Nevins et al., 2003). By combining multiple gene microarray expression patterns and clinical risk factors, work on these strategies can characterize the individual patient profile, leading to the ultimate goal of personalized medicine (Nevins et al., 2003).

4 STATISTICAL CONSIDERATIONS FOR ARRAY DATA ANALYSES

Strategies for high-level analysis for genomic data can be broadly categorized into unsupervised and supervised approaches (Fig. 2). Unsupervised methods are intended for the discovery of hidden taxonomies (class discovery) and require only the gene expression measurements obtained from the experimental data. A number of statistical and computational techniques exist for discovering group patterns in microarray data, which are very useful exploratory and visualization methods for identifying molecular subtypes of tumors and groups of similarly regulated genes associated with a particular state of the transcriptome. In contrast to unsupervised gene expression analyses,

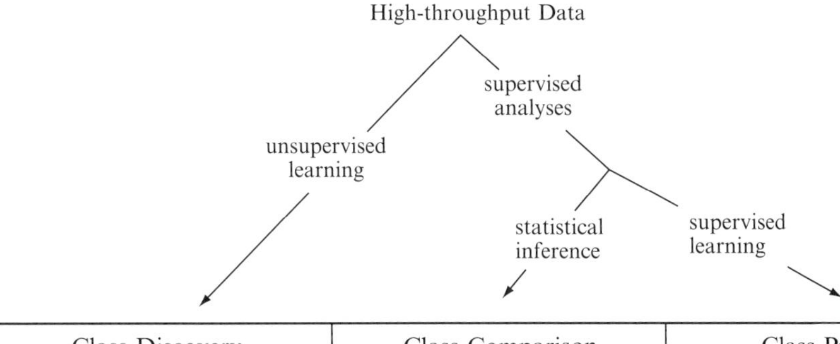

Class Discovery	Class Comparison	Class Prediction
Identification of molecular phenotypes:	Identification of differential gener expression:	Identification of expression signature predictive of a given state or outcome:
•Oncogene-specific classification of mouse cancer models	•Tumor vs.normal tissue gene expression levels	•Cross-species prediction of the tumor subtype
•Discovery of cell-of-origin cancer profiles	•Changes in gene expression level according to treatment	•Prediction of the response to therapeutic targets
•Finding human-like mouse models of cancer	•Gene expression changes during cancer progression (multiple time points)	•Prediction of a functional gene class

Fig. 2. Major types of statistical approaches used to analyze microarray data

supervised methods require a priori information that characterizes a particular class from which the expression profiles are derived (such as the designation "normal" or "tumor" for the tissue sample, or a particular gene ontology for supervising an analysis based upon gene function). Supervised approaches can be utilized to uncover differential gene expression among prespecified classes of interest (class comparison) or used to generate a mathematical model to predict the characteristic of interest in a future specimen based solely on its gene expression profile (class prediction).

The unsupervised learning approach has been used to obtain a simplifying abstraction of high-dimensional microarray data. Various clustering techniques are the most commonly applied for this purpose. The analytical goal of clustering is to find homogeneous groups of samples or genes that share common expression patterns within the same cluster that are more similar to one another than to those contained in different clusters. The clustering process involves feature selection and/or transformation for the appropriate pattern representation, considering proximity measure and an algorithm to group the samples or genes. Selection of the subset of array features for unsupervised learning is supposed to filter out genes that are not meaningful to the clustering, but allow for "natural" grouping in the data to emerge. The selection improves the classification performance by reducing uninformative noise and dimensionality in the data, but does not define any ad hoc classification. This is usually done by excluding those genes whose normalized expression is flat across the clustered samples. However, specific computational algorithms that identify noisy or redundant features in an unsupervised classification context have also been developed for this purpose (Roth and Lange, 2004).

Generally, gene expression is median or mean centered across the arrays prior to clustering to remove dominant effects from the most highly expressed genes on the proximity metric and the clustering result. In a dual channel setting, gene centering additionally eliminates the dependence of the ratios on the amount of expression of a gene in the reference sample. A concept of distance (or similarity as its inverse) is fundamental to clustering and is usually measured by pairwise distances between objects in the data. The Euclidean distance or one minus the Pearson correlation is the proximity metrics usually used for clustering microarray data. As the Euclidean distance measures the absolute distance, a stringent proximity of two profiles is required for them to cluster together. However, with a correlation-based metric, two profiles can be grouped together as long as their expression profiles can be approximated by a linear function (scalar multiple and shift) even though the absolute distance between the expression profiles remains large. The main division of grouping algorithms for producing clusters distinguishes partitional clustering resulting in a single partition of the profiles (*K*-means clustering, self-organizing maps) and hierarchical clustering resulting in a set of nested partitions (agglomerative or divisive). Further descriptions of the many clustering methods available are beyond the scope of this chapter and can be found elsewhere (Jain, 1999; McShane et al., 2003; Simon, 2003).

An example of a class discovery approach to the study of mammary cancer is illustrated by the results depicted in the agglomerative hierarchical clustering analysis in Fig. 3a, b. Hierarchical clustering can be presented graphically as a tree structure referred to as a dendrogram. Building the dendrogram using hierarchical clustering requires, in addition to the proximity distance specified for pairs of profiles, a linkage method to find the distance between merging clusters. Single linkage defines the distance

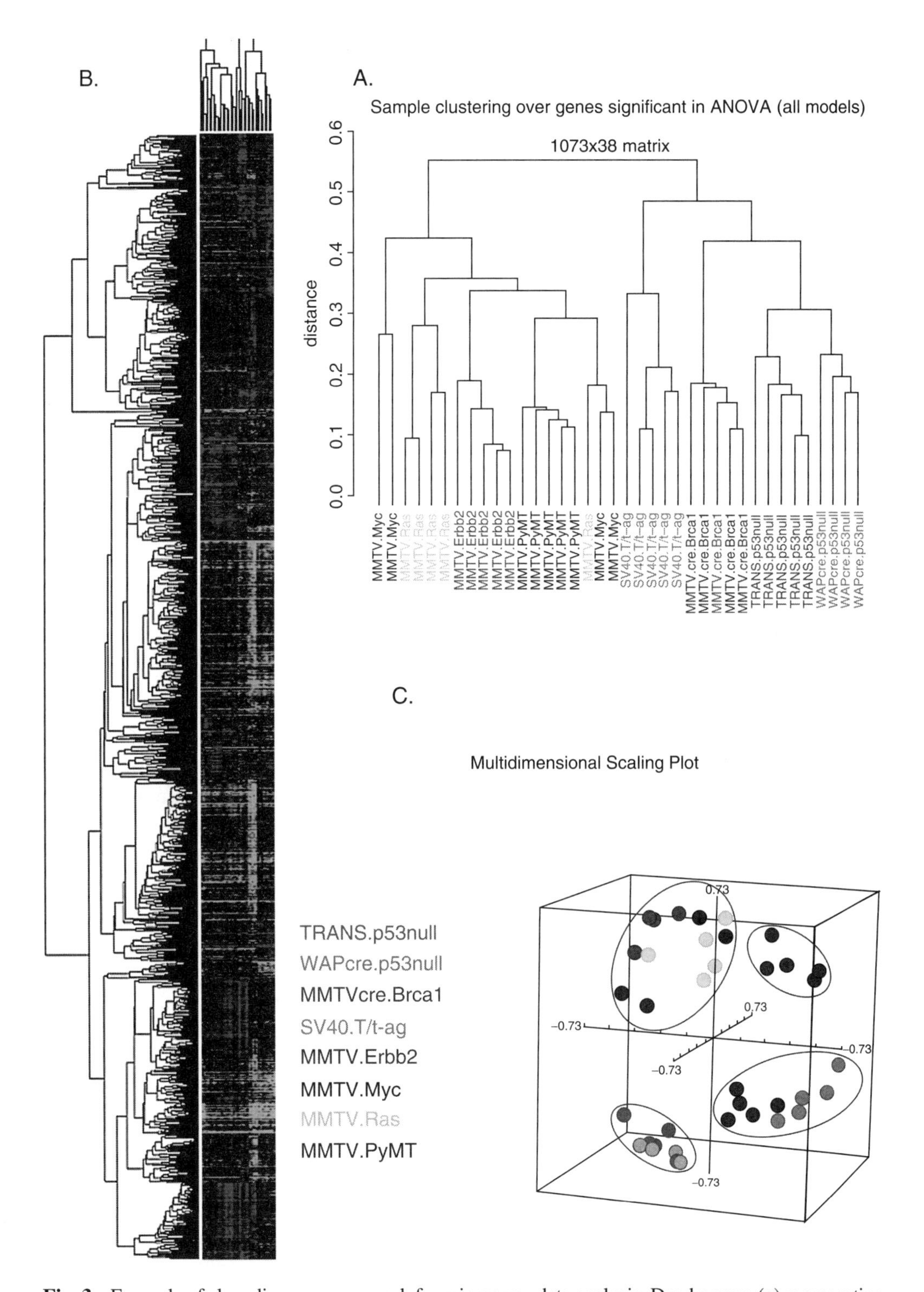

Fig. 3. Example of class discovery approach for microarray data analysis. Dendrogram (**a**) representing agglomerative hierarchical cluster analysis (HCA) (**b**) and multidimensional scaling (MDS) (**c**) of median centered normalized log 2 expression ratios for 1,073 genes selected by univariate *F* test ($p >$ 0.001) for mammary tumor RNA samples from 38 mice representative of eight different mouse models of breast cancer. The Pearson correlation distance metric was used for the HCA and MDS. Average linkage was used to construct the dendrogram: (**a**) sample clustering; (**b**) heatmap representation of the two-way HCA, *rows* representing genes, *columns* representing tumors; *red/green* indicates overexpressed/downexpressed genes in mammary tumor compared to the normal mouse mammary RNA; and (**c**) MDS display for the same samples. BRB-array tools: http://linus.nci.nih.gor/pilot/index.html

between any two clusters as the minimum distance between them; complete linkage is the opposite of single linkage in that it defines the distance between any two clusters as the maximum distance between them. Average linkage uses the mean distance between all possible pairs of members of the two clusters to be merged. The agglomerative algorithm starts with placing each object in its own single cluster and proceeds by merging objects or groups of objects in the order of greatest similarity until all objects and clusters are merged into a single cluster. Figure 3a, b depicts the results from hierarchical clustering analysis, which identifies distinctions between mammary models attributable to oncogene-specific patterns of gene expression.

Another method often used to uncover relationships in microarray data is multidimensional scaling (MDS). MDS represents a set of mathematical techniques used to visualize graphically the structure of multivariate data projected into low-dimensional space. MDS is a method of reducing the dimensionality of data while preserving pairwise distances from the original proximity metric between objects. Principal components (weighted linear combinations of the original variables that capture maximum variance and are independent of each other) may be used to determine the distances between the points in the three-dimensional display. Figure 3c presents an MDS plot for the same data used to generate the hierarchical clusters/dendrogram in Fig. 3a, b. The MD representation further confirms the data structure obtained with the sample hierarchical dendrogram.

GEM models are especially apt for microarray data studies, which aim to uncover genes differentially expressed among the experimental groups (class comparison). Commonly, univariate parametric or nonparametric statistical tests are performed with an account for multiple null hypotheses being tested at a time. As one example Student *t* tests or Wilcoxon tests might be used to identify genes differentially expressed between the tumor and normal mammary tissue in a transgenic mouse model. The individual *p* value measures the probability that the observed difference in gene expression arises by chance only and was designed to control for false positive rate of a test (say $\alpha = 0.05$). When testing multiple hypotheses there is always the possibility that some of the tests have appeared significant just by chance. If in our example a univariate *t* statistic was applied separately to $N = 20{,}000$ genes with $\alpha = 0.05$ (the number of array probes often available), 1,000 genes would be expected to be declared differentially expressed even if no true differences existed between the mammary tumor and normal tissue (αN false positives). Various procedures have been in use to adjust the *p* values or to otherwise correct for multiple testing issues to avoid an abundance of spurious findings. Some of them guard against any false positives, like the Bonferroni correction (usually too conservative for microarray data) or the multivariate permutation adjustment (Callow et al.,). Most often the expected number or proportion of false discoveries among the declared positives is controlled, like in the SAM procedure (Tusher et al., 2001). Multivariate permutation approaches exist that allow for a chosen number or proportion of false discoveries to be controlled with high confidence, e.g., to have 95% confidence that among the genes identified as significant there is no more than 10% of false positives (Korn, 2004).

Microarray technology generates massive amounts of data in terms of the number of variables, but with a relatively very small sample size of biological replicates. The consequence of this is often low statistical power for detecting genes differentially expressed and high false discovery rates. The standard power calculation for the required sample size of an experiment takes into account the desired detectable change in expression, magnitude of population variability, the chosen power to detect the expression change and an acceptable error rate (significance level). Due to genetic background the variability in gene expression is typically much lower in animal models and cell lines compared to that for human subjects where genetic diversity is quite high. Additionally, the variance component alternates across genes and in general low expressers are the most variable. Wei et al. (2004) showed that the number of subjects needed to gain the same power as with inbred animals is much smaller compared to groups of unrelated human individuals. Here we present an example of the variability in a few mouse and human microarray datasets, each profiling two groups of primary mammary tumors (see Table 2 for the data sets description). Figure 4 depicts the cumulative distribution of common standard deviation estimated from the analyzed data sets showing the shift of variability between the mouse and human profiles. In Table 2, we present the estimated power and sample size required for the mouse and human data to detect differential expression in microarray studies for 25, 50, and 75% of the least variable genes with the two-sample *t* test. A sample size of only five independent biological replicates per group from the mouse ensures very high power (97—98%) to detect twofold differences with a 0.001 false positive rate for 50% of the least variable genes, while using five human

Table 2
Power and sample size calculations to detect twofold change in expression with a given false positive rate (α = 0.001) for 25, 50, and 75% least variable genes in the mouse and human microarray datasets

Data set	A			B			C			D		
Percentile	25th	50th	75th	25th	50th	75th	25th	50th	75th	25th	50th	75th
Standard deviation (σ)	0.12	0.19	0.33	0.14	0.20	0.32	0.27	0.34	0.46	0.22	0.29	0.42
Sample size (n=5) **Power ($1 - \beta$)**	0.99	0.98	0.48	0.99	0.97	0.52	0.73	0.45	0.18	0.92	0.65	0.24
Power ($1 - \beta = 0.95$) **Sample size (n)**	3	5	9	4	5	8	7	9	14	6	7	12

The power and sample calculations were performed using R version 2.1 for two-sample two-tailed *t* test and equal number of subjects per group. The expression data were obtained with Affymetrix oligonucleotide arrays and in the preprocessing the robust multiarray average (RMA) method (Irizarry et al., 2003) and quantile normalization were applied to each data set to produce normalized log base 2 gene summary measures

Data set A: mouse model: $p53^{-/-}$;transplant (Jerry et al., 2000); classes: metastatic (n = 14) and nonmetastatic (n = 10) primary mammary tumors; Affymetrix chip: U74Av2

Data set B: mouse model: $p53^{fp/fp}$;WAP*cre* (Lin et al., 2004); classes: ERα positive (n = 5) and ERα negative (n = 5) primary mammary tumors; Affymetrix chip: U74Av2

Data set C: human data: West et al. (2001); classes: ERα positive (n = 25) and ERα negative (n = 24) primary mammary tumors; Affymetrix chip: HuGeneFL

Data set D: human data: Huang et al. (2003a,b); classes: ERα positive (n = 73) and ERα negative (n = 14) primary mammary tumors; Affymetrix chip: U95Av2

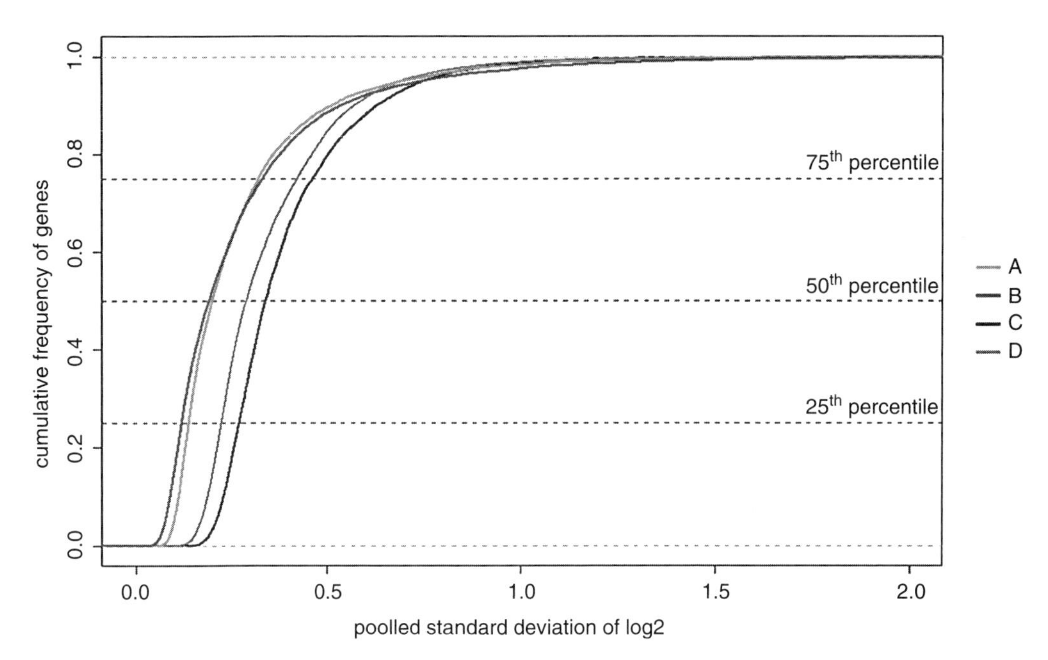

Fig. 4. Cumulative distribution of common standard deviation in the mouse and human mammary tumor groups. This graph demonstrates the empirical distribution function (cdf) for the common standard deviation of log 2 intensity pooled from mammary tumor groups within each of the mouse and human datasets. The empirical cdf $F(x)$ is defined as the proportion of x values less than or equal to x. The x-axis is the standard deviation and the y-axis is the proportion of genes that has the standard deviation below the value of x. The estimated distributions of the standard deviation from the human data sets are shifted towards larger values in relation to those from the mouse data sets. For example, in the mouse data 75% of the genes have the standard deviation of log 2 intensities no greater than 0.32–0.33, while in the human data set no greater than 0.42–0.46. Data sets A–D as described in Table 2

samples results in a tremendous loss of statistical power (45 and 65%). Importantly, new approaches for calculating the statistical power and sample size for microarray data have been developed, which take into account multiple testing and false discovery rates (Pawitan et al., 2005; Gadbury, 2004).

Supervised learning with microarray data involves feature selection, class predictor design and validation. For building a predictor, usually the subset of differentially expressed genes between the classes is selected with the assumption that they are potentially relevant to the class distinction. Sometimes, dimension reduction methods (e.g., principal components) are used to identify the predictive genes instead of a univariate statistic (Khan et al., 2001; West et al., 2001) The next step is to identify an algorithm that will correctly associate a gene expression profile with the prespecified class label based on a set of samples used for training. Numerous algorithms for classifier training have been developed for microarray data (Simon, 2003), including the Fisher linear discriminant analysis, weighted method, compound covariate predictor, nearest neighbor classification, support vector machine and neural networks. These methods for multivariate classifier discovery were available before the advent of microarray technology and designed to model a situation where the number of samples substantially exceeds the number of variables in the classifier. As a consequence, genomic classifiers are very prone to biased overestimation of their predictive accuracy — they can easily predict class membership extremely well in the training dataset, but perform poorly when applied to independent data. To properly estimate the accuracy of a predictor for future samples, split-set validation or crossvalidation are most often used to separate the process of classifier building and validation of its performance (Simon, 2003, 2004).

Class prediction methods are widely used in biomedical and clinical studies involving human microarray data, where development of prognostic and diagnostic genomic classifiers are of great interest. The use of supervised classification of gene expression profiles in mouse models is less frequent, yet has been used in cross-species comparisons as a tool for validation of the cancer models and the identification of expression signatures conserved across species and deregulated in cancer (Lee et al., 2005; Sweet-Cordero et al., 2005). The use of class prediction to forecast the response to drug treatment in mouse models is likely to be of increasing importance in the future (Huang et al., 2003b; Bild et al., 2006).

5 COMPARING MOUSE AND HUMAN EXPRESSION DATA AND NETWORK ANALYSIS

In order to accurately compare mouse models of human cancer with the human disease they are intended to represent, genomic data from each species has to be appropriately translated. This remains an important, but largely unexplored aspect of determining whether the molecular changes observed in mouse models actually reflect the disease process in humans. Generating mouse models based solely on a genetic mutation found in human cancer will not necessarily produce a similar outcome in the model. Likewise, although certain mouse models reproduce similar morphological disease, it is not clear whether these models fully recapitulate the molecular pathways that are altered in human cancer (Callahan and Smith, 2000; Lee et al., 2005; Sweet-Cordero et al., 2005).

It has been hypothesized that gene expression signatures representing similar phenotypes could be conserved in evolutionarily related species (Lee et al., 2005; Sweet-Cordero et al., 2005). By aligning orthologous genomic sequences from different species, evolutionarily conserved sequences have been identified that may represent functional regulatory elements (Eddy, 2002; Cooper, 2003; Ureta-Vidal et al., 2003; Lee et al., 2005). Similarly, expression profiling can help identify conserved molecular pathways operating in the transcriptome during development of cancer in mice and human (Lee et al., 2005). Interspecies comparison of global gene expression such as this is extremely valuable in identifying expression signatures that may be obscured by direct analysis of either a human or a mouse tumor data set alone (Sweet-Cordero et al., 2005). Additionally, such analyses are a fundamental aspect of model validation (Lee et al., 2005; Sweet-Cordero et al., 2005).

Following similar principles in which human breast tumors may be subclassified based on gene expression patterns into groups corresponding to biologic behavior and prognosis (van de Vijver et al., 2002; Sorlie et al., 2003), mouse models of mammary cancer can be categorized with subgroups of human breast cancer based on gene expression data (A.M. Michalowski et al., unpublished data; Herschkowitz et al., 2007). Patterns of gene expression generated from cDNA microarrays have been found to be associated with ER status and tumor grade in humans (Sotiriou et al., 2003). For example, a group of genes differentially expressed in ER positive and negative tumors has the ability to discriminate tumors based on ER status (Gruvberger et al., 2001; Sotiriou et al., 2003). Similarly, mouse tumors can be clustered with human ER breast cancer subgroups; the basal human/mouse ER negative group, luminal type A human/mouse ER positive group, and luminal type B human ER positive/mouse ER negative group (A.M. Michalowski et al., unpublished data). Based on luminal and basal phenotype markers, ER+ WAP-cre;p53fp/fp tumors fall into a subgroup representing ER+ human luminal type A class of tumors, ER negative mouse tumors with dysfunctional p53 cluster in the basal group, and MMTV-oncogene driven tumors cluster with the human luminal type B phenotype. This method of classification allows a more accurate means of comparing mouse models with subgroups of human breast cancer. This is a powerful tool to identify conserved molecular mechanisms of oncogenesis, potential therapeutic agents, and useful prognostic profiles (Lee et al., 2005; Sweet-Cordero et al., 2005; Herschkowitz et al., 2007).

Interspecies comparative gene expression analysis is also valuable when using the mouse gene expression data sets to identify biologically important gene networks in human tumor samples. For example, common expression patterns can be found between mouse and human data sets based upon estrogen receptor expression status using a supervised analytic approach. The integration of human breast cancer sample data sets with that of mouse tumors from a genetically defined background resulted in the identification of a gene set that distinguishes ER positive from ER negative tumors in humans (A.M. Michalowski et al., unpublished data). Interestingly, due to the genetic homogeneity of the mouse background, the predictor for ER status generated by combining the mouse and human expression data had a 95% accuracy rate of human tumor classification and was significantly more robust than the predictors generated from only the mouse (55% accuracy) or human (85% accuracy) data, strongly suggesting functionally conserved expression pathways related to ER status between the two species. This unique cross-species classifier of ER status was 100% accurate for classifying the mouse tumors.

In addition, the mouse/human classifier generates significantly more robust genetic network interactions than does the human classifier. These results demonstrate that mouse models of human breast cancer can be classified with subtypes of human breast cancer, and that a specific group of genes that distinguishes ER positive and negative tumors is conserved between mice and humans. The fact that the mouse/human classifier worked best demonstrates the added value of using mouse data from a relevant mouse model to help define certain biologic properties.

6 USING MOUSE MODELS TO STUDY CANCER EVOLUTION

Changes seen in gene expression through tumor progression such as from early preinvasive lesions such as ductular carcinoma in situ (DCIS) and atypical ductular hyperplasia (ADH) to invasive carcinoma and metastatic disease, have been studied with DNA microarray technology in an attempt to characterize the changes that occur throughout the initiation and progression of the disease (Porter et al., 2001; Ye et al., 2004). In this way, researchers can identify important genes and pathways of oncogenesis that may result in development of advanced diagnostic and treatment methodologies.

6.1 Preinvasive Lesions vs. Primary Breast Cancer

Patterns of coexpressed clusters of genes have been identified in breast cancer cell lines and tumors that relate to specific features, such as histologic subtype and biological behavior (Perou et al., 1999). Historically, DCIS lesions have been characterized by histopathology, and epidemiologic data is used to estimate prognosis (Aubele et al., 2002). However, gene expression analysis of DCIS lesions identifies differences in gene expression compared to normal mammary epithelium, as well as invasive ductal carcinoma (Aubele et al., 2002; Ma et al., 2003; Seth et al., 2003). For example, overexpression of the c-erbB2 oncogene has been found in DCIS, but not in ductal hyperplasia or ADH (Aubele et al., 2002). Interferon inducible genes have also been found to be overexpressed in DCIS compared to other stages of development (Seth et al., 2003). However, although differences in gene expression have been reported for DCIS in comparison to other lesions, there do not appear to be genes or gene profiles specific for DCIS (Porter et al., 2001).

Using laser capture microdissection (LCM), T7-based RNA amplification, and cDNA microarrays, the gene expression profiles of premalignant (ADH), preinvasive (DCIS), and invasive ductal carcinoma (IDC) have been generated in humans as well as mouse models of the disease (Fuller et al., 2003; Ma et al., 2003; Ye et al., 2004). Remarkably, analysis of microarray data obtained in these studies showed that epithelial-specific expression profiles of all stages of human breast cancer, including preinvasive and invasive stages shared significant similarities, suggesting that a majority of the genetic alterations contributing to the initiation and progression of breast cancer are inherently present at the early stages of the disease, and persist throughout its progression (Fuller et al., 2003; Ma et al., 2003; Ye et al., 2004). Additionally, compared to normal breast epithelium, there are significant gene expression changes present in early preinvasive ADH lesions as well as later invasive stages of the disease (Sgroi et al., 1999; Fuller et al., 2003). Although the three distinct stages of breast cancer (ADH, DCIS, and IDC) are very similar at the level of their gene expression, different tumor grades are correlated with specific transcriptional signatures, and tumor grade is linked with the DCIS—IDC stage transition (Ma et al., 2003).

6.2 Metastasis Gene Expression Signatures and Intrinsic Host Factors

Important information can be acquired through gene expression arrays comparing metastatic lesions to primary breast cancer. Although no genes appear to be specific for a metastatic phenotype (Porter et al., 2001), it has been shown by gene expression profiling that primary breast carcinomas are very similar to their distant metastases (Weigelt et al., 2003; Hao et al., 2004; Lahdesmaki et al., 2004; Weigelt and van't Veer, 2004; Huang et al., 2005). Although a commonly suggested theory on metastasis is that metastasis occurs late in tumor progression as a result of accumulation of mutations, the expression profile of metastases suggests that inherent mechanisms may already be in place early in the course of the disease to determine whether a tumor exhibits metastatic behavior or not (Weigelt and van't Veer, 2004). A set of 117 genes has been found to predict metastatic potential in breast carcinomas using microarray gene expression patterns (van de Vijver et al., 2002; Hunter et al., 2003). More recently, a small set of 17 genes was reported to predict metastatic potential for a variety of tumors (Hunter et al., 2003; Ramaswamy et al., 2003; Qiu et al., 2004). Therefore, in contrast to the classic model of metastasis in which only a small population of tumor cells accumulate the required number of mutations for metastasis, these observations suggest that primary tumors may harbor a metastasis signature (Hunter et al., 2003; Weigelt et al., 2003), and that metastatic potential is determined by combinations of early oncogenic alterations rather than a predominance of late stage metastasis promoting events (Bernards and Weinberg, 2002). Significantly, differential gene expression patterns may reflect host genetic profiles that influence and affect metastatic potential.

For example, global expression analysis of tumors from the MMTV-PyMT transgenic model was performed comparing parental homozygous FVB strain with F1 hybrids generated by crossing with several other strains. While all strains expressed the transgene and resultant upregulation of genes associated with cell growth and downregulation of cell adhesion molecules, three classifications of tumors were identified based on global gene expression (1) a group with suppressed tumor growth and dissemination, (2) a group with very aggressive and highly metastatic biological behavior, and (3) a group with intermediate behavior (Hunter et al., 2003; Qiu et al., 2004). Additionally, the group of tumors classified as highly metastatic tended to show altered expression of the same 17 genes previously described as a metastasis predictor in humans. Since all of these tumors were initiated by the same oncogenic effect, the difference in metastatic potential of these tumors is likely related to host genetic factors that modulate metastatic potential (Hunter et al., 2003; Qiu et al., 2004). This indicates that the genetic background of the host can modulate transcriptional profiles, and thus, phenotypic behavior of different mammary cancers. This has important implications for the human population, since it suggests that there may be subpopulations of people based upon their genetic constitution that are more susceptible to metastatic dissemination (Qiu et al., 2004). Thus, carcinogenesis and metastasis are complex phenotypes involving both innate genetic responses and extrinsic factors influencing cellular responses.

Recently bioluminescence imaging techniques were used to identify the organ-specific metastasis signatures of human breast cancer cells introduced into immunodeficient mice (Barnes et al., 1993). Selection of single-cell-derived progenies (SCPs) in vivo from MDA-MB-231 human breast cancer cell line established in culture from the pleural effusion of a breast cancer patient (Cailleau et al., 1974) showed distinct patterns of organ-specific metastasis. SCPs derived from the parental MDA-MB-231 exhibited markedly different abilities to metastasize to bone, lung, or adrenal medulla, which suggests that metastases to different organs require different molecular signatures (Kang et al., 2003; Minn et al., 2005a,b). Transcriptomic profiling analysis revealed that the different SCPs derived from MDA-MB-231 similarly express a previously described "poor-prognosis" gene signature (van de Vijver et al., 2002). Thus, MDA-MB-231 cells express a typical poor-prognosis profile. MDS and hierarchical clustering of the unsupervised transcriptomic data from different SCPs supported the hypothesis that organ-specific metastasis by breast cancer cells is controlled by metastasis-specific genes and these genes are separate from a general poor-prognosis gene expression signature. Furthermore, the 50 bone metastasis genes expressed among the bone-metastatic SCPs could be used to distinguish the primary breast carcinomas that preferentially metastasized to bone from those preferentially metastasized elsewhere (Minn et al., 2005b). These results suggest that the bone-specific metastatic phenotypes and gene expression signature identified in a mouse model may be clinically relevant.

A set of 54 genes that mediates breast cancer metastasis to lung was identified in the SCPs derived from MDA-MB-231 by transcriptomic analysis (Minn et al., 2005a). Several of the lung-organ metastatic genes were functionally verified in vitro through stable short hairpin (sh) RNA interference in the lung-metastatic cells, and the knockdown cell lines were injected through the tail vein to assess lung-metastatic activity. Hierarchical clustering of primary breast carcinomas from a cohort of 82 breast patients was performed using the 54 lung metastasis signature genes. Tumors from patients who developed lung metastasis clustered separately from the tumors that metastasized at nonpulmonary sites. This analysis clinically validated that the lung-metastatic signature genes could be used as classifiers to predict the outcome of the human primary breast cancer (Minn et al., 2005a).

Bone and lung are the most frequent targets of breast cancer metastasis in humans. These results provide evidence that bone and lung impose different requirements for the establishment of metastasis by the circulating cancer cells. These findings shed new light on the biology of breast metastasis.

7 FUTURE DIRECTIONS

Gene expression profiling is a powerful tool that promises to aid in the development of advanced techniques for improving and individualizing the diagnosis and treatment of cancer. Unique gene expression patterns have been reported that predict therapeutic response and prognosis (Sorlie et al., 2003; Jeffrey et al., 2005; van der Pouw Kraan et al., 2005). Breast cancers may be categorized according to molecular classifiers that predict clinical outcome (Perou et al., 1999; Desai et al., 2002a; Sorlie et al., 2003).

Many patients with the same apparent stage of disease based upon anatomic location and histopathology respond very differently to treatment, and these differences in therapeutic response have been ascribed to disease severity and pathogenesis, age and gender, as well as the health and nutritional status of the patient (Wajapeyee and Somasundaram, 2004). Gene expression profiling provides a large amount of information regarding prognostic and predictive factors (West et al., 2001; Huang et al., 2003b). A "poor-prognosis" gene expression signature has been reported that is strongly predictive of a short interval to metastasis in patients with breast cancer without regional lymph node involvement (van't Veer et al., 2002). This type of signature would be useful as a strategy both to select patients who would benefit from chemotherapy, as well as those who would not, thus using chemotherapeutic drugs more appropriately (van't Veer et al., 2002; Nevins et al., 2003; Cleator and Ashworth, 2004). Additionally, a gene expression profile has recently been identified in patients with inflammatory breast cancer (IBC), a clinically aggressive form of locally advanced breast cancer with a poor prognosis (Van Laere et al., 2005). Using unsupervised hierarchical clustering, IBC samples were accurately identified from non-IBC samples using a set of 50 discriminator genes with an accuracy of 88% (Van Laere et al., 2005). These studies show that the correlation of gene expression analysis and risk factors will improve clinical prediction of prognosis, metastasis, and relapse for individual patients, as well as identify potential molecular targets and provide specific tailored treatment, in order to provide more "personalized" medicine based on a patient's individual needs (Huang et al., 2003a; Nevins et al., 2003; van der Pouw Kraan et al., 2005; Van Laere et al., 2005).

Mouse models will undoubtedly continue to play a critical role in understanding mechanisms of cancer development, especially as genomic technologies and bioinformatics tools are applied to these systems. The combined use of bioinformatics applied to GEM models of breast cancer provides great power to the study and intervention of this complex disease. These applications will be invaluable in developing advanced diagnostic and treatment modalities, including the identification of unique biomarkers and therapeutic targets, as well as developing a more personalized diagnostic and therapeutic system for individuals with the complex and heterogeneous disease of breast cancer.

REFERENCES

Abe, K., Hazama, M., Katoh, H., Yamamura, K., and Suzuki, M. 2004. Establishment of an efficient BAC transgenesis protocol and its application to functional characterization of the mouse Brachyury locus. Exp Anim *53*: 311–320

Aubele, M., Werner, M., and Hofler, H. 2002. Genetic alterations in presumptive precursor lesions of breast carcinomas. Anal Cell Pathol *24*: 69–76

Balmain, A., and Nagase, H. 1998. Cancer resistance genes in mice: models for the study of tumour modifiers. Trends Genet *14*: 139–144

Barkan, D., Montagna, C., Ried, T., Green, J. E. 2004. Mammary gland cancer. In: Mouse Models of Human Cancer, ed. E. C. Holland, pp. 103–131. Hoboken: Wiley

Barnes, D., Dublin, E., Fisher, C., Levison, D., and Millis, R. 1993. Immunohistochemical detection of p53 protein in mammary carcinoma: an important new independent indicator of prognosis? Human Pathol *24*: 469–476

Bernards, R., and Weinberg, R. A. 2002. A progression puzzle. Nature *418*: 823

Bild, A. H., Yao, G., Chang, J. T., Wang, Q., Potti, A., Chasse, D., Joshi, M. B., Harpole, D., Lancaster, J. M., Berchuck, A., Olson, J. A., Jr., Marks, J. R., Dressman, H. K., West, M., and Nevins, J. R. 2006. Oncogenic pathway signatures in human cancers as a guide to targeted therapies. Nature *439*: 353–357

Cailleau, R., Young, R., Olive, M., and Reeves, W. J., Jr. 1974. Breast tumor cell lines from pleural effusions. J Natl Cancer Inst *53*: 661–674

Callahan, R., and Smith, G. H. 2000. MMTV-induced mammary tumorigenesis: gene discovery, progression to malignancy and cellular pathways. Oncogene *19*: 992–1001

Callow, M. J., Dudoit, S., Gong, E. L., Speed, T. P., and Rubin, E. M. 2000. Microarray expression profiling identifies genes with altered expression in HDL-deficient mice. Genome Res *10*: 2022–2029

Cardiff, R. 2001. Validity of mouse mammary tumour models for human breast cancer: comparative pathology. Microsc Res Tech *52*: 224–230

Cardiff, R. D. 2003. Mouse models of human breast cancer. Comp Med *53*: 250–253

Cardiff, R. D., and Wellings, S. R. 1999. The comparative pathology of human and mouse mammary glands. J Mammary Gland Biol Neoplasia *4*: 105–122

Cardiff, R. D., Anver, M. R., Gusterson, B. A., Hennighausen, L., Jensen, R. A., Merino, M. J., Rehm, S., Russo, J., Tavassoli, F. A., Wakefield, L. M., Ward, J. M., and Green, J. E. 2000. The mammary pathology of genetically engineered mice: the consensus report and recommendations from the Annapolis meeting. Oncogene *19*: 968–988

Cleator, S., and Ashworth, A. 2004. Molecular profiling of breast cancer: clinical implications. Br J Cancer *90*: 1120–1124

Cooper, G. M., and Sidow, A. 2003. Genomic regulatory regions: insights from comparative sequence analysis. Curr Opin Genet Dev *13*: 604–610

Copeland, N. G., Jenkins, N. A., and Court, D. L. 2001. Recombineering: a powerful new tool for mouse functional genomics. Nat Rev Genet *2*: 769–779

Desai, K. V., Kavanaugh, C. J., Calvo, A., and Green, J. E. 2002a. Chipping away at breast cancer: insights from microarray studies of human and mouse mammary cancer. Endocr Relat Cancer *9*: 207–220

Desai, K. V., Xiao, N., Wang, W., Gangi, L., Greene, J., Powell, J. I., Dickson, R., Furth, P., Hunter, K., Kucherlapati, R., Simon, R., Liu, E. T., and Green, J. E. 2002b. Initiating oncogenic event determines gene-expression patterns of human breast cancer models. Proc Natl Acad Sci USA *99*: 6967–6972

Eddy, S. 2002. Computational genomics of noncoding RNA genes. Cell *217*: 137–140

Fargiano, A. A., Desai, K. V., and Green, J. E. 2003. Interrogating mouse mammary cancer models: insights from gene expression profiling. J Mammary Gland Biol Neoplasia *8*: 321–334

Fuller, A. P., Palmer-Toy, D., Erlander, M. G., and Sgroi, D. C. 2003. Laser capture microdissection and advanced molecular analysis of human breast cancer. J Mammary Gland Biol Neoplasia *8*: 335–345

Gadbury, G. L., Page, G. P., Edwards, J., Kayo, T., Weindruch, R., Permana, P. A., Mountz, J., Allison, D. B. 2004. Power analysis and sample size estimation in the age of high dimensional biology. Stat Methods Med Res *13*: 325–338

Gossen, M., Freundlieb, S., Bender, G., Muller, G., Hillen, W., and Bujard, H. 1995. Transcriptional activation by tetracyclines in mammalian cells. Science *268*: 1766–1769

Green, J. E., and Hudson, T. 2005. The promise of genetically engineered mice for cancer prevention studies. Nat Rev Cancer *5*: 184–198

Green, J. E., Desai, K., Ye, Y., Kavanaugh, C., Calvo, A., and Huh, J. I. 2004. Genomic approaches to understanding mammary tumor progression in transgenic mice and responses to therapy. Clin Cancer Res *10*: 385S–390S

Grisendi, S., and Pandolfi, P. P. 2004. Germline modification strategies. In: Mouse Models of Human Cancer, ed. E. C. Holland, pp. 43–65. Hoboken: Wiley

Gruvberger, S., Ringner, M., Chen, Y., Panavally, S., Saal, L. H., Borg, A., Ferno, M., Peterson, C., and Meltzer, P. S. 2001. Estrogen receptor status in breast cancer is associated with remarkably distinct gene expression patterns. Cancer Res *61*: 5979–5984

Gunther, E. J., Belka, G. K., Wertheim, G. B., Wang, J., Hartman, J. L., Boxer, R. B., and Chodosh, L. A. 2002. A novel doxycycline-inducible system for the transgenic analysis of mammary gland biology. FASEB J *16*: 283–292

Hao, X., Sun, B., Hu, L., Lahdesmaki, H., Dunmire, V., Feng, Y., Zhang, S. W., Wang, H., Wu, C., Wang, H., Fuller, G. N., Symmans, W. F., Shmulevich, I., and Zhang, W. 2004. Differential gene and protein expression in primary breast malignancies and their lymph node metastases as revealed by combined cDNA microarray and tissue microarray analysis. Cancer *100*: 1110–1122

Hennighausen, L. 2000. Mouse models for breast cancer. Oncogene *19*: 966–967

Herschkowitz, J. I., Simin, K., Weigman, V. J., Mikaelian, I., Usary, J., Hu, Z., Rasmussen, K. E., Jones, L. P., Assefnia, S., Chandrasekharan, S., Backlund, M. G., Yin, Y., Khramtsov, A. I., Bastein, R., Quackenbush, J., Glazer, R. I., Brown, P. H., Green, J. E., Kopelovich, L., Furth, P. A., Palazzo, J. P., Olopade, O. I., Bernard, P. S., Churchill, G. A., Van Dyke, T., and Perou, C. M. 2007. Identification of conserved gene expression features between murine mammary carcinoma models and human breast tumors. Genome Biol *8*: R76

Huang, E., Ishida, S., Pittman, J., Dressman, H., Bild, A., Kloos, M., D’Amico, M., Pestell, R. G., West, M., and Nevins, J. R. 2003a. Gene expression phenotypic models that predict the activity of oncogenic pathways. Nat Genet *34*: 226–230

Huang, E., West, M., and Nevins, J. R. 2003b. Gene expression profiling for prediction of clinical characteristics of breast cancer. Recent Prog Horm Res *58*: 55–73

Huang, J., Li, X., Hilf, R., Bambara, R. A., and Muyan, M. 2005. Molecular basis of therapeutic strategies for breast cancer. Curr Drug Targets Immune Endocr Metabol Disord *5*: 379–396

Hunter, K., Welch, D. R., and Liu, E. T. 2003. Genetic background is an important determinant of metastatic potential. Nat Genet *34*: 23–24; author reply 25

Hutchinson, J. N., and Muller, W. J. 2000. Transgenic mouse models of human breast cancer. Oncogene *19*: 6130–6137

Irizarry, R. A., Hobbs, B., Collin, F., Beazer-Barclay, Y. D., Antonellis, K. J., Scherf, U., and Speed, T. P. 2003. Exploration, normalization, and summaries of high density oligonucleotide array probe level data. Biostatistics *4*: 249–264

Jager, R., Friedrichs, N., Heim, I., Buttner, R., and Schorle, H. 2005. Dual role of AP-2gamma in ErbB-2-induced mammary tumorigenesis. Breast Cancer Res Treat *90*: 273–280

Jain, A. K., Murty, M. N, and Flynn, P. J. 1999. Data Clustering: A Review, Vol. 3, pp. 265–322

Jeffrey, S. S., Lonning, P. E., and Hillner, B. E. 2005. Genomics-based prognosis and therapeutic prediction in breast cancer. J Natl Compr Canc Netw *3*: 291–300

Jerry, D. J., Kittrell, F. S., Kuperwasser, C., Laucirica, R., Dickinson, E. S., Bonilla, P. J., Butel, J. S., and Medina, D. 2000. A mammary-specific model demonstrates the role of the p53 tumor suppressor gene in tumor development. Oncogene *19*: 1052–1058

Jolicoeur, P., Bouchard, L., Guimond, A., Ste-Marie, M., Hanna, Z., and Dievart, A. 1998. Use of mouse mammary tumour virus (MMTV)/neu transgenic mice to identify genes collaborating with the c-erbB-2 oncogene in mammary tumour development. Biochem Soc Symp *63*: 159–165

Kang, Y., Siegel, P. M., Shu, W., Drobnjak, M., Kakonen, S. M., Cordon-Cardo, C., Guise, T. A., and Massague, J. 2003. A multigenic program mediating breast cancer metastasis to bone. Cancer Cell *3*: 537–549

Kavanaugh, C., and Green, J. E. 2003. The use of genetically altered mice for breast cancer prevention studies. J Nutr *133*: 2404S–2409S

Kavanaugh, C. J., Desai, K. V., Calvo, A., Brown, P. H., Couldrey, C., Lubet, R., and Green, J. E. 2002. Pre-clinical applications of transgenic mouse mammary cancer models. Transgenic Res *11*: 617–633

Khan, J., Wei, J. S., Ringner, M., Saal, L. H., Ladanyi, M., Westermann, F., Berthold, F., Schwab, M., Antonescu, C. R., Peterson, C., and Meltzer, P. S. 2001. Classification and diagnostic prediction of cancers using gene expression profiling and artificial neural networks. Nat Med *7*: 673–679

Korn, E. L., Troendle, J. F., McShane, L. M., and Simon, R. 2004. Controlling the number of false discoveries: application to high-dimensional genomic data. J Stat Plan Infer *124*: 379–398

Lahdesmaki, H., Hao, X., Sun, B., Hu, L., Yli-Harja, O., Shmulevich, I., and Zhang, W. 2004. Distinguishing key biological pathways between primary breast cancers and their lymph node metastases by gene function-based clustering analysis. Int J Oncol *24*: 1589–1596

Lakhani, S. R., Jacquemier, J., Sloane, J. P., Gusterson, B. A., Anderson, T. J., van de Vijver, M. J., Farid, L. M., Venter, D., Antoniou, A., Storfer-Isser, A., Smyth, E., Steel, C. M., Haites, N., Scott, R. J., Goldgar, D., Neuhausen, S., Daly, P. A., Ormiston, W., McManus, R., Scherneck, S., Ponder, B. A., Ford, D., Peto, J., Stoppa-Lyonnet, D., Bignon, Y. J., Struewing, J. P., Spurr, N. K., Bishop, D. T., Klijn, J. G., Devilee, P., Cornelisse, C. J., Lasset, C., Lenoir, G., Barkardottir, R. B., Egilsson, V., Hamann, U., Chang-Claude, J., Sobol, H., Weber, B., Stratton, M. R., and Easton, D. F. 1998. Multifactorial analysis of differences between sporadic breast cancers and cancers involving BRCA1 and BRCA2 mutations. J Natl Cancer Inst *90*: 1138–1145

Lee, J. S., Grisham, J. W., and Thorgeirsson, S. S. 2005. Comparative functional genomics for identifying models of human cancer. Carcinogenesis *26*: 1013–1020

Lin, S. C., Lee, K. F., Nikitin, A. Y., Hilsenbeck, S. G., Cardiff, R. D., Li, A., Kang, K. W., Frank, S. A., Lee, W. H., and Lee, E. Y. 2004. Somatic mutation of p53 leads to estrogen receptor alpha-positive and -negative mouse mammary tumors with high frequency of metastasis. Cancer Res *64*: 3525–3532

Liu, E. T. 2003. Classification of cancers by expression profiling. Curr Opin Genet Dev *13*: 97–103

Luzzi, V., Holtschlag, V., and Watson, M. A. 2001. Expression profiling of ductal carcinoma in situ by laser capture microdissection and high-density oligonucleotide arrays. Am J Pathol *158*: 2005–2010

Ma, X. J., Salunga, R., Tuggle, J. T., Gaudet, J., Enright, E., McQuary, P., Payette, T., Pistone, M., Stecker, K., Zhang, B. M., Zhou, Y. X., Varnholt, H., Smith, B., Gadd, M., Chatfield, E., Kessler, J., Baer, T. M., Erlander, M. G., and Sgroi, D. C. 2003. Gene expression profiles of human breast cancer progression. Proc Natl Acad Sci USA *100*: 5974–5979

McShane, L. M., Shih, J. H., and Michalowska, A. M. 2003. Statistical issues in the design and analysis of gene expression microarray studies of animal models. J Mammary Gland Biol Neoplasia *8*: 359–374

Medina, D., Sivaraman, L., Hilsenbeck, S. G., Conneely, O., Ginger, M., Rosen, J., and Omalle, B. W. 2001. Mechanisms of hormonal prevention of breast cancer. Ann NY Acad Sci *952*: 23–35

Minn, A. J., Gupta, G. P., Siegel, P. M., Bos, P. D., Shu, W., Giri, D. D., Viale, A., Olshen, A. B., Gerald, W. L., and Massague, J. 2005a. Genes that mediate breast cancer metastasis to lung. Nature *436*: 518–524

Minn, A. J., Kang, Y., Serganova, I., Gupta, G. P., Giri, D. D., Doubrovin, M., Ponomarev, V., Gerald, W. L., Blasberg, R., and Massague, J. 2005b. Distinct organ-specific metastatic potential of individual breast cancer cells and primary tumors. J Clin Invest *115*: 44–55

Nandi, S., Guzman, R. C., and Yang, J. 1995. Hormones and mammary carcinogenesis in mice, rats, and humans: a unifying hypothesis. Proc Natl Acad Sci USA *92*: 3650–3657

Nevins, J. R., Huang, E. S., Dressman, H., Pittman, J., Huang, A. T., and West, M. 2003. Towards integrated clinico-genomic models for personalized medicine: combining gene expression signatures and clinical factors in breast cancer outcomes prediction. Hum Mol Genet *12(Spec No. 2)*: R153–R157

Nikitin, A. Y., Alcaraz, A., Anver, M. R., Bronson, R. T., Cardiff, R. D., Dixon, D., Fraire, A. E., Gabrielson, E. W., Gunning, W. T., Haines, D. C., Kaufman, M. H., Linnoila, R. I., Maronpot, R. R., Rabson, A. S., Reddick, R. L., Rehm, S., Rozengurt, N., Schuller, H. M., Shmidt, E. N., Travis, W. D., Ward, J. M., and Jacks, T. 2004. Classification of proliferative pulmonary lesions of the mouse: recommendations of the mouse models of human cancers consortium. Cancer Res *64*: 2307–2316

Pawitan, Y., Michiels, S., Koscielny, S., Gusnanto, A., and Ploner, A. 2005. False discovery rate, sensitivity and sample size for microarray studies. Bioinformatics *21*: 3017–3024

Perou, C. M., Jeffrey, S. S., van de Rijn, M., Rees, C. A., Eisen, M. B., Ross, D. T., Pergamenschikov, A., Williams, C. F., Zhu, S. X., Lee, J. C., Lashkari, D., Shalon, D., Brown, P. O., and Botstein, D. 1999. Distinctive gene expression patterns in human mammary epithelial cells and breast cancers. Proc Natl Acad Sci USA *96*: 9212–9217

Porter, D. A., Krop, I. E., Nasser, S., Sgroi, D., Kaelin, C. M., Marks, J. R., Riggins, G., and Polyak, K. 2001. A SAGE (serial analysis of gene expression) view of breast tumor progression. Cancer Res *61*: 5697–5702

van der Pouw Kraan, C. T., Dijkstra, C. D., and Verweij, C. L. 2005. Molecular unraveling of disease by means of DNA-microarrays. Ned Tijdschr Geneeskd *149*: 626–631

Qiu, T. H., Chandramouli, G. V., Hunter, K. W., Alkharouf, N. W., Green, J. E., and Liu, E. T. 2004. Global expression profiling identifies signatures of tumor virulence in MMTV-PyMT-transgenic mice: correlation to human disease. Cancer Res *64*: 5973–5981

Ramaswamy, S., Ross, K. N., Lander, E. S., and Golub, T. R. 2003. A molecular signature of metastasis in primary solid tumors. Nat Genet *33*: 49–54

Ried, T., Just, K. E., Holtgreve-Grez, H., du Manoir, S., Speicher, M. R., Schrock, E., Latham, C., Blegen, H., Zetterberg, A., Cremer, T., et al. 1995. Comparative genomic hybridization of formalin-fixed, paraffin-embedded breast tumors reveals different patterns of chromosomal gains and losses in fibroadenomas and diploid and aneuploid carcinomas. Cancer Res *55*: 5415–5423

Roth, V., and Lange, T. 2004. Bayesian class discovery in microarray datasets. IEEE Trans Biomed Eng *51*: 707–718

Sauer, B. 1998. Inducible gene targeting in mice using the Cre/lox system. Methods *14*: 381–392

Seth, A., Kitching, R., Landberg, G., Xu, J., Zubovits, J., and Burger, A. M. 2003. Gene expression profiling of ductal carcinomas in situ and invasive breast tumors. Anticancer Res *23*: 2043–2051

Sgroi, D. C., Teng, S., Robinson, G., LeVangie, R., Hudson, J. R., Jr., and Elkahloun, A. G. 1999. In vivo gene expression profile analysis of human breast cancer progression. Cancer Res *59*: 5656–5661

Shappell, S. B., Thomas, G. V., Roberts, R. L., Herbert, R., Ittmann, M. M., Rubin, M. A., Humphrey, P. A., Sundberg, J. P., Rozengurt, N., Barrios, R., Ward, J. M., and Cardiff, R. D. 2004. Prostate pathology of genetically engineered mice: definitions and classification. The consensus report from the Bar Harbor meeting of the Mouse Models of Human Cancer Consortium Prostate Pathology Committee. Cancer Res *64*: 2270–2305

Simon, R. 2003. Using DNA microarrays for diagnostic and prognostic prediction. Expert Rev Mol Diagn *3*: 587–595

Simon, R. 2004. When is a genomic classifier ready for prime time? Nat Clin Pract Oncol *1*: 4–5

Simon, R. M., Korn, E. L., McShane, L. M., Radmacher, M. D., Wright, G. W., Zhao, Y. 2003. Design and Analysis of DNA Microarray Investigations. New York: Springer

Sorlie, T., Tibshirani, R., Parker, J., Hastie, T., Marron, J. S., Nobel, A., Deng, S., Johnsen, H., Pesich, R., Geisler, S., Demeter, J., Perou, C. M., Lonning, P. E., Brown, P. O., Borresen-Dale, A. L., and Botstein,

D. 2003. Repeated observation of breast tumor subtypes in independent gene expression data sets. Proc Natl Acad Sci USA *100*: 8418–8423

Sotiriou, C., Neo, S. Y., McShane, L. M., Korn, E. L., Long, P. M., Jazaeri, A., Martiat, P., Fox, S. B., Harris, A. L., and Liu, E. T. 2003. Breast cancer classification and prognosis based on gene expression profiles from a population-based study. Proc Natl Acad Sci USA *100*: 10393–10398

Sweet-Cordero, A., Mukherjee, S., Subramanian, A., You, H., Roix, J. J., Ladd-Acosta, C., Mesirov, J., Golub, T. R., and Jacks, T. 2005. An oncogenic KRAS2 expression signature identified by cross-species gene-expression analysis. Nat Genet *37*: 48–55

Tusher, V. G., Tibshirani, R., and Chu, G. 2001. Significance analysis of microarrays applied to the ionizing radiation response. Proc Natl Acad Sci USA *98*: 5116–5121

Ureta-Vidal, A., Ettwiller, L., and Birney, E. 2003. Comparative genomics: genome-wide analysis in metazoan eukaryotes. Nat Rev Genet *4*: 251–262

Van Laere, S., Van der Auwera, I., Van den Eynden, G. G., Fox, S. B., Bianchi, F., Harris, A. L., van Dam, P., Van Marck, E. A., Vermeulen, P. B., and Dirix, L. Y. 2005. Distinct molecular signature of inflammatory breast cancer by cDNA microarray analysis. Breast Cancer Res Treat *93*: 237–246

van't Veer, L. J., Dai, H., van de Vijver, M. J., He, Y. D., Hart, A. A., Mao, M., Peterse, H. L., van der Kooy, K., Marton, M. J., Witteveen, A. T., Schreiber, G. J., Kerkhoven, R. M., Roberts, C., Linsley, P. S., Bernards, R., and Friend, S. H. 2002. Gene expression profiling predicts clinical outcome of breast cancer. Nature *415*: 530–536

van de Vijver, M. J., He, Y. D., van't Veer, L. J., Dai, H., Hart, A. A., Voskuil, D. W., Schreiber, G. J., Peterse, J. L., Roberts, C., Marton, M. J., Parrish, M., Atsma, D., Witteveen, A., Glas, A., Delahaye, L., van der Velde, T., Bartelink, H., Rodenhuis, S., Rutgers, E. T., Friend, S. H., and Bernards, R. 2002. A gene-expression signature as a predictor of survival in breast cancer. N Engl J Med *347*: 1999–2009

Wajapeyee, N., and Somasundaram, K. 2004. Pharmacogenomics in breast cancer: current trends and future directions. Curr Opin Mol Ther *6*: 296–301

Weaver, Z., Montagna, C., Xu, X., Howard, T., Gadina, M., Brodie, S. G., Deng, C. X., and Ried, T. 2002. Mammary tumors in mice conditionally mutant for Brca1 exhibit gross genomic instability and centrosome amplification yet display a recurring distribution of genomic imbalances that is similar to human breast cancer. Oncogene *21*: 5097–5107

Weigelt, B., and van't Veer, L. J. 2004. Hard-wired genotype in metastatic breast cancer. Cell Cycle *3*: 756–757

Weigelt, B., Glas, A. M., Wessels, L. F., Witteveen, A. T., Peterse, J. L., and van't Veer, L. J. 2003. Gene expression profiles of primary breast tumors maintained in distant metastases. Proc Natl Acad Sci USA *100*: 15901–15905

Weiss, W. A., Israel, M., Cobbs, C., Holland, E., James, C. D., Louis, D. N., Marks, C., McClatchey, A. I., Roberts, T., Van Dyke, T., Wetmore, C., Chiu, I. M., Giovannini, M., Guha, A., Higgins, R. J., Marino, S., Radovanovic, I., Reilly, K., and Aldape, K. 2002. Neuropathology of genetically engineered mice: consensus report and recommendations from an international forum. Oncogene *21*: 7453–7463

West, M., Blanchette, C., Dressman, H., Huang, E., Ishida, S., Spang, R., Zuzan, H., Olson, J. A., Jr., Marks, J. R., and Nevins, J. R. 2001. Predicting the clinical status of human breast cancer by using gene expression profiles. Proc Natl Acad Sci USA *98*: 11462–11467

Xu, X., Wagner, K. U., Larson, D., Weaver, Z., Li, C., Ried, T., Hennighausen, L., Wynshaw-Boris, A., and Deng, C. X. 1999. Conditional mutation of Brca1 in mammary epithelial cells results in blunted ductal morphogenesis and tumour formation. Nat Genet *22*: 37–43

Ye, Y., Qiu, T. H., Kavanaugh, C., and Green, J. E. 2004. Molecular mechanisms of breast cancer progression: lessons from mouse mammary cancer models and gene expression profiling. Breast Dis *19*: 69–82

5 Significance of Aberrant Expression of miRNAs in Cancer Cells

George A. Calin, Chang-gong Liu, Manuela Ferracin, Stefano Volinia, Massimo Negrini, and Carlo M. Croce*

ABSTRACT

Alterations in miRNA genes and other non-coding RNAs play a critical role in the pathophysiology of all human cancers: cancer initiation and progression can involve miRNAs (miRNAs) – small non-coding RNAs that can regulate gene expression. At present, the main mechanism of microRNoma (defined as the full complement of miRNAs present in a genome) alteration in cancer cells seems to be represented by aberrant gene expression, characterized by abnormal levels of expression for mature and/or precursor miRNA sequences in comparison with the corresponding normal tissues. miRNA expression profiling has been exploited to identify miRNAs that are potentially involved in the pathogenesis of human cancers. miRNAs and other non-coding RNAs profiling achieved by various methods has allowed the identification of signatures associated with diagnosis, staging, progression, prognosis, and response to treatment of human tumors.

Key Words: Cancer, miRNAs, Oncogene, Tumor-suppressor

1 WHAT miRNAS ARE AND HOW THEY WORK

1.1 miRNAs Are Small Non-coding Regulatory RNAs

miRNAs were first described in 1993 in *C. elegans* by Ambros group at Harvard University (Lee et al., 1993). At present , over 6,000 members of a new class of small non-coding RNAs (ncRNAs), named miRNAs (miRNAs) (Ambros, 2003; Bartel, 2004), have been identified in the last seven years in vertebrates, flies, worms and plants, and also in viruses (Griffiths-Jones et al., 2006). In humans, according to miRBase (http://miRNA.sanger.ac.uk/cgi-bin/sequences/), the miRNoma (defined as the full spectrum of miRNAs) contains over 650 experimentally or "in silico" cloned miRNAs and the total number is expected to surpass the one thousand mark (Bentwich et al., 2005).

G.J. Gordon (ed.), *Cancer Drug Discovery and Development: Bioinformatics in Cancer and Cancer Therapy*,
DOI: 10.1007/978-1-59745-576-3_5, © Humana Press, a part of Springer Science + Business Media, LLC 2009

Compared to the protein-coding genes (PCGs), miRNA genes behave like strangers in the genomic galaxy. MiRNAs are less than 1% of the size of usual PCGs, a reason why they "escaped" cloning for so long a time. Naturally occurring miRNAs are 19–25 nt transcripts cleaved from 70–100 nt hairpin precursor (fold-back) RNA (named pre-miRNA) structure that is transcribed from a larger primary transcript (named pri-miRNA) (Ambros, 2003). No open reading frame (ORF) can be identified in the small piece of genome codifying for miRNAs. While they can be located both inter- and extragenically, more than a half of all known miRNAs are in introns or exons of PCGs or other ncRNAs (Rodriguez et al., 2004). Functionally, it was shown that miRNAs reduce the levels of many of their target transcripts as well as the amount of protein encoded by these transcripts (Lim et al., 2005). About one third of the genes are supposed to interact with various miRNAs as proved by microarray experiments of miRNA-transfected cells (Lim et al., 2005). Combinatorial effects of multiple miRNAs on specific target mRNAs (Krek et al., 2005; Cimmino et al., 2005), as well as redundancy of targets for specific miRNAs (Yekta et al., 2004) have been described. The effects on targets are mainly inhibitory, but positive effects were also identified in the case of *miR-122*, that acts as an enhancer of hepatitis C virus replication by binding to the 5 noncoding region of the virus (Jopling et al., 2005).

1.2 Fine-Tuning Gene Regulation by miRNAs

The antisense single-stranded miRNAs can bind specific mRNA transcripts through sequences that are significantly, though not completely, complementary to the target mRNA. This process is also known as post-transcriptional gene regulation (PTGS). miRNAs seem to be responsible for fine regulation of gene expression, "tuning" the cellular phenotype during delicate processes like development and differentiation in all organisms, from plants to mammals (Sevignani et al., 2006). Many miRNAs are conserved in sequence between distantly related organisms, suggesting that these molecules participate in the essential processes (Pasquinelli, 2002; Ambros, 2004). For example, the members of a cluster on chromosome 13 involved in human lymphomas, *miR-17–92*, are highly conserved in all analyzed primate species (Berezikov et al., 2005).

Target identification has been hampered by the fact that in animals, in contrast to plants, miRNAs do not bind perfectly to their targets. A few nucleotides typically remain unbound, yielding complex secondary structures. Mammalian genes can have more than one miRNA target site in their 3′ UTRs and one miRNA can target more than one mRNA. Bioinformatics approaches have been developed to search for the most thermodynamically favorable miRNA :: mRNA duplex interactions. Several computational procedures are available to predict miRNA targets, such as TargetScan (http://genes.mit.edu/targetscan/) (Lewis et al., 2005), DianaMicroT (http://www.diana.pcbi.upenn.edu) (Kiriakidou et al., 2004), miRanda (http://www.miRNA.org/) (John et al., 2004) and PicTar (http://pictar.bio.nyu.edu/) (Krek et al., 2005). Limited numbers of target mRNAs have been experimentally proven and studied *in vitro*, but the number of confirmed interactions is expected to increase sharply, as prediction tools become more sophisticated and accurate every day. As a proof of the advances in understanding the miRNA :: mRNA interaction, is the recent development of a database named

Tarbase, that provides a means of searching through a comprehensive set of experimentally supported miRNA targets in various organisms, including humans and mice (Sethupathy et al., 2006). The present version includes 570 PCG proved as targets for 128 miRNAs from eight different organisms.

2 ABNORMAL EXPRESSION OF miRNAS IS A HALLMARK OF HUMAN CANCERS

2.1 miRNAs Expression Is Important for Many Cellular Processes

The functions of miRNA are ubiquitous, ranging from the control of leaf and flower development in plants to the modulation of hematopoietic lineage differentiation in mammals (Chen et al., 2004). Several groups have uncovered roles for miRNAs in the coordination of cell proliferation and cell death during development, and in stress resistance and fat metabolism (Ambros, 2003). For example, the *Drosophila* miRNA gene (*miR-14*) suppresses cell death and is required for normal fat metabolism (Xu et al., 2003), while the *bantam* locus encodes a developmentally regulated miRNA that controls cell proliferation and regulates the pro-apoptotic gene *hid* in *Drosophila* (Brennecke et al., 2003). Other examples of miRNAs participating in essential biological processes include *miR-125b* and *let-7* (cell proliferation control), *miR-181* (hematopoietic B-cell lineage fate), *miR-15a* and *miR-16–1* (B-cell survival), *miR-430* (brain patterning), *miR-375* (pancreatic cell insulin secretion) and *miR-143* (adipocyte development) (for reviews see Harfe, (2005), Miska (2005), and Hwang and Mendell (2006)).

2.2 Deregulation of miRNAs Expression in Cancer Cells

Little was known about the expression levels of miRNA genes in normal and neoplastic cells until 2002. The assessment of cancer-specific expression levels for hundreds of miRNAs by traditional techniques, is time-consuming, requires a large amount of total RNA, and the use of radioactive isotopes. cDNA microarrays are useful tools to identify different patterns of expression in a large number of samples. The first developed oligonucleotide miRNA microarray chips (Liu et al., 2004), containing hundreds of human precursor and mature miRNA probes, identified distinct patterns of miRNA expression in human and mouse tissues (tissue-specific miRNAs expression signatures). In the last few years more than a dozen microarray platforms have been developed worldwide to investigate the genome-wide expression of miRNAs. The reliability of this technique was almost always confirmed by other traditional methods for RNA expression quantification such as Northern-blot analysis and Real time RT-PCR. Another method to determine miRNA expression levels involves the use of a bead-based flow cytometric technique (Lu et al., 2005). Since the development of these two methods of assessing global miRNA expression, several commercially available platforms have been developed for miRNA gene expression profiling (Table 1).

Global analysis of miRNA gene expression has shown that these profiles can be used to classify the developmental lineage and differentiation state of tumors; in addition, miRNA profiles appear to provide a distinct signature for tumors which is more

Table 1
Examples of high-throughput methods for miRNA expression

Type	*Principle*	*Advantages*	*References*
miRNA Microarray microchips	Thousands of oligonucleotide probes used to hybridize with mature/precursor miRNA cDNA probe	Screen in a parallel fashion for the expression of a large number of miRNAs through extensive sample collections	Liu et al. (2004), review in Esquela-Kerscher and Slack (2004)
miRNA Oligonucleotide arrays	Membrane filters macroarrays with tens of miRNA oligonucleotides spotted	Simplicity, lab bench oriented	Krichevsky et al. (2003) and Sioud and Rosok (2004)
Bead-based technology	Single miRNA oligos coating polystyrene beads hybridized with biotin labeled ds DNA target, followed by flow cytometry signal detection	Higher specificity and accuracy in respect with microarray microchips	Lu et al. (2005)
Stem-loop qRT-PCR for mature product	Stem-loop RT primer cDNA synthesis followed by quantitative conventional TaqMan PCR	Specific quantification of the mature miRNA	Chen et al. (2005)
Primer-extension quantitative PCR	Tailed miRNA-specific primer cDNA synthesis followed by LNA-primer dependent qPCR with SYBR green	Specific quantification of the mature miRNA	Raymond et al. (2005)
qRT-PCR for precursor miRNA	Hairpin-specific primers used to amplify cDNA	Specific quantification of the precursor miRNA	Schmittgen et al. (2004)
Multiplex RT-PCR	cDNA synthesis followed by pre-PCR amplification and then by TaqMan reactions for each miRNA	Detection of multiple miRNA species in very small samples	Lao et al. (2006)
Invader miRNA assay	Enzymatic cleavage by Cleavase a structure-specific 5 nuclease of a synthetic oligo probe that is in an appropriate overlap-flap structure	Detection of multiple miRNA species in very small samples	Allawi et al. (2004)
In situ detection using LNA-modified oligonucleotide probes	In situ detection using locked nucleic acid LNA-modified DNA probes	In situ spatial and temporal visualization of hybridization	Kloosterman et al. (2006)
Single-molecule method	Hybridization of two spectrally distinguishable fluorescent LNA–DNA oligonucleotides probes to miRNA of interest	Sensitive to femtomolar concentrations of miRNA	Neely et al. (2006)

Note: *ds DNA* double stranded DNA; *LNA* fluorescent locked nucleic acid precise than that provided by protein-coding genes (Lu et al., 2005). These findings lead to a unique opportunity to use these miRNA profiles for tumor diagnostics.

The first report linking miRNAs and cancer (Calin et al., 2002) involved chronic lymphocytic leukemia (CLL), the most common form of adult leukemia in the Western world. Hemizygous and/or homozygous loss at chromosome 13q14 occurs in more than half the CLL cases. Loss of chromosome 13q14 is also found in more than 50% of mantle cell lymphomas, ~30% of multiple myeloma, and about two-thirds of prostate cancers, suggesting that one or more tumor suppressor genes at 13q14 are involved in the pathogenesis of human tumors. However, detailed genetic analysis, including extensive loss of heterozygosity (LOH), mutation, and expression studies have failed to demonstrate the consistent involvement of any of the 12 protein-coding genes located in or close to the deleted region. A cluster of two miRNAs, *mir-15a* and *mir-16–1*, were found to be located in the minimal region of deletion (~30 kb) at 13q14, and to be deleted or down-regulated in ~70% of CLL samples. Furthermore, these miRNAs play essential roles in regulating the apoptotic program in B cells (Cimmino et al., 2005), and therefore can be considered as the first examples of tumor suppressor miRNAs. After the identification of specific miRNA signatures in CLL, miRNAs differentially expressed between tumors and normal counterparts were identified for several tumor types like breast cancer, gliob-

Table 2
Examples of cancer specific miRNA fingerprints identified by high-throughput methods for miRNA expression

Cancer type	*Clinical significance of miRNA fingerprints*	*Reference*
Chronic lymphocytic leukemia	miRNA Signature associated with disease prognosis and progression	Calin et al. (2005a)
Breast carcinomas	miRNA Fingerprints correlated with biopathologic features: estrogen and progesterone receptor expression, tumor stage, vascular invasion, and proliferation index	Iorio et al. (2005)
Hepatocellular carcinomas	miRNA Expression levels correlated with the degree of HCC diferentiation	Murakami et al. (2006)
Lung carcinomas	Molecular signatures correlated with tumor histology; miRNA expression profiles correlated with survival of lung adenocarcinomas	Yanaihara et al. (2006)
Leukemias and carcinomas	miRNA Profiles can differentiate solid cancers and miRNA fingerprints better classify poorly differentiated tumours as messenger RNA profiles	Lu et al. (2005)
Human solid cancers	Common miRNA are differentially expressed in various solid cancers and are targeting cancer specific protein coding genes	Volinia et al. (2006)

lastoma, hepatocellular carcinoma, colorectal carcinoma, lung cancer, and pituitary tumors (Table 2 and included references). Common patterns of miRNA modulation were identified, such as *miR-16–1* downregulation in CLL, breast cancer and pituitary tumors (Calin et al., 2004; Iorio et al., 2005; Bottoni et al., 2005). Deregulated miRNA may be considered interesting targets for cancer-specific therapies.

3 THE TARGETS OF DEREGULATED miRNAS ARE MAIN PLAYERS IN THE MALIGNANT PROCESS

The first important interaction of miRNA::mRNA with clear cancer connection was elegantly proved by the Slack group at Yale University (Johnson et al., 2005). It was for a long time known that, in lung cancers, activation of *RAS* genes by point mutations represents early events (Malumbres and Barbacid, 2003). RAS protein is significantly higher in lung tumors than in normal lung tissue, while *let-7* expression is lower in lung cancer cells. This correlation led to the identification of a direct regulation of RAS by the *let-7* miRNA family (Johnson et al., 2005). Exogenous delivery of *let-7* to the lung might either prevent the formation of lung tumors (from premalignant lesions) or shrink tumors by activating RAS mutations (Slack and Weidhans, 2006).

Another interesting molecular dissection of a miRNA::mRNA interaction important for the cancer phenotype was done by the Mendell group at Johns Hopkins University (O'Donnell et al., 2005) and by the Hammond group at North Carolina University at Chapel Hill (He et al., 2005b). The oncogene c-MYC encodes a transcription factor that regulates, via several targets including E2F1 transcription factor, cell proliferation and survival (Dang et al., 2005). A feedback regulatory loop in which MYC directly binds and activates transcription of the cluster *miR-17–92* that consequently negatively regulates E2F1 by direct interaction, while c-MYC directly induces expression of the E2F1 that in turn induces c-MYC, has been described (O'Donnell et al., 2005). This fine molecular dissection of an important cellular pathway has cancer implications, as it has been shown that *c-MYC* and *miR-17–92* cooperate and such cooperation accelerates B cell tumorigenesis (He et al., 2005b). Such results offer a rational basis for targeted therapy, for examples by using antisense miRNAs against the clustered miRNAs, a decision that will overload the regulatory loop, with the acceleration of the c-MYC – E2F1 feedback with consequent cell death by ARF-p53 pathway (Hammond, 2006).

miRNAs are natural antisense interactors with players in the eukaryotic survival and cell cycle programs. The overexpression of antiapoptotic protein BCL2 is a main genetic event in human tumorigenesis, including follicular lymphoma, lung cancer and B-cell CLL (Cory et al., 2003). This activation, except in all cases of follicular lymphomas where a translocation t(14;18) is responsible (Tsujimoto et al., 1984, 1985), has an unknown mechanism. Loss of *miR-15a/miR-16–1* in CLL results in BCL2 overexpression and it was recently proved by our group that restoration of *mir-15/miR-16* in leukemia cells induces apoptosis by directly interacting with BCL2 mRNA (Cimmino et al., 2005). These results are encouraging in the light of new promising results regarding the therapeutic potential of antisense Bcl2 as chemosensitizer for cancer therapy (Kim et al., 2004).

4 THE "miRNA CASCADE": A MODEL OF miRNA INVOLVEMENT IN HUMAN CANCERS

The classical tumorigenesis model postulates the need of bi-allelic alterations for a TSG and cooperation between heterozygous alterations of OGs. Recently haploinsuficiency, defined as alterations of one allele with loss-of-function, was proposed as an important mechanism for TSG inactivation (Fodde and Smits, 2002). MiRNAs are contributors to oncogenesis functioning as tumor suppressors (TS) (as is the case of *miR-15a/miR-16–1* cluster) or as oncogenes (OG) (as is the case of *miR-155* or *miR17–92 cluster*) (Fig. 1) (Calin et al., 2004; Cimmino et al., 2005; Chen, 2005; Berezikov and Plasterk, 2005; Gregory and Shiekhattar, 2005). We have proved by

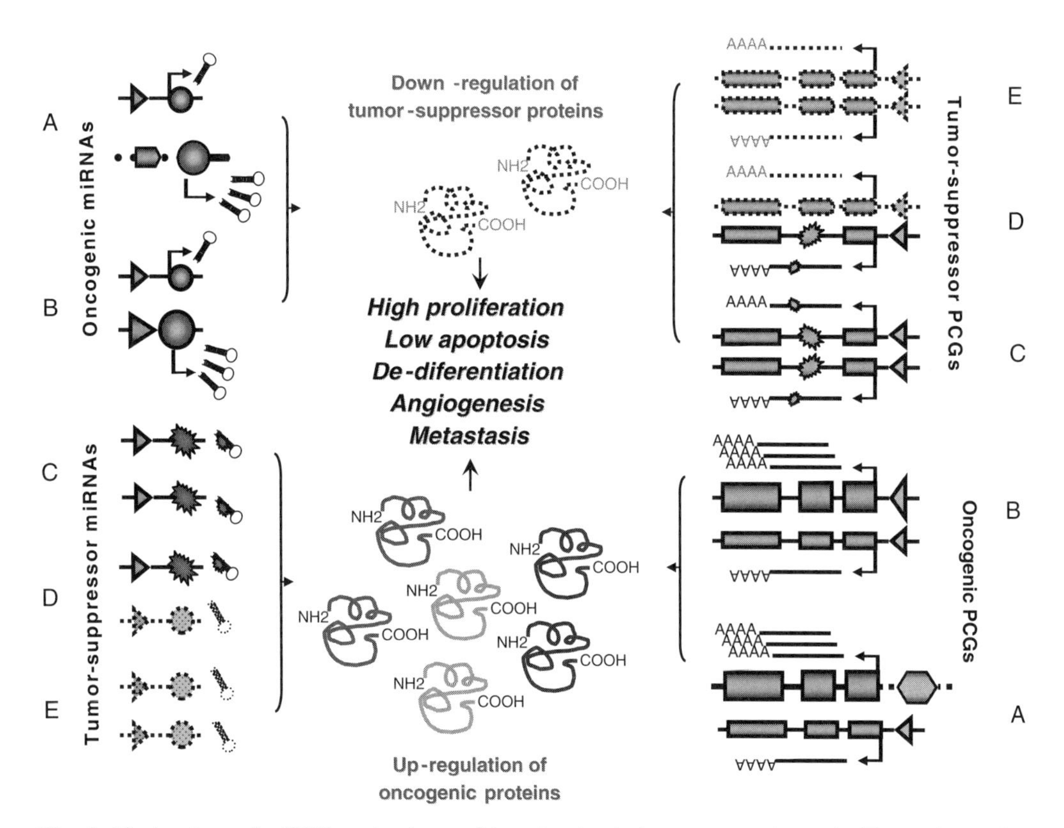

Fig. 1. *Mechanisms of miRNA activation and inactivation in human tumorigenesis.* The main mechanisms common for miRNAs and cancer-specific PCGs are represented by chromosomal translocation/rearrangement, genomic amplification, biallelic mutations, deletion/promoter methylation plus mutation and biallelic deletions/promoter methylations or combination. The effects of oncogenic miRNAs activation are the same as of inactivation of tumor-suppressor PCGs. Conversely, the effects of suppressor miRNAs inactivation are the same as of activation of oncogenic PCGs. For example, effects of t (14;18), (q32;q21), or del13q13.4 in leukemic cells are the same – overexpression of the antiapoptotic BCL2 protein, in the former case by juxtaposition of oncogene Bcl2 to Ig enhancers, while in the latter by downregulation of suppressor *miR-16–1* and *miR-15a* that negatively regulate BCL2 production. The promoter regions are presented as *triangles* and the structural genes as *circles* in miRNAs or *rectangles* in PCGs

our genome-wide miRNA expression profiling that relatively minor variations in the levels of expression of a miRNA or mutations that affect moderately the conformation of miRNA::mRNA pairing could have disastrous consequences for the cell. The explanation is represented by the large number of targets of each miRNA and the relatively large number of altered miRNAs, making it very probable that two or more PCGs from the same molecular pathway/interacting pathways are disturbed. It was shown that the miRNAs differentially expressed in human solid cancers have as targets important PCGs such as *RB1* (Retinoblastoma 1) and *TGFBR2* (transforming growth factor beta receptor II) (Volinia et al., 2006). The downregulation of the suppressor *miR-15a/miR-16–1* induces overexpression of BCL2 and possibly other genes that may be important for tumorigenesis, while the overexpression of oncogenic *miR-17–92* cooperates with *c-MYC* that stimulates proliferation. Therefore, the miRNAs may act "in cascade" over several cancer specific PCGs, that in turn could influence the transcription or function of several other PCGs and ncRNAs including miRNAs. If such alterations occur in the somatic cells, the miRNA alterations initiate or contribute to tumorigenesis, while if present in the germline this could represent cancer predisposing events. Furthermore, specific mutations or very rare polymorphisms in miRNAs or interacting regions of miRNAs could represent predisposing events for cancer initiation (Calin et al., 2005a; He et al., 2005a).

The abnormalities found to influence the activity of miRNAs are the same as those described to target PCGs, including chromosomal rearrangements, genomic amplifications or deletions, and mutations (Fig. 1). In a specific tumor both abnormalities in PCGs and miRNAs can be identified. Inactivation of tumor suppressor PCGs and activation of oncogenic miRNAs have the same molecular consequences – reduced levels of proteins blocking proliferation and activating apoptosis. By contrast, activation of oncogenic PCGs and inactivation of suppressor miRNAs is followed by accumulation of proteins that stimulates proliferation and decrease apoptosis. A paradigm for this model is human B-cell CLL, in which *miR-15a* and *miR-16–1* are located in the most frequently deleted genomic region, are down-regulated in the majority of cases, harbor mutations in familial cases and induce apoptosis in a leukemia model by targeting the ubiquitous over expressed antiapoptotic BCL2 gene.

5 THE miRNA STORY FROM SCIENTIST BENCH-SIDE TO PATIENT BEDSIDE

5.1 miRNAs as a New Class of Biomakers Used in Human Cancer Diagnosis

If the miRNAs are active players in human oncogenesis, then they will have an impact on diagnosis and prognosis of cancer (Table 2). In fact the evidence that miRNAs represent new diagnostic and prognostic factors in human cancers is rapidly accumulating (Croce and Calin, 2005; Calin et al., 2005b). In B-cell CLL, a unique miRNA signature is associated with prognostic factors (such as the levels of expression of the 70-kD Zeta-associated protein (ZAP-70) and the presence or absence of mutations in the immunoglobulin heavy-chain variable-region gene (*IgVH*)) and with the time from diagnosis to initiation of therapy in B cell CLL (Calin et al., 2005a). In diffuse large B cell lymphoma (DLBCL), independent studies revealed that significantly higher levels of *miR-155* are identified in cases with poorer prognosis (an activated B cell phenotype) than with the germinal center phenotype (Eis et al., 2005; Kluiver et al., 2005). Expression of members of *let-7* family correlates with postoperative survival in lung cancer, the group of patients with reduced expression showing significantly shorter survival after potentially curative resection (Takamizawa et al., 2004). In lung adenocarcinomas, high *miR-155* and low *let-7a-2* expression correlate with poor survival (Yanaihara et al., 2006). In breast carcinomas, expression of miRNAs was correlated with specific breast cancer bio-pathologic features, such as estrogen and progesterone receptor expression (the members of *miR-30* family), tumor stage (*miR-213* and *miR-203*), vascular invasion (*miR-9–3* and *miR-10b*) or proliferation index (members of *let-7* family) (Iorio et al., 2005). Comprehensive analysis of miRNA expression patterns in hepatocellular carcinoma found a set of miRNAs, *miR-92*, *miR-20*, and *miR-18* that were inversely correlated with the degree of differentiation (Murakami et al., 2006). Such evidence strongly suggests that quantification of miRNAs may be diagnostically useful for cancer patients.

5.2 miRNAs as New Potential Agents for Cancer Therapy

If abnormal miRNA expression in cancer cells represents important genetic abnormalities, then miRNAs could be used as potential targets for therapeutic intervention. One such approach could involve the oncogene c-KIT pathway, targeted by the small molecule drug, Gleevec (Imatinib mesylate). C-KIT is overexpressed by activating point mutations in gastrointestinal stromal tumors (GIST), adult mastocytosis, small cell lung cancer and testicular germ cell cancer. In GIST, the effects of Gleevec are dramatic (Sattler and Salgia, 2003), but as in every type of single drug chemotherapy, resistance is a dangerous drawback. It was proved that *miR-221* and *miR-222* directly target c-KIT (Felli et al., 2005). It would be of interest to device a possible combined therapy of Gleevec and *miR-221–miR-222* for refractory GIST cases.

The development of modified miRNA molecules with higher in vivo efficiency, such as the locked nucleic acid (LNA)-modified oligonucleotides (Orom et al., 2006), the anti-miRNA oligonucleotides (AMOs) (Weiler et al., 2005), or the "antagomirs" (Krutzfeldt et al., 2005) represent interesting steps for bringing these fundamental research advances into medical practice. The development of knockout or transgenic miRNA mouse models and the prophylactic or curative studies in case they develop cancer will be of help to better understand the therapeutic potential of miRNAs.

ACKNOWLEDGEMENTS

Dr Calin is supported by the CLL Global Research Foundation and by an MD Anderson Trust grant.

REFERENCES

Allawi, H.T., Dahlberg, J.E., Olson, S., Lund, E., Olson, M., Ma, W.P., Takova, T., Neri, B.P., and Lyamichev, V.I. 2004. Quantitation of microRNAs using a modified Invader assay. RNA *10*, 1153–1161.

Ambros, V. 2003. MicroRNA pathways in flies and worms: growth, death, fat, stress, and timing. Cell *113*, 673–676.

Ambros, V. 2004. The functions of animal microRNAs. Nature *431*, 350–355.

Bartel, D.P. 2004. MicroRNAs: genomics, biogenesis, mechanism, and function. *Cell 116*, 281–297.

Bentwich, I., Avniel, A., Karov, Y., Aharonov, R., Gilad, S., Barad, O., Barzilai, A., Einat, P., Einav, U., Meiri, E., . 2005. Identification of hundreds of conserved and nonconserved human microRNAs. *Nat Genet 37*, 766–770.

Berezikov, E., and Plasterk, R.H. 2005. Camels and zebrafish, viruses and cancer: a microRNA update. *Hum Mol Genet 14*, R183–R190.

Berezikov, E., Guryev, V., van de Belt, J., Wienholds, E., Plasterk, R.H., and Cuppen, E. 2005. Phylogenetic shadowing and computational identification of human microRNA genes. *Cell 120*, 21–24.

Bottoni, A., Piccin, D., Tagliati, F., Luchin, A., Zatelli, M.C., and degli Uberti, E.C. 2005. miR-15a and miR-16–1 down-regulation in pituitary adenomas. *J Cell Physiol 204*, 280–285.

Brennecke, J., Hipfner, D.R., rk, A., Russell, R.B., Cohen, S.M.S., A., Russell, R.B., and Cohen, S.M. 2003. Bantam encodes a developmentally regulated microRNA that controls cell proliferation and regulates the proapoptotic gene hid in Drosophila. *Cell 113*, 25–36.

Calin, G.A., Dumitru, C.D., Shimizu, M., Bichi, R., Zupo, S., Noch, E., Aldler, H., Rattan, S., Keating, M., Rai, K., . 2002. Frequent deletions and down-regulation of micro- RNA genes miR15 and miR16 at 13q14 in chronic lymphocytic leukemia. *Proc Natl Acad Sci USA 99*, 15524–15529.

Calin, G.A., Sevignani, C., Dumitru, C.D., Hyslop, T., Noch, E., Yendamuri, S., Shimizu, M., Rattan, S., Bullrich, F., Negrini, M., . 2004. Human microRNA genes are frequently located at fragile sites and genomic regions involved in cancers. *Proc Natl Acad Sci USA 101*, 2999–3004.

Calin, G.A., Ferracin, M., Cimmino, A., Di Leva, G., Shimizu, M., Wojcik, S.E., Iorio, M.V., Visone, R., Sever, N.I., Fabbri, M., . 2005a. A MicroRNA signature associated with prognosis and progression in chronic lymphocytic leukemia. *N Engl J Med 353*, 1793–1801.

Calin, G.A., Garzon, R., Cimmino, A., Fabbri, M., and Croce, C.M. 2005b. MicroRNAs and leukemias: How strong is the connection? *Leuk Res 30*, 653–655.

Chen, C.Z. 2005. MicroRNAs as oncogenes and tumor suppressors. *N Engl J Med 353*, 1768–1771.

Chen, C.Z., Li, L., Lodish, H.F., and Bartel, D.P. 2004. MicroRNAs modulate hematopoietic lineage differentiation. *Science 303*, 83–86.

Chen, C., Ridzon, D.A., Broomer, A.J., Zhou, Z., Lee, D.H., Nguyen, J.T., Barbisin, M., Xu, N.L., Mahuvakar, V.R., Andersen, M.R., . 2005. Real-time quant ification of microRNAs by stem-loop RT-PCR. *Nucleic Acids Res 33*, e179.

Cimmino, A., Calin, G.A., Fabbri, M., Iorio, M.V., Ferracin, M., Shimizu, M., Wojcik, S.E., Aqeilan, R., Zupo, S., Dono, M., . 2005. miR-15 and miR-16 induce apoptosis by targeting BCL2. *Proc Natl Acad Sc USA 102*, 13944–13949.

Cory, S., Huang, D.C.S., and Adams, J.M. 2003. The BCL2 family: roles in cell survival and oncogenesis. *Oncogene 22*, 8590–8607.

Croce, C.M., and Calin, G.A. 2005. miRNAs, Cancer, and Stem Cell Division. *Cell 122*, 6–7.

Dang, C.V., O'Donnell, K.A., and Juopperi, T. 2005. The great MYC escape in tumorigenesis. *Cancer Cell 8*, 177–178.

Eis, P.S., Tam, W., Sun, L., Chadburn, A., Li, Z., Gomez, M.F., Lund, E., and Dahlberg, J.E. 2005. Accumulation of miR-155 and BIC RNA in human B cell lymphomas. *Proc Natl Acad Sci USA 102*, 3627–3632.

Esquela-Kerscher, A., and Slack, F.J. 2004. The age of high-throughput microRNA profiling. *Nat Methods 1*, 106–107.

Felli, N., Fontana, L., Pelosi, E., Botta, R., Bonci, D., Facchiano, F., Liuzzi, F., Lulli, V., Morsilli, O., Santoro, S., . 2005. MicroRNAs 221 and 222 inhibit normal erythropoiesis and erythroleukemic cell growth via kit receptor down-modulation. *Proc Natl Acad Sci USA 102*, 18081–18086.

Fodde, R., and Smits, R. 2002. Cancer biology. A matter of dosage. *Science 298*, 761–763.

Gregory, R.I., and Shiekhattar, R. 2005. MicroRNA biogenesis and cancer. *Cancer Res 65*, 3509–3512.

Griffiths-Jones, S., Grocock, R.J., van Dongen, S., Bateman, A., and Enright, A.J. 2006. miRBase: microRNA sequences, targets and gene nomenclature. NAR *Database Issue*, D140–D144.

Hammond, S.M. 2006. MicroRNAs as oncogenes. *Curr Opin Genet Dev 16*, 4–9.

Harfe, B.D. 2005. MicroRNAs in vertebrate development. *Curr Opin Genet Dev 15*, 410–415.

He, H., Jazdzewski, K., Li, W., Liyanarachchi, S., Nagy, R., Volinia, S., Calin, G.A., Liu, C.G., Franssila, K., Suster, S., . 2005a. The role of microRNA genes in papillary thyroid carcinoma. Proc Natl Acad Sci USA *102*, 19075–19080.

He, L., Thomson, J.M., Hemann, M.T., Hernando-Monge, E., Mu, D., Goodson, S., Powers, S., Cordon-Cardo, C., Lowe, S.W., Hannon, G.J., . 2005b. A microRNA polycistron as a potential human oncogene. *Nature 435*, 828–833.

Hwang, H.W., and Mendell, J.T. 2006. MicroRNAs in cell proliferation, cell death, and tumorigenesis. *Br J Cancer 94*, 776–780.

Iorio, M.V., Ferracin, M., Liu, C.G., Veronese, A., Spizzo, R., Sabbioni, S., Magri, E., Pedriali, M., Fabbri, M., Campiglio, M., 2005. microRNA gene expression deregulation in human breast cancer. *Cancer Res 65*, 7065–7070.

John, B., Enright, A.J., Aravin, A., Tuschl, T., Sander, C., and Marks, D.S. 2004. Human MicroRNA targets. *PLoS Biol 2*, e363.

Johnson, S.M., Grosshans, H., Shingara, J., Byrom, M., Jarvis, R., Cheng, A., Labourier, E., Reinert, K.L., Brown, D., and Slack, F.J. 2005. RAS is regulated by the let-7 microRNA family. *Cell 120*, 635–647.

Jopling, C.L., Yi, M., Lancaster, A.M., Lemon, S.M., and Sarnow, P. 2005. Modulation of hepatitis C virus RNA abundance by a liver-specific MicroRNA. *Science 309*, 1577–1581.

Kim, H., Emi, M., Tanabe, K., and Toge, T. 2004. Therapeutic potential of antisense Bcl-2 as a chemosensitizer for cancer therapy. *Cancer 101*, 2491–2502.

Kiriakidou, M., Nelson, P.T., Kouranov, A., Fitziev, P., Bouyioukos, C., Mourelatos, Z., and Hatzigeorgiou, A. 2004. A combined computational-experimental approach predicts human microRNA targets. *Genes Dev 18*, 1165–1178.

Kloosterman, W.P., Wienholds, E., de Bruijn, E., Kauppinen, S., and Plasterk, R.H. 2006. In situ detection of miRNAs in animal embryos using LNA-modified oligonucleotide probes. *Cell 3*, 27–29.

Kluiver, J., Poppema, S., de Jong, D., Blokzijl, T., Harms, G., Jacobs, S., Kroesen, B.J., and van den Berg, A. 2005. BIC and miR-155 are highly expressed in Hodgkin, primary mediastinal and diffuse large B cell lymphomas. *J Pathol 207*, 243–249.

Krek, A., Grun, D., Poy, M.N., Wolf, R., Rosenberg, L., Epstein, E.J., MacMenamin, P., da Piedade, I., Gunsalus, K.C., Stoffel, M., 2005. Combinatorial microRNA target predictions. *Nat Genet 37*, 495–500.

Krichevsky, A.M., King, K.S., Donahue, C.P., Khrapko, K., and Kosik, K.S. 2003. A microRNA array reveals extensive regulation of microRNAs during brain development. *RNA 9*, 1274–1281.

Krutzfeldt, J., Rajewsky, N., Braich, R., Rajeev, K.G., Tuschl, T., Manoharan, M., and Stoffel, M. 2005. Silencing of microRNAs in vivo with 'antagomirs'. *Nature 438*, 685–689.

Lao, K., Xu, N.L., Yeung, V., Chen, C., Livak, K.J., and Straus, N.A. 2006. Multiplexing RT-PCR for the detection of multiple miRNA species in small samples. *Biochem Biophys Res Commun 343*, 85–89.

Lee, R.C., Feinbaum, R.L., and Ambros, V. 1993. The C. elegans heterochronic gene lin-4 encodes small RNAs with antisense complementarity to lin-14. *Cell 75*, 843–854.

Lewis, B.P., Burge, C.B., and Bartel, D.P. 2005. Conserved seed pairing, often flanked by adenosines, indicates that thousands of human genes are microRNA targets. *Cell 120*, 15–20.

Lim, L.P., Lau, N.C., Garrett-Engele, P., Grimson, A., Schelter, J.M., Castle, J., Bartel, D.P., Linsley, P.S., and Johnson, J.M. 2005. Microarray analysis shows that some microRNAs downregulate large numbers of target mRNAs. *Nature 433*, 769–773.

Liu, C.G., Calin, G.A., Meloon, B., Gamliel, N., Sevignani, C., Ferracin, M., Dumitru, C.D., Shimizu, M., Zupo, S., Dono, M., 2004. An oligonucleotide microchip for genome-wide microRNA profiling in human and mouse tissues. *Proc Natl Acad Sci USA 101*, 9740–9744.

Lu, J., Getz, G., Miska, E.A., Alvarez-Saavedra, E., Lamb, J., Peck, D., Sweet-Cordero, A., Ebert, B.L., Mak, R.H., Ferrando, A.A., 2005. MicroRNA expression profiles classify human cancers. *Nature 435*, 834–838.

Malumbres, M., and Barbacid, M. 2003. RAS oncogenes: the first 30 years. *Nat Rev Cancer 3*, 459–465.

Miska, E.A. 2005. How microRNAs control cell division, differentiation and death. *Curr Opin Genet Dev 15*, 563–568.

Murakami, Y., Yasuda, T., Saigo, K., Urashima, T., Toyoda, H., Okanoue, T., and Shimotohno, K. 2006. Comprehensive analysis of microRNA expression patterns in hepatocellular carcinoma and non-tumorous tissues. *Oncogene 25*, 2537–2545.

Neely, L.A., Patel, S., Garver, J., Gallo, M., Hackett, M., McLaughlin, S., Nadel, M., Harris, J., Gullans, S., and Rooke, J. 2006. A single-molecule method for the quantitation of microRNA gene expression. *Nat Methods 3*, 41–46.

O'Donnell, K.A., Wentzel, E.A., Zeller, K.I., Dang, C.V., and Mendell, J.T. 2005. c-Myc-regulated microRNAs modulate E2F1 expression. *Nature 435*, 839–843.

Orom, U.A., Kauppinen, S., and Lund, A.H. 2006. LNA-modified oligonucleotides mediate specific inhibition of microRNA function. *Gene 372*, 137–141.

Pasquinelli, A.E. 2002. MicroRNAs: deviants no longer. *Trends Genet 18*, 171–173.

Raymond, C.K., Roberts, B.S., Garrett-Engele, P., Lim, L.P., and Johnson, J.M. 2005. Simple, quantitative primer-extension PCR assay for direct monitoring of microRNAs and short-interfering RNAs. *RNA 11*, 1737–1744.

Rodriguez, A., Griffiths-Jones, S., Ashurst, J.L., and Bradley, A. 2004. Identification of mammalian microRNA host genes and transcription units. *Genome Res 14*, 1902–1910.

Sattler, M., and Salgia, R. 2003. Targeting c-Kit mutations: basic science to novel therapies. *Leuk Res 28*, S11–S20.

Schmittgen, T.D., Jiang, J., Liu, Q., and yang, L. 2004. A high-throughput method to monitor the expression of microRNA precursor. *Nucleic Acid Res 32*, 43–53.

Sethupathy, P., Corda, B., and Hatziegeorgiou, A.G. 2006. TarBase: A comprehensive database of experimentally supported animal microRNA targets. *RNA 12*, 192–197.

Sevignani, C., Calin, G.A., Siracusa, L.D., and Croce, C.M. 2006. Mammalian microRNAs: a small world for fine-tuning gene expression. *Mamm Genome 17*, 189–202.

Sioud, M., and Rosok, O. 2004. Profiling microRNA expression using sensitive cDNA probes and filter arrays. *Biotechniques 37*, 574–576, 578–580.

Slack, F.J., and Weidhans, J.B. 2006. MicroRNAs as a potential magic bullet in cancer. *Future Med 2*, 73–82.

Takamizawa, J., Konishi, H., Yanagisawa, K., Tomida, S., Osada, H., Endoh, H., Harano, T., Yatabe, Y., Nagino, M., Nimura, Y., . 2004. Reduced expression of the let-7 microRNAs in human lung cancers in association with shortened postoperative survival. *Cancer Res 64*, 3753–3756.

Tsujimoto, Y., Finger, L.R., Yunis, J., Nowell, P.C., and Croce, C.M. 1984. Cloning of the chromosome breakpoint of neoplastic B cells with the t 14;18 chromosome translocation. *Science 226*, 1097–1099.

Tsujimoto, Y., Cossman, J., Jaffe, E., and Croce, C.M. 1985. Involvement of the bcl-2 gene in human follicular lymphoma. *Science 228*, 1440–1443.

Volinia, S., Calin, G.A., Liu, C.G., Ambs, S., Cimmino, A., Petrocca, F., Visone, R., Iorio, M., Roldo, C., Ferracin, M., 2006. A microRNA expression signature of human solid tumors defines cancer gene targets. *Proc Natl Acad Sci USA 103*, 2257–2261.

Weiler, J., Hunziker, J., and Hall, J. 2005. Anti-miRNA oligonucleotides AMOs. ammunition to target miRNAs implicated in human disease? *Gene Ther 13*, 496–502.

Xu, P., Vernooy, S.Y., Guo, M., and Hay, B.A. 2003. The drosophila MicroRNA Mir-14 suppresses cell death and is required for normal fat metabolism. *Curr Biol 13*, 790–795.

Yanaihara, N., Caplen, N., Bowman, E., Kumamoto, K., Okamoto, A., Yokota, J., Tanaka, T., Calin, G.A., Liu, C.G., Croce, C.M., 2006. microRNA Signature in lung cancer diagnosis and prognosis. *Cancer Cell 9*, 189–198.

Yekta, S., Hshih, I., and Bartel, D.P. 2004. MicroRNA-directed cleavaga of HOXB8 mRNA. *Science 304*, 594–596.

6 Proteomic Methods in Cancer Research

Scot Weinberger and Egisto Boschetti

ABSTRACT

Recent advancements and progress in proteomics technologies and research protocols have made a demonstrable impact upon clinical investigations, particularly in the area of cancer research. This chapter reviews the overall requirements and approaches involved in clinical proteomics research with particular emphasis on and review of accomplishments in the field of cancer research and therapy. A detailed discussion of the challenges in clinical proteomic research is presented along with a valuable review of protein purification and protein analytical platforms. Extensive discussion on the use of various clinical proteomic mass spectrometric approaches is provided.

Key Words: Translational proteomics, Clinical proteomics, Proteomics, Protein purification, Mass spectrometry, Electrophoresis, Protein microarrays

1 INTRODUCTION

Ever since the term proteome, and its associate, proteomics, was coined (Wasinger et al., 1995; Wilkins et al., 1996a–c; Williams et al., 1996), we have witnessed an explosion of interest and research in the areas formerly referred to as protein or enzyme analysis and protein biochemistry. Not surprisingly, the evolving enthusiasm and promise of modern proteomics techniques and technologies have made a demonstrable impact upon cancer research. A recent search of PubMed citations for proteomic based cancer research publications has indicated a total of about 1,250 publications from 2001 to May of 2006 (see Fig. 1). Moreover, the demonstrated growth in proteomics based cancer research indicates that in the year 2008, about 800 published works will be released. The majority of this work has focused upon applying proteomics technologies to the study of cancer in a variety of organ systems (see Fig. 2), while only 30% of all noted cancer proteomics publications were directed towards the development and testing of proteomic technologies.

G.J. Gordon (ed.), *Cancer Drug Discovery and Development: Bioinformatics in Cancer and Cancer Therapy*,
DOI: 10.1007/978-1-59745-576-3_6, © Humana Press, a part of Springer Science + Business Media, LLC 2009

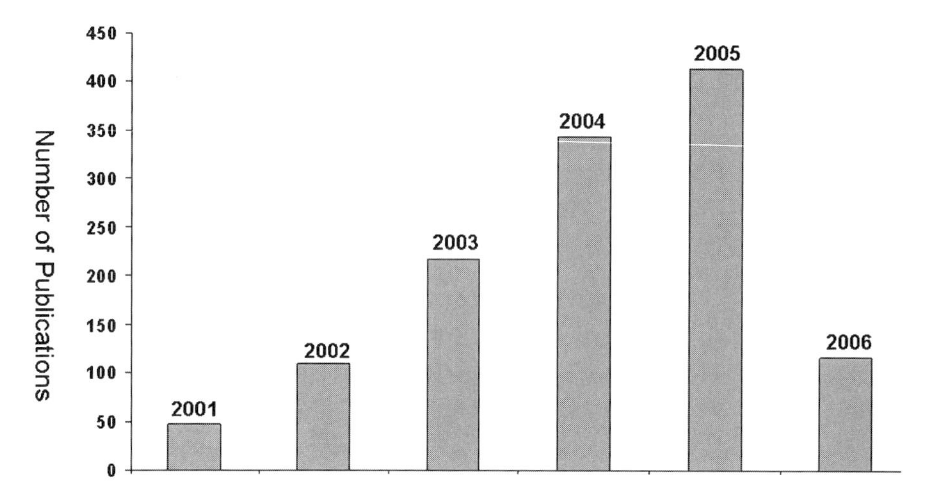

Fig.1. *Proteomic cancer publications.* This figure depicts the total number of PubMed citations that contain the term proteomics and cancer. A growing trend is indicated from 2001 to 2005. 2006 data are incomplete, as it represents total publications as of May, 2006. See text for further details

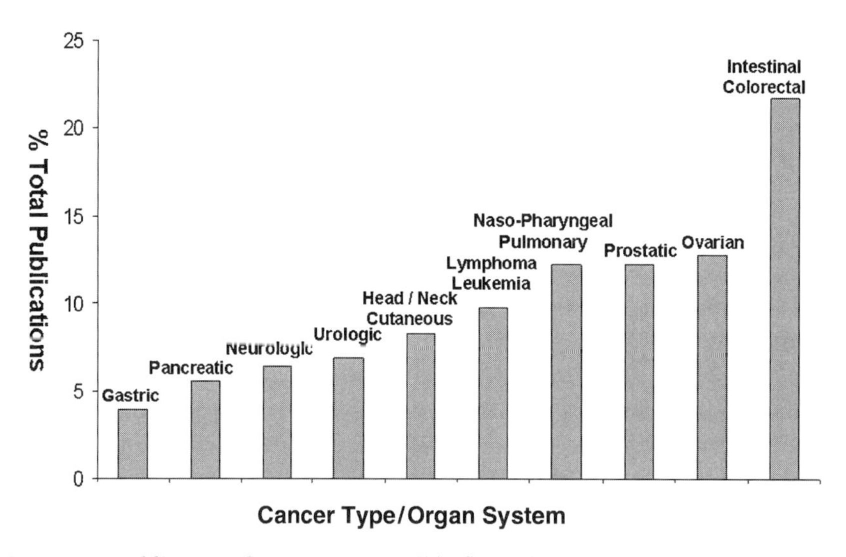

Fig. 2. *Proteomic publications by cancer type.* This figure illustrates the percentage of proteomic studies by cancer type and organ system for all abstracts noted in Fig. 1. As can be seen, proteomic cancer research has been broadly adapted, studying urogenital, integument, nervous, and digestive systems almost in unity. It is somewhat surprising that breast cancer studies represented less than 1% of all proteomic based cancer abstracts, and are not captured by this figure

Roughly compared to genomics research, clinical proteomics approaches are less mature and are far more challenged by the inherent analytical demands placed upon them by the study of proteins in authentic systems. Proteomic studies require tackling an extensive range of protein abundance. Furthermore, there is no direct or easily achievable means by which one can isolate and purify proteins or amplify their abundance when originally present at trace levels. Incompatible practices and methods of clinical specimen collection, processing, and storage compound the challenges. In many cases, established clinical sample collection protocols duly serve their original assay or histological assessment, but introduce unwanted artifacts such as proteolytic cleavage, protein fixation or chemical modification, as well as undesirable levels of electrolytes or media stabilizing proteins. When one considers these challenges, it is of little surprise that proteomic research techniques remain at an early stage of evolution, without a clearly dominant approach (see Fig. 3).

Among today's most popular clinical proteomic research activities is differential protein display or expression monitoring. Differential protein display is a comparative technique that contrasts protein profiles between different organisms, individuals, pathogenic and/or metabolic conditions, and phenotypic response to environmental or chemical challenges. Almost universally, protein phenotypic studies are employed as a first approach in generating a phenomenological model that can be correlated with the clinical question at hand. In cancer research, these phenomenological studies most frequently attempt to differentiate normal cells from transformed or transitional cells, and healthy individuals from afflicted cohorts.

Differential expression analysis performed in basic proteomic research and clinical proteomic studies differ in terms of their overarching goals and analytical requirements. For example, basic research activities often involve studies performed upon pooled biological samples such as lysates or secretions from harvested cells, lysates or products of

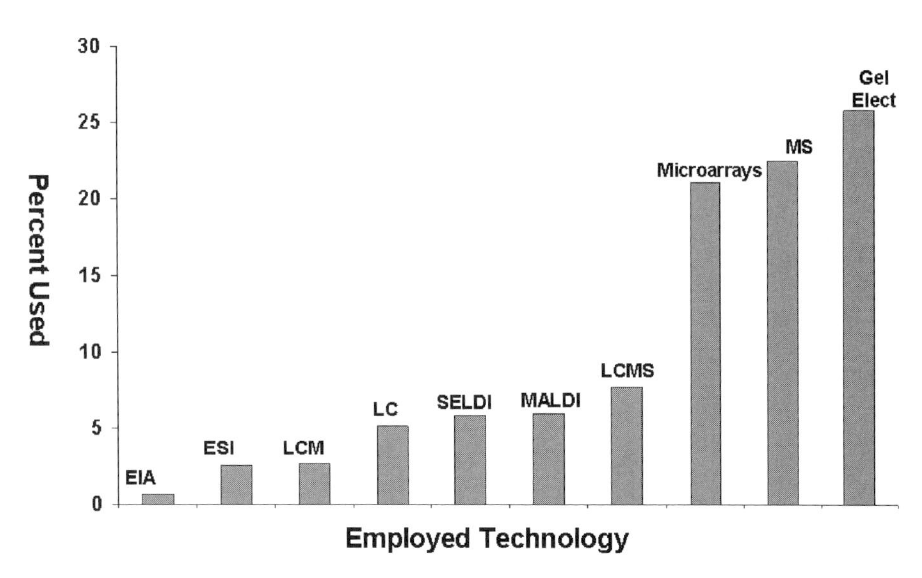

Fig. 3. *Proteomic technologies employed in cancer research.* The figure describes the employment rate of various proteomic research technologies among all citations noted in Fig. 1. Microarray, Mass Spectrometry, and Gel Electrophoresis remain the techniques of choice. Mass Spectrometry values represent the composite of all MS technologies, including SELDI, MALDI, ESI, LC-MS, and Fourier Transform Ion Cyclotron Resonance MS. Abbreviations: EIA, Enzyme Immuno Assay; ESI, Electrospray Ionization; LCM, Laser Capture Micro-dissection; LC, Liquid Chromatography; SELDI, Surface Enhanced Laser Desorption/Ionization; MALDI, Matrix-Assisted Laser Desorption/ Ionization; LCMS, Liquid Chromatography Mass Spectrometry; MS, Mass Spectrometry; Gel Elect, Gel Electrophoresis

cultivated bacteria, or combined pools of biological fluids and tissues from many laboratory animals or human subjects. Under these conditions, basic research studies are not hampered by sample limitations, and protein purification often relies on established techniques such as low- and high-pressure liquid chromatography, serial chromatography, and various electrophoretic approaches. Often, the goal in basic research studies is to catalogue and identify every protein associated with a particular sample. In this case, proteins are frequently reduced to peptides for the purpose of protein identification and characterization typically by electrospray ionization or matrix-assisted laser desorption/ionization mass spectrometry.

In contrast, clinical proteomic studies endeavor to follow the progress of the disease in an individual or small population with the ultimate aim of finding biomarkers potentially useful as diagnostic agents or new drug targets. Under such circumstances, sample or tissue availability is limited and the dependence upon highly efficient, small-scale techniques is high. Typically, protein populations between groups are compared using univariate or multivariate analysis schemes with the ultimate aim of elucidating a protein or groups of proteins whose expression levels correlate with a given clinical condition (Weinberge et al., 2002).

This chapter reviews the various proteomic methods and technologies employed in modern cancer research. The challenges of proteomic analysis are explored and a review of various analytical means is presented.

2 PROTEOME COMPLEXITY AND THE NEED FOR FRACTIONATION

It is well known that the protein composition of tissue extracts and biological fluids is extremely complex, not only because of the numerous encoding genes, but also because of the extremely large number of possible post-translational modifications. The human genome comprises more than 25,000 different genes but proteins are more numerous because of splice variants, regulated and dysregulated proteolysis, and of course all possible modifications occurring during maturation. Furthermore, proteins are diversely expressed. Within clinical samples, some proteins are highly concentrated while others are present only as a few copies and only at certain stages of the cell cycle. It is acknowledged that serum proteins have a difference in concentration that can reach ten or even 12 orders of magnitude (Adkins et al., 2002; Castagna et al., 2005; Thadikkaran et al., 2005).

Because of these factors, a prudently designed clinical proteomic study must begin with fractionation of the clinical sample. It is essential that the selected sample collection and fractionation scheme do not introduce unwanted artifacts, which may be erroneously construed as potential biomarkers. If the initial protocol is long or performed at room temperature, the unwanted activation of nascent proteases is a true concern, resulting in the modification of the initial protein mixture. The latter is particularly troublesome for plasma or serum samples. To mitigate unwanted proteolysis, collected samples should be cooled to 4°C as soon as it is practical to do so, and sample fractionation is best performed under similar conditions. An alternative approach is the immediate addition of anti-proteases, with the acknowledged risk of modifying the nascent proteome and introducing artifacts of different origins.

3 SAMPLE PRE-FRACTIONATION METHODS

Beyond initial sample collection and storage challenges, the key to successful sample fractionation lies within prudent exploitation of the physicochemical properties of proteins, to separate them under the best conditions. A clinical sample is typically a biological fluid or tissue that may need to be a solubilized. Techniques for tissue fractionation, cell separation and sub-cellular fractionation, as well as solubilization, to produce working protein solutions, will not be described in this section. Rather a discussion of separation techniques starting from protein solutions using modern methods of fractionation or isolation will be reviewed.

3.1 Chromatographic Techniques Applied to Proteomics

For years, liquid chromatographic (LC) separation methods have been used for protein separation and purification. There are few restrictions to the use of LC in the clinical proteomics field, with the possible exception of sample availability and load. The scarcity of clinical samples frequently demands the use of small or very small columns with potential complications for the collection of numerous small fractions. Further the typical goal in proteomic studies is not the isolation of one single protein but rather a fractionation, making subsequent protein detection more universal and data analysis simpler in its interpretation. LC separations are applied to native proteins, denatured proteins, and to protein fragments after total or partial hydrolysis by proteases. Chromatographic methods can be implemented using a very large choice of solid phase adsorbents. Among those of low specificity are ion exchangers and hydrophobic interaction sorbents. Medium specificity sorbents that target groups of proteins sharing a common moiety include IMAC sorbents (proteins capable to interact with metal ions), boronic acids for glycoproteins, immobilized lectin (for sub-groups of glycoproteins), hydroxyapatite, and immobilized enzyme inhibitors (e.g., benzamidine for the capture of serine proteases). Highly specific sorbents are also available such as immobilized Protein A or Protein G, well-known for the selective extraction and immobilization of antibodies or FC fusion products.

3.1.1 Ion Exchange Chromatography

Ion exchange (IEX) is the most widely used chromatographic method for protein separation. It is based on the interaction that occurs between a charged protein at a given pH and the complementary charge of a solid phase resin at the same pH. Its separation efficiency depends on a number of factors which have been studied in depth at theoretical and practical levels (Fernandez et al., 1996). A wide variety of functional groups have been described, but the most popular are weak and strong cation and weak and strong anion exchangers. Resins for different applications, in different particle sizes, different ligand densities and pore sizes are easily available (Boschetti, 1994).

Human serum is one of the most widely described protein mixtures fractionated by IEX chromatography. For proteomics research, IEX has been extensively described to simplify the initial sample and then analyze the fractions with the aim of identifying new species (Tirumalai et al., 2003). Throughout the available literature, many examples of proteome fractionation methods have been reported based on ion exchange chromatography (Link et al., 1997; Lopez, 2000; Washburn et al., 2001; Wagner et al., 2002). Ion exchange chromatography is used in combination with other separation methods, such as size exclusion (Opiteck et al., 1998), chromatofocusing (Wall et al., 2000), affinity capture (Geng et al., 2001), or reverse phase chromatography (Tomlinson and Chicz, 2003). However, multidimensional chromatography as used in proteomics fractionation generally does not exceed two dimensions due to the high number of fractions to manage (pH-adjustment, desalting, re-injection in following dimension) and analyze.

3.1.2 Hydrophobic Interaction Chromatography

Hydrophobic interaction chromatography (HIC) is another means to fractionate a protein mixture. HIC is based on the capability of non-polar residues of proteins (clusters of hydrophobic amino acids) to associate in aqueous solutions with hydrophobic chains of the solid phase. The formation of these complexes is characterized by a high entropy contribution to the free energy of the whole aggregation and the enthalpy contribution is low or even negative. The association between proteins and hydrophobic resins is generally promoted by the presence of high concentrations of salts of strong lyotropic effect. Several mechanistic retention models have been described to explain retention strength and the capacity factor of a given protein at a particular lyotropic salt concentration (Melander et al., 1984; Staby and Mollerup, 1996). In typical examples, HIC pre-fractionation allowed detecting novel proteins from cytosolic soluble fraction of *H. influenzae* on a phenyl column. Approximately 150 proteins, bound to the column, were identified, but only 30 for the first time (Fountoulakis et al., 1999).

3.1.3 Affinity Chromatography and Immunoprecipitation

Affinity chromatography, in its various forms, provides specific means to extract targeted groups of proteins or even a single protein by adsorption onto a specifically designed solid phase. It is based on the ability of a protein or another biopolymer to recognize a natural or synthetic partner. Affinity chromatography sorbents consist of a porous matrix on which a selected ligand (bait molecule) is chemically attached directly or by means of a spacer arm. On a mechanistic level, affinity chromatography is performed during the use of protein microarrays, where the bait is selectively immobilized upon an appropriately conditioned planar surface, and subsequently used to pan biological mixtures for proteins of corresponding affinity (often termed the prey).

Affinity chromatography provides higher selectivity of the bait-prey interaction than those existing for ion exchangers and hydrophobic sorbents. The specificity of this interaction is basically governed by the law of mass action with the involvement of association and dissociation constants. In affinity chromatography, a high association rate during loading and washing and a high dissociation rate during elution are needed. High dissociation constant rates can be achieved either by addition of chaotropic agents, or deforming agents into the elution buffer or even by selective elution such as EDTA, for Ca^{++} binding proteins. Competition with a free ligand is another mode to force the protein to desorb to the benefit of the competing molecule as it is frequently the case for lectin-glycorpotein complexes.

In affinity chromatography, high selectivity means that only one protein or one well-defined category binds to the immobilized ligand with high affinity in the presence of a large number of other proteins. Affinity chromatography is a term that embraces a variety of different types of chromatographic recognition; in some cases single proteins can be selectively bound on the solid phase (e.g., immunosorbents), in other cases homogeneous groups of proteins are selectively captured. Groups in this case share one or few common properties such as the recognition for a sugar, an amino acid, a metal ion, or even a dye molecule.

Affinity chromatography has been extensively used as a mean to separate proteins. In most of the cases this technique is used to separate proteins as groups for further analysis. However, well-known applications are used for the removal of high abundance protein as described later in this chapter. Lectin affinity chromatography for group protein binding has been extensively used for the separation of glycoproteins (Geng et al., 2001). As reported, lectin affinity adsorption was used for the separation of glycoconjugates from a variety of expressed protein mixtures prior to proteomics analysis (Corthals et al., 2000b; Lopez et al., 2000; Geng et al., 2001; Brzeski et al., 2003; Ghosh et al., 2004).

Immunoprecipitation uses antibodies that are selective for one or a group of proteins sharing a similar epitope. This approach is commonly used for the separation/analysis of phosphorylated proteins (Ramamoorthy et al., 2004) and protein isoforms such as tumor necrosis factor (Watts et al., 1997). In practice the selected antibody is mixed with the protein extract and incubated for the time that is necessary to form an immune-complex. Then the immune-complex is separated using a Protein A column. This approach is not largely reported in the literature; however, it deserves attention as a pre-fractionation method prior to analysis by two-dimensional electrophoresis or mass spectrometry because of its high level of specificity. By using very specific antibodies the purities of serine/threonine-phosphorylated proteins were significantly enhanced with consequent better specificity of their substrate for kinases (Gronborg et al., 2002).

Immunoprecipitation used alone or in association with other separation methods, was also extensively described as a means for analyzing phosphotyrosyl proteins in cerebrospinal fluid (Yuan and Desiderio, 2003) and the mapping of phosphorylation sites from human T-cells (Brill et al., 2004). Moreover the immunoprecipitation principle is also interesting for investigating the formation of protein–protein complexes and therefore contributing to the elucidation of some pathways (Figeys et al., 2001; Ren et al., 2003; Schulze and Mann, 2004). Immunoprecipitation methods also embrace immunoadsorption using solid phases where the antibodies are covalently attached. The benefit found in this approach is to prevent the contamination of the captured proteins by the antibody. This technology derives directly from immunoadsorption as extensively used in the last three decades for preparative protein purification in biopharmaceutical applications.

3.1.4 Immobilized Metal Affinity Chromatography

Immobilized metal affinity chromatography (IMAC) has been repeatedly reported as an effective means to separate histidine-exposed proteins (Ren et al., 2003; Smith et al., 2004). IMAC chromatography has also been applied to calcium binding proteins (Lopez et al., 2000) and to the separation of phospho-proteins (Ficarro et al., 2002). In Surface

Enhanced Laser Desorption analysis, IMAC targets have frequently been used in the discovery of putative cancer biomarkers (Fung et al., 2001; Wilson et al., 2004; Zhang et al., 2004a).

3.2 *Electrophoresis Based Methods*

Among the various electrophoretic methods routinely used in clinical proteomic analysis, are SDS-polyacrylamide gel electrophoresis, two-dimensional electrophoresis, and isoelectric focusing. Mini-preparative gel electrophoresis and continuous electrophoresis in free liquid films are two specific versions of these electrophoretic schemes. The former has recently been applied to enrich low-abundance brain proteins, by eluting dozens of fractions sequentially (Fountoulakis and Juranville, 2003).

3.2.1 Continuous Electrophoresis

Continuous electrophoresis in free liquid films (also called free-flow electrophoresis: FFE) is based on the continuous flow of an electrolyte in a direction normal to the line of forces of the electrical field and the protein mixture to be separated is added contiguously at a small spot in the flowing medium. Components of the initial mixture are deflected in diagonal trajectories and thus separated according to their electrophoretic mobility and can be collected at the bottom of the chamber as distinct fractions. This technique has the advantage that relatively large samples can be processed. FFE suffers from its high diffusion coefficients; however, to limit this phenomenon, possible solutions have been reported, such as the micro-fabricated FFE device useful for continuous separation of proteins (Kobayashi et al., 2003).

3.2.2 Isoelectric Point Approaches

Applications based on isoelectric point have been reported for protein fractionation. Built-in forces inherent to this kind of protein migration prevent entropic peak dissipation. A significant advantage of this method is immediately evident in the separation of proteins of large size that had always been problematic while using standard methods of isoelectric focusing. This latter has been proposed as the first dimension of a two-dimensional map, the eluted fractions being directly analyzed by orthogonal SDS-polyacrylamide gel electrophoresis (Hoffman et al., 2001). Individual bands from the second dimension can then be eluted and analyzed by electrospray ionization, tandem MS allowing for the identification of a large number of proteins.

Another preparative isoelectric separation method is the Rotofor. The device is assembled from 20 sample chambers, separated by liquid-permeable nylon screens, except at the extremities, where cation- and anion-exchange membranes are placed. It can be used as first dimension in a two-dimensional electrophoresis-like process, in which each fraction can be further fractionated using chromatography as reported (Zhu and Lubman, 2004). Modifications to the original principle have been described such as miniaturized devices accommodating isoelectric membranes in replacement of carrier ampholytes with significant advantages (Zuo and Speicher, 2002; Shang et al., 2003).

Another miniprep electromigration technique for the separation of proteins based on their isoelectric point is "Off-gel IEF" (Ros et al., 2002). Likewise in a multi-compartment separation technique, the system is composed of liquid chambers positioned on top of an IPG continuous gel slab. Upon application of an electric field, perpendicularly to the liquid chamber, charged species (those having pI values above and below the pH of the IPG gel) are extracted away from the chamber's electric field. After separation, only the globally neutral species (pI = pH of the IPG gel) remain in the solution. A practical extension of this initial principle uses a multi-well device, composed of a series of compartments of small volumes compatible with current instruments for separation (Michel et al., 2003).

Preparative isoelectric focusing as an electrophoresis-based method for preparative protein separation made significant progress when multi-compartment electrolyzers using isoelectric membranes was conceived (Wenger et al., 1987; Righetti et al., 1990). The advantages of such a device were immediately apparent. First, it allowed separating proteins by groups of well-defined isoelectric points in solution with very limited danger of protein precipitation. Second. this approach was directly compatible for additional sub-fractionation with the added advantage of enhancing protein concentration.

Finally, another preparative electrically-driven protein separation is based on the use of a thin bed of neutral dextran beads embedded with carrier's ampholytes where proteins move according to their isoelectric point when submitted to an electric field (Gorg et al., 2002). The focusing process is induced by the presence of carrier ampholytes. Dextran beads are used as an anti-convective medium where proteins move freely to reach their isoelectric point. Separated proteins are collected with the beads and then used for further analysis. However, some down-stream analyses are complicated; this is the case of mass spectrometry because ampholyte carriers, which are very numerous and not compatible with mass spectrometry, are difficult to remove.

3.3 Depletion of High Abundance Proteins and Compression of Expression Dynamic Range

3.3.1 Depletion Technologies

Albumin, immunoglobulins, transferrin and a few other very high-abundance proteins in plasma, serum, or even CSF, represent a challenge for the proper detection of many low concentration proteins. For instance, albumin is a source of trouble in mass spectrometry as it suppresses the detection of numerous other species. Similarly two-dimensional electrophoresis cannot reveal species that are covered by the signal of albumin. To resolve the situation it has been proposed to remove one or more high-abundance species prior to mass spectrometry and two-dimensional electrophoresis analysis.

Serum albumin is removed by either using Cibacron Blue dye or anti-albumin antibodies attached to chromatographic beads. Immunoglobulins G are removed by using immobilized Protein A or Protein G or even anti-IgG antibodies. Albumin and IgG can also be removed simultaneously by mixing two different sorbents. Recently it has been proposed to remove several proteins at a time, using a mixture of immunosorbents against the major plasma proteins such as albumin, IgG, IgA, transferrin, haptoglobin and α1-anti-trypsin (Martosella et al., 2005; Zolotarjova et al., 2005).

While seemingly advantageous, all these depletion methods, including immunoadsorption, can be problematic because they remove target proteins that may be closely associated with the removed species while simultaneously resulting in unwanted dilution (Zhou et al., 2004b; Zolotarjova et al., 2005). It has been recently reported that depletion methods may produce artifacts in terms of disease significance (Mehta et al., 2003; Tirumalai et al., 2003).

3.3.2 Compression Technologies

As previously noted, clinical sample proteomes are highly complex in their diversity and expression levels. The most representative example is the human serum where albumin alone represents 60% of the total protein load and the nine most abundant species (albumin, immunoglobulins G, haptoglobin, transferrin, transthyretin, α_1-antitrypsin, α_1-acidic glycoprotein, hemopexin and α_2-macroglobulin) constitute about 90% of the entire serum proteome. It is commonly admitted that the 50 first proteins represent 99% of the protein mass and the remaining 1% comprises more than 100,000 other proteins.

One possible solution to the abundance challenge is to selectively enrich the abundance of low copy number proteins, by capturing them using a diverse library of affinity ligands. The principle of solid phase adsorption would be advantageous since it selectively concentrates the target protein instead of diluting it, as it is the case with depletion. Based on the principle of over-saturating affinity beads and extending the principle to all serum proteins at a time, one could concentrate a large number of low abundance species, hence compressing the dynamic range of abundance. This method necessitates a very large number of highly selective ligands, each of them attached to a distinct bead in a number exceeding the number of target proteins. If such diversity of beads is mixed together and a large excess volume of protein mixture (with large differences in protein concentration such as in the serum) is loaded, high abundance proteins very rapidly saturate the corresponding beads while low abundance species continue to adsorb as long as the sample is available. Based on this principle, a novel approach has been described using solid phase ligand libraries (Righetti et al., 2005; Thulasiraman et al., 2005). The library is comprised of discrete beads, each of them carrying a relatively large number of copies of the same ligand, and ligands are different from one bead to another.

The use of such a highly diverse combinatorial library of affinity ligands in overloading conditions results in a large reduction of protein concentration difference. High-abundance proteins such as albumin, IgG and others are partially eliminated while low abundance proteins are concomitantly and progressively concentrated as the solid phase is continuously fed. Retained proteins are then eluted in bulk or sequentially from the affinity library using buffer modifiers such as ionic strength, pH, chaotropic agents or organic solvents with subsequent analysis by any number of analytical methods.

When applied to clinical samples with down-stream 2-D Gel and/or MS analysis, this compression approach has resulted in the novel discovery of proteins present in biological fluids. For example, human urine was analyzed before and after treatment using the same ligand library with the discovery of 251 proteins never described before (Castagna et al., 2005).

4 CLINICAL PROTEOMIC ANALYTICAL METHODS

This section describes the most important analytical methods applied in the realm of clinical proteomics studies. For the most part, clinical proteomic analysis relies upon electrophoretic, liquid chromatographic, mass spectrometric, protein array, and bioinformatic technologies and products, many of which are discussed herein.

4.1 Top-Down vs. Bottom-Up Approaches

In clinical proteomic analysis, basic analytical approaches can be roughly grouped into two different schemes: those that directly analyze nascent proteins as they present themselves in living systems (Top-Down) and those that directly examine their proteolytic fragments (Bottom-Up). Top-down methodologies include: Surface Enhanced Laser Desorption/Ionization (SELDI) analysis (Merchant and Weinberger, 2000), 2-D gel electrophoresis (Jain, 2002); differential gel electrophoresis (Friedman et al., 2004; Alfonso et al., 2005); virtual 2-D gel electrophoresis (Loo et al., 1996, 2001); hyphenated LC combined with mass spectrometry (MS) (Chong et al., 2001; Kachman et al., 2002); protein arrays (Geho et al., 2004; Alessandro et al., 2005; Clarke and Chan, 2005), and electrospray ionization (ESI) Fourier transform ion cyclotron resonance mass spectrometry (FT ICRMS) (Ge et al., 2002; Meng et al., 2004).

Top-Down approaches benefit from the ability to directly study the protein in question. As such, investigators can correlate putative peptide sequence, measured protein physical characteristics (such as isoelectric point and hydrophobic index) with observed protein molecular weight, providing additional avenues towards protein identification as well as facile detection of posttranslational modifications. For high throughput schemes, such as SELDI MS, Top-Down biomarker discovery protocols can be immediately translated into assays for the purpose of diagnostic validation. However, the benefits of Top-Down analysis are accompanied by several demands, particularly in the area of MS detection.

Because proteomic samples are complex solutions that often thwart the generation of useful MS signals, MS detection is almost universally hyphenated with at least one separation scheme. Even so, most protein separation schemes fail to provide sufficient resolution, often creating complex protein sub-populations in each fraction. Ionization potentials of intact proteins can be vastly different, making confident qualitative and quantitative analysis somewhat problematic. The latter is particularly true for amphiphobic and hydrophobic proteins as well as for proteins with post-translational modifications (PTMs, i.e., glycosylation and phosphorylation). Regardless of ion yield, analysis of complex protein fractions often places a high demand upon MS resolving power, especially for Top-Down studies using Electrospray Ionization (ESI) as an ionization source. For proteins with molecular weights in excess of about 10 kDa, confident deductions regarding protein identity and PTMs become increasingly dependent upon mass accuracy, often requiring the use of sophisticated MS platforms.

Because Bottom-Up studies analyze peptide fragments of nascent proteins, many of the previously described challenges encountered in the Top-Down approach are circumvented. Most commonly used proteolytic approaches generate protein hydrosylates with molecular weights below 5 kDa, reducing mass accuracy demands for determining protein identity or peptide PTMs. Compared to their parent proteins, peptides are more uniformly ionized, somewhat mitigating the challenge of qualitative and quantitative detection.

Most frequently, Bottom-Up MS analysis is hyphenated with at least one on-line peptide separation scheme. A typical platform consists of a reverse phase (RP) HPLC system and is joined to an ESI MS. When more complete proteome coverage is desired, Bottom-Up MS analyses rely upon tandem peptide separation schemes, such as combined ion exchange (IEX) and RP chromatography. Typically 5–10 fractions for the first dimension are collected and then subjected to RP HPLC MS analysis.

While Bottom-Up approaches are useful in cataloguing protein populations, they do not scale well in terms of creating practical assays. Further, the end product of a Bottom-Up study is a data base identification of the peptide or peptides detected during the purification scheme. Unless 100% peptide coverage of the parent protein is obtained, the analyst remains uncertain as to true nature of the biomarker in the living system. The detection of important PTM's may be overlooked, this is particularly troublesome when abnormal proteolysis is observed as a direct consequence of pathologic or metabolic alterations. Often, the analyst is compelled to search for the nascent protein to address PTM and assay development issues. Clearly in the clinical proteomics world, de novo Bottom-Up studies eventually lead to directed Top-Down research (see Fig. 4).

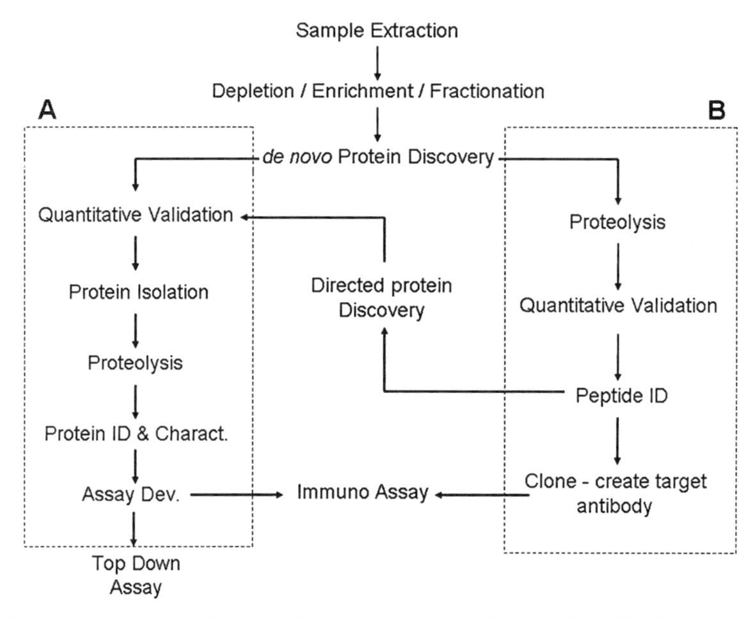

Fig. 4. *Clinical proteomic workflow.* (**a**) Top-down proteomics studies; (**b**) Bottom-up proteomics studies. Clinical proteomic analysis is often first performed by using a non-directed or de novo protein discovery approach. Both top-down and bottom-up methodologies can be applied. Once protein identification has been established, undirected studies frequently give way to directed analysis, particularly during the phase of assay development or biomarker validation

4.2 Top-Down Methodologies

4.2.1 Two-Dimensional Gel Electrophoresis–Mass Spectrometry

The most widely used Top-Down protein discovery approach is two-dimensional gel electrophoresis/mass spectrometry (2-D MS) (Goerg et al., 2004). While 2-D gels do not directly study proteins by mass spectrometry, they do provide a means to generate protein profiles of authentic samples. Briefly, proteins are firstly separated by their isoelectric point via immobilized pH gradients and then further fractionated using sodium dodecyl sulfate polyacrylamide gel electrophoresis (SDS PAGE). PAGE slabs are then stained creating a two-dimensional array of spots. Slabs are digitally imaged using an optical scanner, and scanned images can be differentially compared, with particular focus upon spot location and stain intensity. As such, three-dimensional patterns can be generated for each sample (protein pI, protein molecular weight, and protein abundance).

Early attempts at interpreting 2-D gel profiles were facilitated by using artificial intelligence and machine learning programs. One particular program termed MELANIE (Medical Electrophoresis Analysis Interactive Expert System) was created to automatically classify 2-D profiles using heuristic clustering analysis and hierarchical classification (Appel et al., 1988; Pun et al., 1988). The overarching goal of this work was to create a means to determine disease-associated patterns with the intent of creating a computer based diagnostic regimen. It was successfully applied towards the diagnosis of liver cirrhosis and the distinction of a variety of cancer types from cancerous biopsies (Appel et al., 1991). Later the work was extended towards the comparative analysis of plasma/serum obtained from apparently healthy individuals and from patients with a few selected, known diseases. Despite their apparent complexity, the patient electropherograms revealed readily detectable modifications of the reference protein profile for the selected diseases. Several disease associated spot patterns were elucidated from patients with monoclonal gammopathies, hypogammaaglobulinemia, hepatic failure, chronic renal failure, and hemolytic anemia (Tissot et al., 1991).

While initially successful, the 2-D approach failed to translate to the clinic; it was inherently troubled in terms of its limited reproducibility, restricted dynamic range of detection, and laboriously slow throughput. More recently, a new 2-D approach termed Difference Gel Electrophoresis (DIGE) has been introduced as a possible means to improve reproducibility and throughput (Unlu et al., 1997). DIGE is a modification of the classical 2-D approach in which multiple samples can be analyzed within the same gel, allowing for the simultaneous analysis of experimental and control cohorts. DIGE is performed by fluorescently tagging multiple samples with different amine reactive dyes, running them on the same 2-D gel, and then performing post-run fluorescence imaging of the gel, allowing for direct superimposition of groups. In this manner, one sees a reduction in the number of gels processed and imaged. Further, the affects of gel-to-gel irreproducibility are somewhat minimized. DIGE based analysis have been successfully performed in the study of colorectal cancer (Friedman et al., 2004; Alfonso et al., 2005).

While 2-D gel patterns can be instructive, ultimately the identity of unique markers must be established. Towards this end, 2-D gel analysis has been married with ESI and Matrix Assisted Laser Desorption/Ionization (MALDI) mass spectrometry. Protein spots of interest are typically excised from the gel and then de-stained. The excised gel plugs are then digested using proteases with specific proteolytic activity, such as trypsin, endo-Lys-c, and V-8 protease. Liberated peptides diffuse out of the gel plugs lending themselves for subsequent MS analysis. An over-view of this process along with suggested protocols is provided by Corthals et al (Gygi et al., 2000). The specific details of MS based protein identification are subsequently discussed in this chapter.

4.2.2 Virtual 2-D Gel Analysis

To address the reproducibility and throughput issues associated with classical 2-D gel analysis, several researchers hyphenated the direct analysis of polyacrylamide gels with MALDI mass spectrometry, eliminating the step of SDS PAGE (Loo et al., 1996). Ultra thin (less than 10 μm when dry) gels were soaked in MALDI matrix solution and then directly analyzed in a MALDI-TOF mass spectrometer. Initial spectra were acquired from isoelectric focusing, native, and SDS gels. Virtual 2-D gels were created by MALDI scanning isoelectric gels. Virtual 2-D gels were extended to the study of the *E. coli* proteome (Loo et al., 2001). When compared to classical 2-D studies of the same proteome, virtual 2-D analysis allowed for the postulation of protein identities (<50 kDa) based upon improved molecular mass determination and pI (±0.3 pH units).

Data reduction and display algorithms were created to allow for facilitated viewing and studying of virtual 2-D results (Walker et al., 2001). At its most advanced state, virtual 2-D gel analysis demonstrated high sensitivity (analogous to silver stain detection limits) and improved throughput, resolution, and mass accuracy when compared to the classical 2-D analysis. However, complications associated with interfacing gels with MALDI MS systems fundamentally limited broader adoption of this approach.

4.2.3 Surface Enhanced Laser Desorption/Ionization

SELDI protein array technology represents a collection of analytical tools and protocols that address the challenges of protein separation, protein purification, and protein detection by mass spectrometry (Merchant and Weinberger, 2000; Fung et al., 2001; Lin et al., 2003). SELDI array surfaces function as solid phase extraction media that support on-probe isolation and cleanup of analytes prior to mass spectrometric investigation. Analytes with physical and chemical properties that are complimentary to the array surface functional groups are adsorbed, while others are washed away during the sample preparation process.

After adsorption and purification upon the array surface, retained proteins are subsequently desorbed and ionized using matrix assisted laser desorption/ionization and are typically detected by a time-of-flight mass spectrometer. To date, there are over of 220 published studies using SELDI in cancer research. Recently, SELDI technology has demonstrated the ability to translate the fundamental discovery of protein biomarkers into predictive assays for the purpose of diagnosing the presence of epithelial ovarian carcinoma (Zhang et al., 2004b). In a related study, researchers used SELDI to uncover host specific protein PTMs capable of classifying cancer subtypes (Fung et al., 2005).

SELDI is one of the most successful top-down approaches to address qualitative, quantitative, and throughput challenges of pattern recognition diagnostic test development. Significant advancements in automated sample preparation, MS analysis, and statistical analysis affords the investigator the ability to generate extensive protein profiles for the purpose of de novo biomarker discovery, often using less than 100 uL of the total sample. Because proteins are sequestered according to their physico-chemical characteristics, competitive ion suppression is significantly reduced when compared to MALDI or ESI MS analysis, improving assay qualitative and quantitative reliability. Because SELDI is a true top-down MS discovery approach, initial protein discovery conditions can be leveraged to create protein purification or directed protein discovery schemes.

Compared to de novo or directed SELDI protein discovery, SELDI based protein identification is far more laborious and pales by comparison to the throughput achievable in Bottom-Up studies. However, since only statistically meaningful biomarkers need be identified, the additional burden of protein identification does not significantly impact the entire assay development process, and as such, SELDI based diagnostic test development remains one of the most rapid approaches towards the creation of well characterized protein clinical assays.

4.2.4 Liquid Chromatography: Mass Spectrometry (LC-MS)

In a similar vein as virtual 2-D gel analysis, researchers combined protein isoelectric focusing with nonporous silica RP HPLC–ESI Time of Flight (TOF) MS analysis to create a liquid phase, three-dimensional protein separation method (Wall et al., 2001). The fast scanning speed, high resolution/mass accuracy, and sensitivity of the TOF MS analyzer allowed for the detection of several hundred unique proteins in the pI range of 4.8–8.5 from cytosolic fractions of human erythroleukemia cell line lysates. Proteins were identified by combining determined pI (±0.5 units) and intact molecular weight (±150 ppm error). Using molecular weight and peptide fingerprinting results, PTM's and sequence modifications were noted. The combination of pI, RP elution time and MW was also used to create 2-D pI-MS protein maps, where proteins are displayed as bands, whose gray scale was proportional to the intensity of the protein molecular ion peak.

Two-dimensional chromatofocusing liquid separations were also combined with RP HPLC–ESI TOF MS analysis to separate and analyze proteins from human breast epithelial whole cell lysates (Chong et al., 2001). Because isoelectric point separation was now achieved using an LC column, throughput was remarkably improved. Top-Down LC-MS methodologies have been successfully applied to differential studies of premalignant and malignant human breast cell lines (Chong et al., 2001) as well as epithelial ovarian carcinoma (Kachman et al., 2002).

4.2.5 Top-Down Fourier Transform Ion Cyclotron Resonance MS (FT-ICR-MS) Analysis

The advent of high mass resolving power and accuracy afforded by FT-ICR-MS has enabled the direct study of proteins, with or without prior LC separation, with impressive results (Kelleher et al., 1999). ESI has been coupled with FT-ICR to enable accurate mass assignments for intact proteins as well as large fragments of proteins whose masses easily sum to those of the intact parent. Identified proteins or protein fragments of interest can be directly sequenced using low energy collisional induced dissociation (CID), electron capture dissociation (ECD), as well as infrared multi-photon dissociation (IRMPD) (McLafferty, 2001; Reid and McLuckey, 2002). In order to achieve these ends, a number of data processing improvements were required.

While ESI has the ability to produce multiply charged ions, bringing large molecules down to a mass (*m*) to charge (*z*) ratio amenable to FT-ICR analysis, it also provides a fundamental disadvantage because determined mass becomes ambiguous unless charge can be firmly established. For pure compounds or for low mw species that only form up to three multiply charged ions, the latter is easily achieved by using a simple algorithm that uses multiple peaks in the same mass with different charge states. However, for complex mixtures of large peptides and proteins, a great number of multiple charge envelopes will overlap, making this straight forward approach fail. The high resolving power of FT-ICR allows for direct, unambiguous determination of charge state, based upon the incremental m/z difference for a given isotopic distribution. The latter was leveraged to create a pattern recognition based, automatic charge state assignment algorithm from deconvoluted isotopic envelopes, allowing on-the-fly determination of protein profiles, even for complex mixtures (Senko et al., 1995). Later a new computer algorithm known as THRASH (Thorough High Resolution Analysis of Spectra by Horn) was developed to further accelerate protein signal deconvolution and identification. THRASH combines a subtractive peak finding routine to locate possible isotopic clusters with a Fourier transform/ Patterson method for primary charge determination. A least-squares fitting to a theoretically derived isotopic abundance distribution was then used for final m/z determination. Further, a new signal to noise calculation procedure was devised for the accurate determination of baseline and background noise (Horn et al., 2000b). In terms of deriving de novo protein sequence from ESI FT-ICR generated ECD fragmented proteins, McLafferty and coworkers developed an algorithm to convert fragment ion mass values into most probable protein sequence (Horn et al., 2000a). The algorithm has been successful in deriving a sequence from proteins as large as 10 kDa in size.

Top Down FT-ICR-MS measurements have been effectively performed in the study of *Bacillus cerus* T spores (Demirev et al., 2001), proteins of thiamine biosynthesis (Ge et al., 2002), as well as human lung cancer cell lines (Yan et al., 2005). When combined with size exclusion – nano LC separation, a number of biomarkers distinguishing stage III/IV epithelial ovarian carcinoma from postmenopausal age matched controls were highlighted (Bergen et al., 2003).

Among the previously discussed Top Down approaches, FT-ICR arguably represents the most exciting and promising advancement. However, several challenges remain. FT-ICR-MS instrumentation is exquisitely expensive and challenging to competently operate, making the likelihood of broad-based adaptation low without significant burden reduction. Further, because of m/z range restrictions, Top-Down FT-ICR approaches must be inherently linked with electrospray ionization. As such, quantitative results will be invariably affected by competitive ion suppression, placing high requirements upon chromatographic resolution and retention time reproducibility. Further, sequence determination of large ESI formed ions is greatly dependent upon charge state (Reid and McLuckey, 2002; Yan et al., 2005). As such, advanced algorithms for automated charge state selection and distillation of multiple MS/MS analysis to create composite sequence determination are still required.

4.3 Bottom-Up Methodologies

4.3.1 Shotgun LC-MS and LC-MS/MS Analysis

Shotgun LC-MS and LC-MS/MS analysis has emerged as a popular approach to catalogue proteomes of interest as well as a means to discover protein based biomarkers. Protein populations are irreversibly reduced prior to global digestion with an enzyme of known specificity. In most cases, trypsin is employed. In few instances, proteins are initially fractionated using IEX, isoelectric focusing, or affinity chromatography prior to global digestion. The resultant highly complex peptide pool is then subjected to LC-MS/MS analysis. Initial studies often focused upon specific proteomes of modest complexity and employed a single stage of LC separation (RP HPLC). The human urinary proteome was evaluated by combining low flow rate gradient HPLC with ESI quadrupole TOF MS and MS/MS analysis (Davis et al., 2001b). In all about 200 proteins were identified during a 24 h analytical period.

4.3.2 MudPIT: Towards Higher Shotgun Resolution

In order to address the extensive peptide complexity created in global digestion schemes, a group of researchers originated what is to date the most widely adapted method for shotgun proteomics known as multidimensional protein identification technology (MudPIT) (Wolters et al., 2001). MudPIT is a multidimensional LC method that integrates SCX and RP resin in a single, biphasic column. As is the case with many landmark advancements, variations on a common theme emerge, and today many researchers combine a two-phase separation scheme using distinct SCX and RP columns. For example, an automated SCX exchange – gradient RP HPLC–MS/MS system was used to evaluate complex peptide digestion mixtures. Investigators noted greater than 40% increase in the number of peptide and protein detections when compared to a RP LC control (Davis et al., 2001a). For the purpose of this review, we will not distinguish between integrated and distinct MudPIT analysis.

Since its introduction, MudPIT analysis has been extended towards the investigation of whole proteomes, cellular organelles, protein complexes (Paoletti et al., 2004; Washburn, 2004), membrane bound proteins (Wolters, 2004), and plant proteomes (Park, 2004). More recently, MudPIT has been used to monitor changes in global protein expression patterns in cells and tissue as a function of developmental, physiologic and disease processes (Kislinger and Emili, 2005). In terms of cancer research, MudPIT has been applied to the study of epithelial ovarian carcinoma (Somiari et al., 2005), silenced p53 effectors (Benzinger et al., 2005), lung micro-vascular endothelial cells (Durr et al., 2004), serine protease inhibitors (Chen et al., 2005), pancreatic carcinoma (Mauri et al., 2005) and breast cancer(Jessani et al., 2005; Somiari et al., 2005).

Unlike most Top-Down approaches, MudPIT analysis provides a large number of protein identifications as part of its de facto operation. Frequently hundreds to thousands of putative identifications and measured peptide signals are generated in a single analysis, making the task of data interpretation and quality assurance challenging, to say the least. Towards this end, an advanced data management program known as Pep-Miner has been developed (Beer et al., 2004). Pep-Miner functions by clustering similar spectra from multiple LC-MS/MS runs. The major effect is a marked reduction in the huge amounts of data to a more manageable size, allowing for convenient storage and post-processing of acquired spectra. In one study Pep-Miner was applied to a MudPIT analysis of lung cancer cells, reducing an initial 517,000 spectra to 20,900 clusters while identifying 2,518 peptides derived from 830 proteins (Beer et al., 2004).

A major computational effort has been focused upon improving the confidence in reported peptide identifications. SEQUEST (Yates et al., 1995) is one of the most broadly used automated peptide identification tools. SEQUEST functions to correlate un-interpreted tandem mass spectra of modified peptides produced under low-energy collision conditions, with amino acid sequences in a protein database. Observed peptide fragmentation patterns in the tandem mass spectra are used to directly search and fit linear amino acid sequences into the data base. However, confidence in the identified peptides has been shown to be dependent upon spectral quality. Recently, work has been directed towards the creation of two different pre-processing approaches to assess spectral quality prior to SEQUEST identification: binary classification, which predicts whether or not SEQUEST will be able to make an identification; and statistical regression, which predicts a more universal quality metric involving the number of b- and y-ion peaks (Bern et al., 2004). Another algorithm, known as Logistic Identification of Peptides (LIP), has been developed and reportedly achieves high sensitivity and selectivity for peptide classifications when compared to manually verified gold standards. The LIP index is a weighted average of SEQUEST output variables based on logistic regression models (Higdon et al., 2004). Neural networks and specific statistical models have also been applied to SEQUEST mining results in order to normalize reported scores with respect to peptide composition and length. The investigators reported an improved sensitivity and specificity of peptide identification compared to the standard SEQUEST filtering procedure (Razumovskaya et al., 2004).

The overall implications of potentially false positive identifications from database mining experiments of large tandem MS experiments are discussed in Cargile et al. (2004). The authors indicate significant false positive identification rates, even when previously suggested probability score thresholds have been applied. Other researchers have evaluated several protein identification programs with respect to search selectivity and sensitivity (Chamrad et al., 2004).

Even with present software advancements, the analyst is advised to manually verify putative protein identities from shotgun studies. Clearly manual inspection is impractical on a global level, sustaining pressure to improve quality assurance routines for shotgun proteome inventory studies. However, in the realm of biomarker discovery, differential algorithms can be applied between cohorts to identify a finite subset of markers. Under such circumstances, manual verification is not an egregious task.

4.3.3 Quantitative Challenges in Shotgun Analysis

As is the case for any protein differential display regimen, shotgun differential proteomic studies need to address both qualitative and quantitative figures of merit. Towards this end, a computer program called RelEx, which uses a least-squares regression for the calculation of peptide ion current ratios from MS derived ion chromatograms, has been created (MacCoss et al., 2003)

In an effort to constrain quantitative error, quantitative Bottom-Up studies have used specific labeling motifs. Isotope Coded Affinity Tags (ICAT) was introduced as an approach to further improve quantitative accuracy and concurrent sequence identification of individual proteins within complex mixtures (Gygi et al., 1999). Briefly, ICAT tags are biotin containing isotopic mass tags with specific reactivity for reduced cysteine. Globally digested protein pools are labeled with different ICAT tags after which cysteine containing peptides are extracted using avidin beads. Differentially labeled populations are recombined for simultaneous analysis using LC-MS/MS. Initially the approach was used to compare protein expression within *Saccharomyces cerevisiae*. Later ICAT was used to illustrate differences between mRNA abundance and protein expression in the same yeast (Aebersold et al., 2000). Subsequently, ICAT was extended to the analysis of native and campthothecin-treated cortical neurons (Yu et al., 2002), quantitative profiling of LNCap prostate cancer cells (Meehan and Sadar, 2004), breast cancer (Pawlik et al., 2006), and the identification of androgen co-regulated protein networks of human prostate cancer cells (Wright et al., 2004). Several reviews examining stable isotope affinity tags have been authored (Tao and Aebersold, 2003; Wright and Aebersold, 2003; Zhou et al., 2004a).

One complication of the ICAT labeling scheme is that the covalently linked tag causes a measurable difference in ion fragmentation resulting in additional burden for automated sequence interpretation. Studies comparing low energy CID fragmentation patterns of peptides labeled with the ICAT reagent to those of unmodified peptides revealed the formation of ions attributed to the modified Cys peptide as well as those unique to the labeling reagent (Borisov et al., 2002). Further, since labeling is dependent upon the presence of Cys in the studied peptide, the majority of a given protein's coverage is discarded, often with great risk of missing key posttranslational modifications.

Another quantitative labeling motif with more universal peptide coverage than ICAT is Isobaric Tags for Relative and Absolute Quantitation (iTRAQ™) (Chong et al., 2006). The iTRAQ method uses a multiplexed set of four isobaric reagents specifically reactive for primary amines, thus theoretically labeling every peptide from a protein digested by a c-terminal lysine or arginine specific protease (i.e., Trypsin or endo-Lys-C). Resultant peptides are identical in mass and single ms analysis mode, but generate highly specific tandem MS signatures. In terms of cancer research, iTRAQ has been used to study breast cancer(Overall and Dean, 2006) as well as endometrial carcinoma(DeSouza et al., 2005). A specific software program known as I-Tracker has been developed to facilitate relative and absolute quantitative studies (Shadforth et al., 2005).

Recent work has compared the quantitative performance of ICAT, DIGE, and iTRAQ using a six-protein mixture, a reconstituted protein mixture (BSA spiked into human plasma devoid of six abundant proteins), and complex HCT-116 cell lysates as the samples (Wu et al., 2006). All three techniques yielded quantitative results with reasonable accuracy when the six-protein or the reconstituted protein mixture was used. In DIGE, accurate quantification was sometimes compromised due to co-migration or partial co-migration of proteins. The iTRAQ method was found to be more susceptible to errors in precursor ion isolation, which could be aggravated with increasing sample complexity. The quantification sensitivity of each method was estimated by the number of peptides detected for each protein. In this regard, the global-tagging iTRAQ technique was more sensitive than the cysteine-specific cICAT method, which in turn was as sensitive as, if not more sensitive than, the DIGE technique. Protein profiling on HCT-116 and HCT-116 p53 –/– cell lysates displayed limited overlapping among proteins identified by the three methods, suggesting the complementary nature of these methods.

4.4 Protein Identification and Characterization

The analytical and computational challenges of mass spectrometry biomarker discovery and mass spectrometry protein identification are inextricably linked. To fully appreciate this inter-relationship, one needs to consider the basic approaches to pattern-based biomarker discovery and the inherent requirement for protein identification. For regardless of the discovery approach, diagnostic proteins should be well characterized by confidently confirming identity, and establishing precise primary sequence as well as all salient post-translational modifications (PTMs). In this section, we present a modest over-view of MS protein identification methods.

As previously noted, regardless of the biomarker discovery scheme, statistically validated diagnostic candidates must be identified in terms of their primary amino acid sequence and exhibited PTMs. Nowadays, protein identification and characterization is almost universally achieved using single or tandem mass spectrometry in combination with computational algorithms.

4.4.1 Peptide Mass Fingerprinting

With the continued growth of protein sequence data bases as well as the emergence of cDNA databases, it became possible to derive peptide sequence and protein identity by correlating MS measurements with theoretical peptide fragments of a previously known sequence. Henzel *et al.* introduced a computer algorithm, later coined Fragfit, for the automated identity determination of proteins separated by 2-D gels (Henzel et al., 1993; Arnott et al., 1996). Peptides were generated by reduction, alkylation, and tryptic digestion and then analyzed via MALDI TOF. Fragfit functioned by searching an existing protein sequence database for multiple peptides of individual proteins that match the measured masses. In a parallel effort, Mann and coworkers created routines to correlate MS results to protein identities found within the protein databases (Mann et al., 1993)

Today, the process of identifying proteins based upon single MS measurements of specific proteolytic fragments searched against protein or cDNA databases is generically referred to as peptide mass fingerprinting (PMF). High throughput (HT) PMF analysis is frequently performed in hyphenated 2-D gel MALDI MS analysis. Alternatively several in-gel digestions are queued up for automated analysis using LC ESI MS. As was the case in shotgun experiments, HT PMF studies generate a tremendous amount of data, creating corresponding challenges in insuring quality in protein identifications. Accordingly, computer algorithms directed towards improving protein identification from PMF studies have been created.

4.4.2 Sequencing and Protein ID via MS/MS

While PMF can often provide initial protein identification, in cases in which insufficient protein purification is achieved, or in studies with limited peptide coverage, substantial irregular peptide cleavage, and/or PTM's, PMF algorithms generally fail to provide a confident and complete list of all proteins found in the original sample. Consequently, tandem MS analysis is relied upon as the gold standard for establishing the peptide primary sequence, PTM, and protein identification. In 1990, pioneering work for today's modern peptide $MS^{(n)}$ analysis was performed by Cooks and Stafford using a quadrupole ion trap mass spectrometer (QIT MS) (Kaiser et al., 1990). A number of small peptides were ionized using Cs + surface ionization, injected into the trap, mass selected, and then activated by low energy CID, resulting in dissociation. Product ions were mass selectively ejected and then analyzed to determine the primary sequence. Tandem MS data on sub-femtomole levels of gramicidin S was demonstrated. In the same year, Van Berkel and others combined ESI with QIT single and multiple MS analysis to demonstrate low energy CID fragmentation and peptide sequencing (Van Berkel et al., 1990). One year later, one-line capillary RP LC was combined with ESI QIT MS analysis (McLuckey et al., 1991). In addition to ESI, MALDI generated ions were also analyzed using QIT MS (Qin and Chait, 1995). Today, the majority of tandem ms experiments are performed using on-line capillary RP HPLC with ESI ion trap devices. Sequences are automatically processed and protein identification conferred using various algorithms such as SEQUEST.

Other tandem MS schemes currently used for peptide sequence determination include the ESI tandem quadrupole TOF MS analyzers (Shevchenko et al., 1997), ESI FTICR MS (Wu et al., 1995), MALDI post source decay (PSD) analysis (Kaufman et al., 1993), MALDI quadrupole TOF analysis (Krutchinksy et al., 1998), MALDI TOF–TOF MS^2 analysis (Bienvenut et al., 2002; Juhasz et al., 2002; Yergey et al., 2002), and MALDI QIT–TOF MS analysis (Ding et al., 1999).

5 CONCLUSIONS

In spite of all the technological progress made to date, clinical proteomic studies are still challenged in a number of important dimensions. Perhaps most critical, is the ability to effectively detect low copy number proteins in a relatively straight-forward manner. Even with current improvements in protein fractionation and analytical detection, the practical lower limit of detection remains in the attomole range, making detection at trace levels difficult, to say the least. Without substantial advancements in this area, it will remain a true challenge to detect early signs of protein "leakage" into biological fluids such as blood and urine that are pathognomonic of the early onset of disease.

Concomitant advancements in the dynamic range of qualitative and quantitative detection are also required. In fact, in some cases, the goals of maximizing dynamic range and analytical sensitivity can pose some interesting choices in terms of compromise and trade-off. Often, a given analytical platform has a fixed dynamic range of response. The latter is usually due to fundamental limitations in the physical generation of signal, or in the down-stream processing of acquired electronic signatures. As one digs deeper into the dirt, it becomes easy to loose the ability to accurately detect and quantify major constituents. Indeed, the old adage of the "forest through the trees" certainly applies.

Instrument access and automation also pose an interesting dilemma. Modern proteomic research devices have greatly benefited from advancements in engineering and automation, making them accessible and easy to use in the hands of the veritable novice or the aspiring translational researcher. Even today, automated mass spectrometric, protein microarray, electrophoretic, and chromatographic devices still require a reasonable level of savvy so that the devices are appropriately used. Indeed, it is now easier than ever before to generate large volumes of data and results that often evoke feelings of elation, particularly when studying diseases with significant morbidity and mortality, such as cancer. The translational researcher is best served by developing a practical understanding of each platform's operating principles. Moreover, broadly adopted, peer review standard protocols and controls need to be employed to insure that experimental results truly reflect significant clinical findings and are not the direct consequence of unknown or uncontrolled artifacts of the employed analytical regimen.

Clinical specimen collection and processing remains an area of concern in proteomics research. Most clinical samples are acquired using protocols that have effectively evolved since the dawn of clinical chemistry. These protocols have matured in response to existing assay requirements as well as financial pressures exerted by the healthcare industry. In many cases, sample collection and storage introduce buffering and stabilizing media that thwart effective proteomic analysis. Furthermore, the practice of drawing blood or collecting urine is so diverse, that artifactual proteomic signatures can be created, making the differentiation of sample origin possible when employing multivariate analysis. Clearly, standards for proteomic clinical sample collection and storage need to be broadly adapted, if multi-site validation of putative biomarkers can be routinely realized. Moreover, these new proteomic compatible protocols need to be practical in terms of their scope and financial impact to the health care enterprise, for any broadly adapted proteomic assay will need to be financially achievable.

REFERENCES

Adkins, J. N. et al. 2002. Toward a human blood serum proteome: analysis by multidimensional separation coupled with mass spectrometry. Mol Cell Proteomics 1:947–955.

Aebersold, R. et al. 2000. New approaches to quantitative proteome analysis. Biotecnol Apl 17:46–47.

Alessandro, R. et al. 2005. Proteomic approaches in colon cancer: promising tools for new cancer markers and drug target discovery. Clin Colorectal Cancer 4:396–402.

Alfonso, P. et al. 2005. Proteomic expression analysis of colorectal cancer by two-dimensional differential gel electrophoresis. Proteomics 5:2602–2611.

Appel, R. et al. 1988. Automatic classification of two-dimensional gel electrophoresis pictures by heuristic clustering analysis: a step toward machine learning. Electrophoresis 9:136–142.

Appel, R. D. et al. 1991. The MELANIE project: from a biopsy to automatic protein map interpretation by computer. Electrophoresis 12:722–735.

Arnott, D. P. et al. 1996. Identification of proteins from two-dimensional electrophoresis gels by peptide mass fingerprinting. ACS Symp Ser 619:226–243.

Beer, I. et al. 2004. Improving large-scale proteomics by clustering of mass spectrometry data. Proteomics 4:950–960.

Benzinger, A. et al. 2005. Targeted proteomic analysis of 14-3-3s, a p53 effector commonly silenced in cancer. Mol Cell Proteomics 4:785–795.

Bergen, H. R. III et al. 2003. Discovery of ovarian cancer biomarkers in serum using NanoLC electrospray ionization TOF and FT-ICR mass spectrometry. Dis Markers 19:239–249.

Bern, M. et al. 2004. Automatic quality assessment of Peptide tandem mass spectra. Bioinformatics 20(Suppl 1):I49–I54.

Bienvenut, W. V. et al. 2002. Matrix-assisted laser desorption/ionization-tandem mass spectrometry with high resolution and sensitivity for identification and characterization of proteins. Proteomics 2:868–876.

Borisov, O. V. et al. 2002. Low-energy collision-induced dissociation fragmentation analysis of cysteinyl-modified peptides. Anal Chem 74:2284–2292.

Boschetti, E. 1994. Advanced sorbents for preparative protein separation purposes. J Chromatogr 658:207.

Brill, L. M. et al. 2004. Robust phosphoproteomic profiling of tyrosine phosphorylation sites from human T cells using immobilized metal affinity chromatography and tandem mass spectrometry. Anal Chem 76:2763.

Brzeski, H. et al. 2003. Albumin depletion method for improved plasma glycoprotein analysis by two-dimensional difference gel electrophoresis. Biotechniques 35:1128.

Cargile, B. J. et al. 2004. Potential for false positive identifications from large databases through tandem mass spectrometry. J Proteome Res 3:1082–1085.

Castagna, A. et al. 2005. Exploring the hidden human urinary proteome via ligand library beads. J Proteome Res 4:1917–1930.

Chamrad, D. C. et al. 2004. Evaluation of algorithms for protein identification from sequence databases using mass spectrometry data. Proteomics 4:619–628.

Chen, E. I. et al. 2005. Maspin alters the carcinoma proteome. FASEB J 19:1123–1124, 10 1096/fj 04-2970fje.

Chong, B. E. et al. 2001. Differential screening and mass mapping of proteins from premalignant and cancer cell lines using nonporous reversed-phase HPLC coupled with mass spectrometric analysis. Anal Chem 73:1219–1227.

Chong, P. K. et al. 2006. Isobaric tags for relative and absolute quantitation (iTRAQ) reproducibility: Implication of multiple injections. J Proteome Res 5:1232–1240.

Clarke, W., and Chan, D. W. 2005. ProteinChips: the essential tools for proteomic biomarker discovery and future clinical diagnostics. Clin Chem Lab Med 43:1279–1280.

Corthals, G. L. et al. 2000b. The dynamic range of protein expression: a challenge for proteomic research. Electrophoresis 21:1104.

Davis, M. T. et al. 2001a. Automated LC-LC-MS-MS platform using binary ion-exchange and gradient reversed-phase chromatography for improved proteomic analyses. J Chromatogr B: Biomed Sci Appl 752:281–291.

Davis, M. T. et al. 2001b. Towards defining the urinary proteome using liquid chromatography-tandem mass spectrometry. II. Limitations of complex mixture analyses. Proteomics 1:108–117.

Demirev, P. A. et al. 2001. Tandem mass spectrometry of intact proteins for characterization of biomarkers from Bacillus cereus T spores. Anal Chem 73:5725–5731.

DeSouza, L. et al. 2005. Search for cancer markers from endometrial tissues using differentially labeled tags iTRAQ and cICAT with multidimensional liquid chromatography and tandem mass spectrometry. J Proteome Res 4:377–386.

Ding, L. et al. 1999. High-efficiency MALDI-QIT-ToF mass spectrometer. Proc SPIE-Int Soc Opt Eng 3777:144–155.

Durr, E. et al. 2004. Direct proteomic mapping of the lung microvascular endothelial cell surface in vivo and in cell culture. Nat Biotechnol 22:985–992.

Fernandez, M. A. et al. 1996. Characterization of protein adsorption by composite silica-polyacrylamide gel anion exchangers II. Mass transfer in packed columns and predictability of breakthrough behavior. J Chromatogr 746:185–198.

Ficarro, S. B. et al. 2002. Phosphoproteome analysis by mass spectrometry and its application to Saccharomyces cervisiae. Nat Biotechnol 20:301–305.

Figeys, D. et al. 2001. Mass spectrometry for the study of protein–protein interactions. Methods 24:230.

Fountoulakis, M., and Juranville, J. F. 2003. Enrichment of low-abundance brain proteins by preparative electrophoresis. Anal Biochem 313:267.

Fountoulakis, M. et al. 1999. Enrichment of low-copy-number gene products by hydrophobic interaction chromatography. J Chromatogr A 833:157–168.

Friedman, D. B. et al. 2004. Proteome analysis of human colon cancer by two-dimensional difference gel electrophoresis and mass spectrometry. Proteomics 4:793–811.

Fung, E. T. et al. 2001. Protein biochips for differential profiling. Curr Opin Biotechnol 12:65–69.

Fung, E. T. et al. 2005. Classification of cancer types by measuring variants of host response proteins using SELDI serum assays. Int J Cancer 115:783–789.
Ge, Y. et al. 2002. Top down characterization of larger proteins (45 kDa) by electron capture dissociation mass spectrometry. J Am Chem Soc 124:672–678.
Geho, D. H. et al. 2004. Opportunities for nanotechnology-based innovation in tissue proteomics. Biomed Microdevices 6:231–239.
Geng, M. et al. 2001. Proteomics of glycoproteins based on affinity selection of glycopeptides from tryptic digests. J Chromatogr 752:293.
Ghosh, D. et al. 2004. Lectin affinity as an approach to the proteomic analysis of membrane glycoproteins. J Proteome Res 3:841.
Goerg, A. et al. 2004. Current two-dimensional electrophoresis technology for proteomics. Proteomics 4:3665–3685.
Gorg, A. et al. 2002. Sample prefractionation with Sephadex isoelectric focusing prior to narrow pH range two-dimensional gels. Proteomics 2:1652–1657.
Gronborg, M. et al. 2002. A mass spectrometry-based proteomic approach for identification of serine/threonine-phosphorylated proteins by enrichment with phospho-specific antibodies: identification of a novel protein, Frigg, as a protein kinase A substrate. Mol Cell Proteomics 1:517–527.
Gygi, S. P. et al. 1999. Quantitative analysis of complex protein mixtures using isotope-coded affinity tags. Nat Biotechnol 17:994–999.
Gygi et al. 2000. Evaluation of two-dimensional gel electrophoresis-based proteome analysis technology. Proc Natl Acad Sci USA 17:9390–9395.
Henzel, W. J. et al. 1993. Identifying proteins from two-dimensional gels by molecular mass searching of peptide fragments in protein sequence databases. Proc Nat Acad Sci USA 90:5011–5015
Higdon, R. et al. 2004. LIP index for peptide classification using MS/MS and SEQUEST search via logistic regression. OMICS 8:357–369.
Hoffman, P. et al. 2001. Continuous free-flow electrophoresis separation of cytosolic proteins from the human colon carcinoma cell line LIM 1215: a non two-dimensional gel electrophoresis-based proteome analysis strategy. Proteomics 1:807–818.
Horn, D. M. et al. 2000a. Automated de novo sequencing of proteins by tandem high-resolution mass spectrometry. Proc Nat Acad Sci USA 97:10313–10317.
Horn, D. M. et al. 2000b. Automated reduction and interpretation of high resolution electrospray mass spectra of large molecules. J Am Soc Mass Spectrom 11:320–332.
Jain, K. K. 2002. Role of Proteomics in diagnosis of Cancer. Technol Cancer Res Treat 4: 281–286
Jessani, N. et al. 2005. A streamlined platform for high-content functional proteomics of primary human specimens. Nat Methods 2:691–697.
Juhasz, P. et al. 2002. MALDI-TOF/TOF technology for peptide sequencing and protein identification. Mass Spectrom Hyphenated Techn Neuropeptide Res 375–413.
Kachman, M. T. et al. 2002. A 2-D liquid separations/mass mapping method for interlysate comparison of ovarian cancers. Anal Chem 74:1779–1791.
Kaiser, R. E. Jr. et al. 1990. Collisionally activated dissociation of peptides using a quadrupole ion-trap mass spectrometer. Rapid Commun Mass Spectrom 4:30–33.
Kaufman, R. et al. 1993. Mass spectrometric sequencing of linear peptides by product-ion analysis in a felfectron time-of-flight mass spectrometer using matrix-assisted laser desorption ionization. Rapid Commun Mass Spectrom 7:902–910.
Kelleher, N. L. et al. 1999. Top down versus bottom up protein characterization by tandem high-resolution mass spectrometry. J Am Chem Soc 121:806–812.
Kislinger, T., and Emili, A. 2005. Multidimensional protein identification technology: current status and future prospects. Expert Rev Proteomics 2:27–39.
Kobayashi, H. et al. 2003. Free-flow electrophoresis in a microfabricated chamber with a micromodule fraction separator continuous separation of proteins. J Chromatogr A 990:169–178.
Krutchinksy, A. N. et al. 1998. Orthogonal injection of matrix-assisted laser desorption/ionization ions into a time-of-flight spectrometer through a collisional damping interface. Rapid Commun Mass Spectrom 12:508–518.
Lin, S. et al. 2003. Means of hydrolyzing proteins isolated upon ProteinChip array surfaces: Chemical and enzymatic approaches. P. Michael Conn (ed.). Handbook of Proteomic Methods. Totowa, NJ: Humana, pp. 59–72.

Link, A. J. et al. 1997. A strategy for the identification of proteins localized to subcellular spaces: applications to E. coli periplasmic proteins. Int J Mass Spectrom Ion Proc 160:303–316.
Loo, R. R. O. et al. 1996. Interfacing polyacrylamide gel electrophoresis with mass spectrometry. Techniques in Protein Chemistry VII Symposium of the Protein Society, 9th, Boston, July 8–12, 1995, 305–313.
Loo, R. R. O. et al. 2001. Virtual 2-D gel electrophoresis: visualization and analysis of the E. coli proteome by mass spectrometry. Anal Chem 73:4063–4070.
Lopez, M. F. 2000. Better approaches to finding the needle in a haystack: optimizing proteome analysis through automation. Electrophoresis 21:1082–1093.
Lopez, M. F. et al. 2000. High-throughput profiling of the mitochondrial proteome using affinity fractionation and automation. Electrophoresis 21:3427–3440.
MacCoss, M. J. et al. 2003. A correlation algorithm for the automated quantitative analysis of shotgun proteomics data. Anal Chem 75:6912–6921.
Mann, M. et al. 1993. Use of mass spectrometric molecular weight information to identify proteins in sequence databases. Biol Mass Spectrom 22:338–345.
Martosella, J. et al. 2005. Reversed-phase high-performance liquid chromatographic prefractionation of immunodepleted human serum proteins to enhance mass spectrometry identification of lower abundant proteins. J Proteome Res 4:1522–1537.
Mauri, P. et al. 2005. Identification of proteins released by pancreatic cancer cells by multidimensional protein identification technology: a strategy for identification of novel cancer markers. FASEB J 19:1125–1127, doi:10.1096/fj 04–3000fje.
McLafferty, F. W. 2001. Tandem mass spectrometric analysis of complex biological mixtures. Int J Mass Spectrom 212:81–87.
McLuckey, S. A. et al. 1991. Ion spray liquid chromatography/ion trap mass spectrometry determination of biomolecules. Anal Chem 63:375–383.
Meehan, K. L., and Sadar, M. D. 2004. Quantitative profiling of LNCaP prostate cancer cells using isotope-coded affinity tags and mass spectrometry. Proteomics 4:1116–1134.
Mehta, A. I. et al. 2003. Biomarker amplification by serum carrier protein binding. Disease Markers 19:1–10.
Melander, W. R. et al. 1984. Salt-mediated retention of proteins in hydrophobic-interaction chromatography. Application of solvophobic theory. J Chromatogr 317:67–85.
Meng, F. et al. 2004. Molecular-level description of proteins from saccharomyces cerevisiae using Q-FT mass spectrometry for top down proteomics. Anal Chem 76:2852–2858.
Merchant, M., and Weinberger, S. R. 2000. Recent advancements in surface-enhanced laser desoprtion/ionization time-of-flight mass spectrometry. Electrophoresis 21:1164–1177.
Michel, P. et al. 2003. Protein fractionation in a multicompartment device using Off-Gel isoelectric focusing. Electrophoresis 24:3–11.
Opiteck, G. J. et al. 1998. Comprehensive two-dimensional higher-performance liquid chromatography for the isolation of overexpressed proteins and proteome mapping. Anal Biochem 258:349–361.
Overall, C. M., and Dean, R. A. 2006. Degradomics: Systems biology of the protease web. Pleiotropic roles of MMPs in cancer. Cancer Metastasis Rev 25:69–75.
Paoletti, A. C. et al. 2004. Principles and applications of multidimensional protein identification technology. Exp Review of Proteomics 1:275–282.
Park, O. K. 2004. Proteomic studies in plants. J Biochem Mol Biol 37:133–138.
Pawlik, T. M. et al. 2006. Proteomic analysis of nipple aspirate fluid from women with early-stage breast cancer using isotope-coded affinity tags and tandem mass spectrometry reveals differential expression of vitamin D binding protein. BMC Cancer 6:68.
Pun, T. et al. 1988. Computerized classification of two-dimensional gel electrophoretograms by correspondence analysis and ascendant hierarchical clustering. Appl Theor Electrophor 1:3–9.
Qin, J., and Chait, B. T. 1995. Preferential fragmentation of protonated gas-phase peptide ions adjacent to acidic amino acid residues. J Am Chem Soc 117:5411–5412.
Ramamoorthy, S. et al. 2004. Intracellular mechanisms mediating the anti-apoptotic action of gastrin. Biochem Biophys Res Commun 323:44–48.
Razumovskaya, J. et al. 2004. A computational method for assessing peptide- identification reliability in tandem mass spectrometry analysis with SEQUEST. Proteomics 4:961–969.
Reid, G. E., and McLuckey, S. A. 2002.\"Top down\" protein characterization via tandem mass spectrometry. J Mass Spectrom 37:663–675.

Ren, L. et al. 2003. Improved immunomatrix methods to detect protein:protein interactions. J Biochem Biophys Methods 57:143–157.
Righetti, P. G. et al. 1990. Preparative purification of human monoclonal antibody isoforms in a multi-compartment electrolyser with immobiline membranes. J Chromatogr 500:681–696.
Righetti, P. G. et al. 2005. Proteome analysis in the clinical chemistry laboratory: Myth or reality? Clinica Chimica Acta 357:123–139.
Ros, A. et al. 2002. Protein purification by off-gel electrophoresis. Proteomics 2:151–156.
Schulze, W. X., and Mann, M. 2004. Novel proteomic screen for peptide-protein interactions. J Biol Chem 279:10756–10764.
Senko, M. W. et al. 1995. Automated assignment of charge states from resolved isotopic peaks for multiply charged ions. J Am Soc Mass Spectrom 6:52–56.
Shadforth, I. P. et al. 2005. i-Tracker: for quantitative proteomics using iTRAQ. BMC Genomics 6:145.
Shang, T. Q. et al. 2003. Carrier ampholyte-free solution isoelectric focusing as a prefractionation method for the proteomic analysis of complex protein mixtures. Electrophoresis 24:2359–2368.
Shevchenko, A. et al. 1997. Rapid 'de Novo' peptide sequencing by a combination of nanoelectrospray isotopic labeling and a quadrupole/time-of-flight mass spectrometer. Rapid Commun Mass Spectrom 11:1015–1024.
Smith, S. D. et al. 2004. Using immobilized metal affinity chromatography, two-dimensional electrophoresis and mass spectrometry to identify hepatocellular proteins with copper-binding ability. J Proteome Res 3:834–840.
Somiari, R. I. et al. 2005. Proteomics of breast carcinoma. J Chromatogr B Anal Technol Biomed Life Sci 815:215–225.
Staby, A., and Mollerup, J. 1996. Solute retention of lysozyme in hydrophobic interaction perfusion chromatography. J Chromatogr A 734:205–212.
Tao, W. A., and Aebersold, R. 2003. Advances in quantitative proteomics via stable isotope tagging and mass spectrometry. Curr Opin Biotechnol 14:110–118.
Thadikkaran, L. et al. 2005. Recent advances in blood-related proteomics. Proteomics 5:3019–3034.
Thulasiraman, V. et al. 2005. Reduction of the concentration difference of proteins in biological liquids using a library of combinatorial ligands. Electrophoresis 26:3561–3571.
Tirumalai, R. S. et al. 2003. Characterization of the low molecular weight human serum proteome. Mol Cell Proteomics 2:1096–1103.
Tissot, J. D. et al. 1991. High-resolution two-dimensional protein electrophoresis of pathological plasma/serum. Appl Theor Electrophoresis 2:7–12.
Tomlinson, A. J., and Chicz, R. M. 2003. Microcapillary liquid chromatography/tandem mass spectrometry using alkaline pH mobile phases and positive ion detection. Rapid Commun Mass Spectrom 17:909–916.
Unlu, M. et al. 1997. Difference gel electrophoresis. A single gel method for detecting changes in protein extracts. Electrophoresis 18:2071–2077.
Van Berkel, G. J. et al. 1990. Electrospray ionization combined with ion trap mass spectrometry. Anal Chem 62:1284–1295.
Wagner, K. et al. 2002. An automated on-line multidimensional HPLC system for protein and peptide mapping with integrated sample preparation. Anal Chem 74:809–820.
Walker, A. K. et al. 2001. Mass spectrometric imaging of immobilized pH gradient gels and creation of\"virtual\" two-dimensional gels. Electrophoresis 22:933–945.
Wall, D. B. et al. 2000. Isoelectric focusing nonporous RP HPLC: a two-dimensional liquid-phase separation method for mapping of cellular proteins with identification using MALDI-TOF mass spectrometry. Anal Chem 72:1099–1111.
Wall, D. B. et al. 2001. Isoelectric focusing nonporous silica reversed-phase high-performance liquid chromatography/electrospray ionization time-of-flight mass spectrometry: a three-dimensional liquid-phase protein separation method as applied to the human erythroleukemia cell-line. Rapid Commun Mass Spectrom 15:1649–1661.
Washburn, M. P. 2004. Technique review: Utilisation of proteomics datasets generated via multidimensional protein identification technology (MudPIT). Brief Funct Genomic Proteomic 3:280–286.
Washburn, M. P. et al. 2001. Large-scale analysis of the yeast proteome by multidimensional protein identification technology. Nat Biotechnol 19:242–247.

Wasinger, V. C. et al. 1995. Progress with gene-product mapping of the Mollicutes: Mycoplasma genitalium. Electrophoresis 16:1090–1094.

Watts, A. D. et al. 1997. Separation of tumor necrosis factor alpha isoforms by two-dimensional polyacrylamide gel electrophoresis. Electrophoresis 18:1806–1091.

Weinberger, S. R. et al. 2002. Current achievements using ProteinChip array technology. Curr Opin Chem Biol 6:86–91.

Wenger, P. et al. 1987. Amphoteric, isoelectric immobiline membranes for preparative isoelectric focusing. J Biochem Biophys Methods 14:29–43.

Wilkins, M. R. et al. 1996a. From proteins to proteomes: large scale protein identification by two-dimensional electrophoresis and amino acid analysis. Bio/technology 14:61–65.

Wilkins, M. R. et al. 1996b. Progress with proteome projects: why all proteins expressed by a genome should be identified and how to do it. Biotechnol Genetic Eng Rev 13:19–50.

Wilkins, M. R. et al. 1996c. Current challenges and future applications for protein maps and post-translational vector maps in proteome projects. Electrophoresis 17:830–838.

Williams, K. L. et al. 1996. Analytical biotechnology and proteome analysis. Australasian Biotechnol 6:162–164, 166–167.

Wilson, L. L. et al. 2004. Detection of differentially expressed proteins in early-stage melanoma patients using SELDI-TOF mass spectrometry. Ann N Y Acad Sci 1022:317–322.

Wolters, D. A. 2004. Applications of MudPIT technology. BIOspektrum 10:162–164.

Wolters, D. A. et al. 2001. An automated multidimensional protein identification technology for shotgun proteomics. Anal Chem 73:5683–5690.

Wright, M. E., and Aebersold, R. 2003. Differential expression proteomic analysis using isotope coded affinity tags. Proteomic Genomic Anal Cardiovasc Dis 213–233.

Wright, M. E. et al. 2004. Identification of androgen-coregulated protein networks from the microsomes of human prostate cancer cells. Genome Biol 5(1):R4.

Wu, Q. et al. 1995. Characterization of cytochrome c variants with high-resolution FTICR mass spectrometry: Correlation of fragmentation and structure. Anal Chem 67:2498–2509.

Wu, W. W. et al. 2006. Comparative study of three proteomic quantitative methods, DIGE, cICAT, and iTRAQ, using 2D gel- or LC-MALDI TOF/TOF. J Proteome Res 5:651–658.

Yan, B. et al. 2005. A graph-theoretic approach for the separation of b and y ions in tandem mass spectra. Bioinformatics 21:563–574.

Yates, J. R. III et al. 1995. Method to correlate tandem mass spectra of modified peptides to amino acid sequences in the protein database. Anal Chem 67:1426–1436.

Yergey, A. L. et al. 2002. De novo sequencing of peptides using MALDI/TOF-TOF. J Am Soc Mass Spectrom 13:784–791.

Yu, L. R. et al. 2002. Isotope-coded affinity tag analysis of native and camptothecin-treated cortical neurons. Bioforum Int 6:328–331.

Yuan, X., and Desiderio, D. M. 2003. Proteomics analysis of phosphotyrosyl-proteins in human lumbar cerebrospinal fluid. J Proteome Res 5:476–487.

Zhang, Z. et al. 2004a. Three biomarkers identified from serum proteomic analysis for the detection of early stage ovarian cancer. Cancer Res 64:5882–5890.

Zhang, Z. et al. 2004b. Three biomarkers identified from serum proteomic analysis for the detection of early stage ovarian cancer. Cancer Res 64:5882–5890.

Zhou, H. et al. 2004a. Quantitative protein analysis by solid phase isotope tagging and mass spectrometry. Methods Mol Biol 261:511–518.

Zhou, M. et al. 2004b. An investigation into the human serum "interactome". Electrophoresis 25:1289.

Zhu, Y., and Lubman, D. M. 2004. Narrow-band fractionation of proteins from whole cell lysates using isoelectric membrane focusing and nonporous reversed-phase separations. Electrophoresis 25:949–958.

Zolotarjova, N. et al. 2005. Differences among techniques for high-abundant protein depletion. Proteomics 5:3304–3313.

Zuo, X., and Speicher, D. W. 2002. Comprehensive analysis of complex proteomes using microscale solution isoelectricfocusing prior to narrow pH range two-dimensional electrophoresis. Proteomics 2:58–68.

7 Comprehensive Genomic Profiling for Biomarker Discovery for Cancer Detection, Diagnostics and Prognostics

Xiaofeng Zhou, Nagesh P. Rao, Steven W. Cole, and David T. Wong

ABSTRACT

Tumors develop through the combined processes of genetic instability and selection, resulting in clonal expansion of cells that have accumulated the most advantageous set of genetic aberrations. These genetic instabilities manifest themselves as a series of genetic alterations, including discrete mutations and chromosomal aberrations. With the human genome deciphered, high-throughput technologies are advancing studies to link genome-wide gene expression and mutation profiles with biological and disease phenotypes. Recent advances in comprehensive genomic characterization present an unprecedented opportunity for advancing the treatment of cancer, but many challenges remain to be overcome before we can fully exploit genomic markers/targets for cancer prediction, diagnosis, treatment and prognostics. Here, we review recent developments in comprehensive genomic characterization at the DNA level, and consider some of the challenges that remain for defining the precise genomic portrait of tumors. We then offer some potential solutions that may help overcome these challenges.

Key Words: Loss of heterozygosity, Comparative genomic hybridization, Spectral karyotyping, Multicolor fluorescence in situ hybridization, Tiling array, SNP array, Personalized therapies

1 INTRODUCTION

Molecular understanding of genomic aberrations can have significant clinical value in diagnosis, treatment and prognostics of cancer. Four decades ago, there was the milestone discovery of the Philadelphia chromosome (a translocation between chromosome

G.J. Gordon (ed.), *Cancer Drug Discovery and Development: Bioinformatics in Cancer and Cancer Therapy*,
DOI: 10.1007/978-1-59745-576-3_7, © Humana Press, a part of Springer Science + Business Media, LLC 2009

9 and 22, which fuses the Bcr gene and the Abl tyrosine kinase gene) (Nowell and Hungerford, 1960) as the primary genetic change in chronic myelogenous leukemia (CML). This led to one of the first effective targeted therapies for cancer: treatment of CML with the tyrosine kinase inhibitor imatinib (Gleevec). Since then, many exciting clinical advances have been made based on our increasing knowledge of the tumor genome. The completion of the human genome project (Lander et al., 2001; Venter et al., 2001) now makes it possible to systematically query the cancer genome in ways that were hitherto impossible. Microarrays designed to analyze targeted genomic regions relevant to various cancers are being developed. For example, in chronic lymphocytic leukemia (CLL), efforts are being made to optimize therapeutic options based on specific genomic aberrations (Schwaenen et al., 2004; Stilgenbauer and Dohner, 2005). Associations between genomic aberrations with disease prognosis have been found for a variety of tumor types, including prostate cancer (Paris et al., 2004), breast cancer (Callagy et al., 2005), gastric cancer (Weiss et al., 2004), head and neck cancer (Rosin et al., 2000) and lymphoma (Martinez-Climent et al., 2003; Rubio-Moscardo et al., 2005). Many more studies are in progress or near completion. These findings will provide a new paradigm for cancer treatment that is fundamentally guided by a genomic perspective of the disease (reviewed in Futreal et al., 2004; Mundle and Sokolova, 2004; Avivi and Rowe, 2005; Granville and Dennis, 2005; Jeffrey et al., 2005).

A critical biological difficulty confronting the identification and eventual translation of genomic markers/targets for cancer prediction, diagnostics, treatment and prognostics is distinguishing the genomic aberrations driving malignant cell growth from those that are byproducts of abnormal proliferation (Zhou et al., 2006). Among these key processes are germline variations that lead to hereditary cancer predispositions, the acquisition of transforming DNA or RNA sequences from cancer viruses, somatic mutations in the cancer genome, and epigenetic mechanisms (such as DNA methylation or histone modification) that promote oncogenesis by modifying cancer-related genes. Somatic genomic alterations such as point mutations, genomic amplifications or deletions, loss of allelic heterozygosity, and chromosomal translocations are believed to play a central role in the development of most solid tumors (Weir et al., 2004). A variety of high-throughput genetic and molecular technologies have been developed and enable the identification of a broad range of genetic abnormalities, including the analysis of chromosome karyotyping, loss of heterozygosity (LOH), comparative genomic hybridization (CGH), digital karyotyping (DK) (Wang et al., 2002), fluorescence in situ hybridization (FISH), restriction landmark genome scanning (RLGS) (Imoto et al., 1994), representational difference analysis (RDA) (Lisitsyn and Wigler, 1993), and statistical inference of chromosomal changes from gene expression data (Crawley and Furge,2002; Zhou et al., 2004a, 2005). The recent development of multicolor staining-based cytogenetic techniques such as multicolor fluorescence in situ hybridization (M-FISH) and spectral karyotyping (SKY) have further improved the ability to analyze the tumor genome (Liehr et al., 2002). However, none of the existing genomic techniques can capture all these genetic changes in a single analysis (Fig. 1; reviewed in Albertson et al., 2003; Zhou et al., 2006). This represents a major obstacle to the comprehensive analysis of tumor genomes and their relationship to clinical phenotypes.

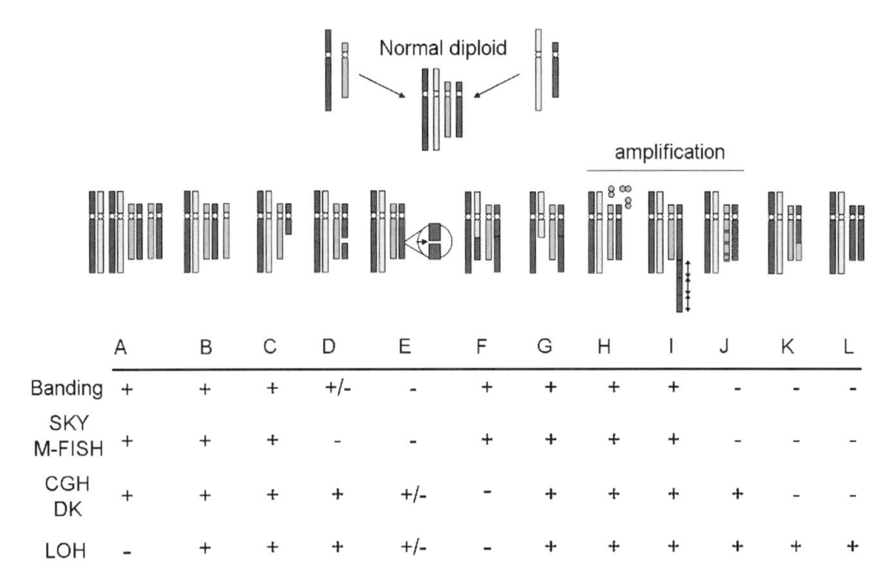

Fig. 1. Detection and mapping of chromosomal abnormalities using different genomic and cytogenetic approaches. *A*, polyploid; *B*, aneuploid; *C*, gross deletion; *D*, interstitial deletion; *E*, microdeletion; *F*, reciprocal translocation; *G*, nonreciprocal translocation; *H*, double minutes; *I*, HSR; *J*, distributed insertion; *K*, somatic recombination; *L*, duplication and loss (adapted from Zhou et al., 2006; with kind permission of Future Drugs Ltd) (*See Color Plates*)

In this chapter, we survey the recent technological advances in the characterization of tumor genomes at the DNA level and consider some of the major challenges that remain in defining genomic markers/targets for cancer prediction, diagnostics, treatment and prognostics. We also identify potential solutions that may overcome these outstanding challenges to improve the diagnosis and treatment of solid tissue malignancies.

2 TECHNICAL OVERVIEW OF THE WIDELY USED GENOMIC APPROACHES

2.1 Systematic Copy Number Analysis

CGH was developed to survey gene copy number abnormalities (amplifications and deletions) across a whole genome (Kallioniemi et al., 1992). In a typical CGH analysis, differentially labeled test/disease and reference genomic DNAs are cohybridized to normal metaphase chromosomes to generate fluorescence ratios along the length of chromosomes that provide a cytogenetic representation of DNA copy number variation. This was the first effective approach to scanning the entire genome for variations in DNA content (Pinkel and Albertson, 2005a,b). However, chromosome-based CGH has a limited mapping resolution (~10–20-Mb). Array-based CGH is a second-generation approach in which fluorescence ratios on microarrayed DNA elements provide a locus-by-locus measure of gene copy number variation (Pinkel et al., 1998; Ishkanian et al., 2004) (Fig. 2). Although this approach can potentially increase mapping resolution, most array CGH methods have utilized large genomic clones (e.g., bacterial artificial

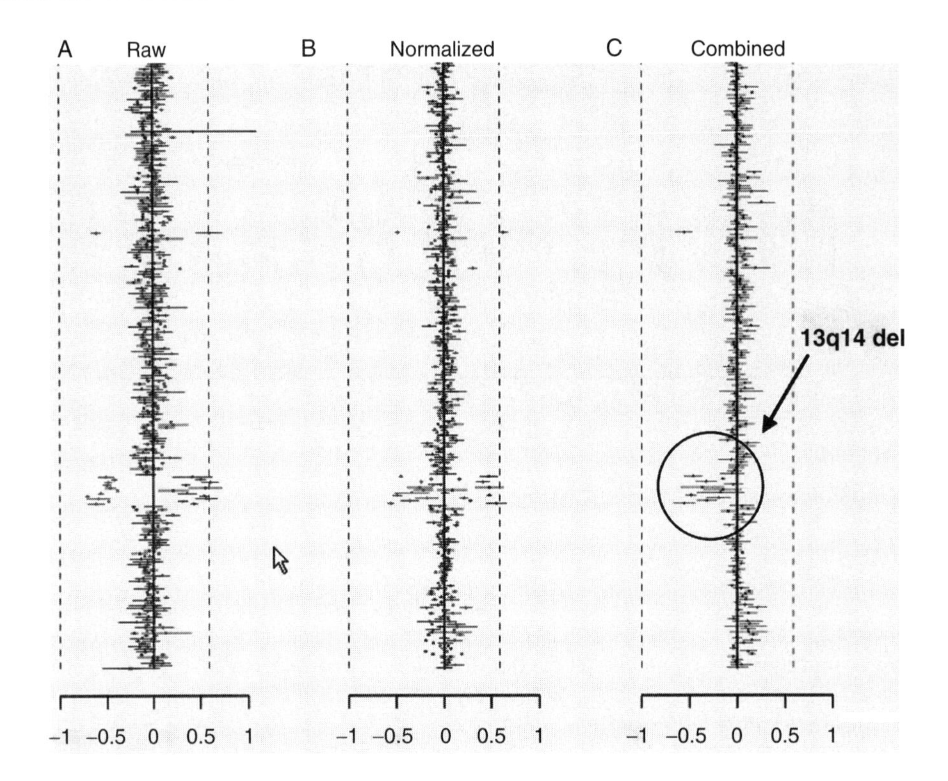

D

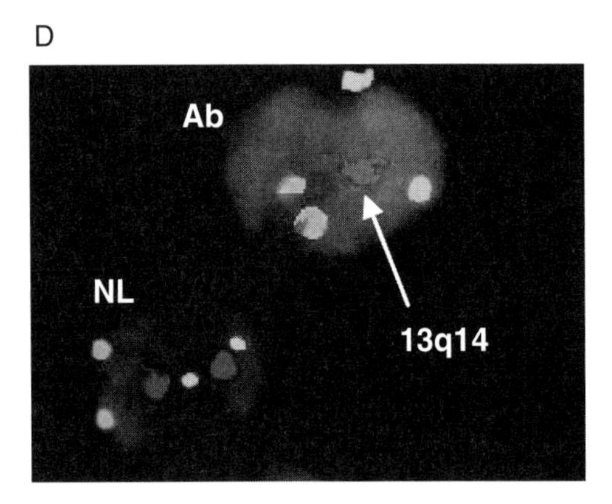

Fig. 2. A typical BAC array-based CGH result on a CLL sample where a dye-swap experiment was performed. CGH was performed using array with BAC clones representing genomic regions that are commonly involved in CLL ((**A**) raw data, (**B**) normalized, and (**C**) combined data). A deletion of the 13q14 region was observed. (**D**) Targeted FISH analysis with DNA probes specific to 13q14 (*red*) confirmed this submicroscopic deletion (abnormal cell (Ab) with loss of one red signal when compared to the normal cell (NL)) (*See Color Plates*)

chromosomes, (BACs)) which limit spatial sensitivity. In addition, large genomic clones also suffer from reduced specificity due to the inclusion of common repeats (e.g., *Alu* and long interspersed nuclear elements – LINEs), redundant sequences (e.g., low copy repeats, LCRs), also known as segmental duplications), and segments of extensive sequence similarity (pseudogenes or paralogous genes) (Mantripragada et al., 2004).

Recently, several additional higher-density tools for CGH analysis have become available with the completion of the human genome sequence. These include cDNA array-based CGH (Pollack et al., 1999; Zhou et al., 2004b), oligonucleotide array-based CGH (Lucito et al., 2003; Brennan et al., 2004), tiling array-based CGH (Ishkanian et al., 2004), and copy number analysis using high-density SNP microarrays (Bignell et al., 2004; Zhao et al., 2004, 2005; Zhou et al., 2004d). Tiling and SNP array-based approaches have drawn the most attention due to their high resolution. Tiling arrays have the potential to resolve small (gene level) gains and losses (resolution 40-kb) that might be missed by marker-based genomic arrays which contain a large number of gaps due to the distance between the targeted probes (Ishkanian et al., 2004; Davies et al., 2005). It is conceivable that even higher-resolution tiling arrays will become available in the future, providing an opportunity to map genomic alterations at close to base pair resolution. The SNP array-based approach provides the unique advantage of concurrent CGH and LOH analysis, which is discussed in further detail below (Zhao et al., 2004; Zhou et al., 2004d).

2.2 *Systematic Allelic Imbalance Analysis*

Chromosomal aberrations include segments of allelic imbalance identifiable by LOH at the polymorphic loci, which can be used to identify regions harboring tumor suppressor genes. Allelic losses, which are caused by mitotic recombination, gene conversion, or nondisjunction cannot be detected by CGH and thus require LOH analysis for their identification. This approach is "favored" by the Knudson two-hit hypothesis (Knudson, 1971, 1996) for hunting tumor suppressor genes. The discovery of the first tumor suppressor gene, RB1 (Friend et al., 1986), followed the Knudson two-hit hypothesis that tumor suppressor genes are inactivated by a recessive mutation in one allele followed by the loss of the other wild-type allele, which can be detected by LOH. Traditionally, polymorphic markers, such as restriction fragment length polymorphisms (RFLPs) and microsatellite markers, have been used to detect LOH through allelotypic comparisons of DNA from a cancer sample and a matched normal sample (Vogelstein et al., 1989). However, this approach is tedious, labor intensive, and requires a large amount of sample DNA, allowing only a modest number of markers to be screened. High-density whole genome allelotyping cannot be readily performed. The mapping of the human genome has allowed for the identification of millions of SNP loci(http://www.ncbi.nlm.nih.gov/SNP/), which makes them ideal markers for various genetic analyses, including LOH. Because of their abundance, even spacing, and stability across the genome, SNPs have significant advantages over RFLPs and microsatellite markers as a basis for high-resolution whole genome allelotyping with accurate copy number measurements. High-density oligonucleotide arrays have recently been generated to support large-scale high-throughput SNP analysis (Wang et al., 1998). It is now possible to genotype over 500,000 SNP markers using the Affymetrix Mapping 500K SNP oligonucleotide array. LOH patterns generated by SNP array analysis have a high degree of concordance with previous microsatellite analyses of the same cancer samples (Lindblad-Toh et al., 2000). Additionally, shared regions of LOH from SNP arrays can cluster lung cancer samples into subtypes (Janne et al., 2004), and distinct patterns of LOH are found to associate with clinical features in primary breast, bladder, head and neck and prostate tumors

(Hoque et al., 2003; Lieberfarb et al., 2003; Wang et al., 2004; Zhou et al., 2004c,d). One unique advantage of this SNP array-based approach is that the intensity of sample hybridization to the array probes can also be used to infer copy number changes (similar to CGH) (Fig. 3) (Bignell et al., 2004; Zhao et al., 2004; Zhou et al., 2004d). This unique feature has been explored by algorithms implemented in several independent bioinformatics/statistical software packages, including dChipSNP (Zhao et al., 2004) and Copy Number Analysis Tool (Huang et al., 2004). The use of these novel analytic tools to analyze data from high-density SNP arrays now allows the analysis of DNA copy number to be combined with LOH analysis to distinguish copy number gains, copy number neutral loss of heterozygosity, and copy number losses, to comprehensively map the configuration of tumor genomes (Zhao et al., 2004).

2.3 Cytogenetic-Based Approaches: Old Techniques with New Twists

Cytogenetics has flourished since the introduction of chromosome-banding techniques in 1969 (Caspersson et al., 1969a,b). One major drawback of these approaches is the requirement for in vitro culture and metaphase preparation of the cells of interest, which limits its application to studies of solid cancers. Nevertheless, cytogenetic approaches continue to play an important role in genomic profiling because they facilitate direct visualization of chromosomal abnormalities. These cytogenetic techniques also complement CGH and LOH by providing information on chromosomal structural rearrangements that are not resolved by DNA copy number analyses. For example, translocations are one of the most common genomic abnormalities in cancer (Futreal et al., 2004), but they cannot be detected by CGH or LOH. An experienced cytogeneticist, however, can readily detect many forms of chromosomal translocations using classical cytogenetic techniques, such as karyotyping (chromosome banding). A karyotype analysis usually involves blocking cells in mitosis, and microscopically viewing condensed chromosomes stained with Giemsa dye, which stains regions of chromosomes that are rich in the base pairs Adenine (A) and Thymine (T) to produce a dark band. Karyotype analysis is performed over 500,000 times per year in the U.S. and Canada as part of the standard clinical test for prenatal and postnatal screening, as well as for the diagnosis of cancers — hematological malignancies in particular. However, many cancer cells have complex karyotypes that are difficult to interpret (as illustrated in Fig. 1). Recently, several new labeling techniques have been introduced in the field of molecular cytogenetics, including spectral karyotyping (SKY), multiple fluorescence in situ hybridization (M-FISH), cross-species color banding (Rx-FISH), color-changing karyotyping (CCK) (Henegariu et al., 1999), and multicolor chromosome banding. These techniques permit the simultaneous viewing of all chromosomes in different colors, and thus considerably improve the detection of subtle rearrangements. For example, both SKY and M-FISH use a combinatorial labeling scheme with spectrally distinguishable fluorochromes. The chromosome-specific probe pools (chromosome painting probes) are generated from flow cytometry-sorted chromosomes and then amplified and fluorescently labeled by degenerate oligonucleotide-primed polymerase chain reaction. With the introduction of these techniques in 1996 (Liyanage et al., 1996; Schrock et al., 1996; Speicher et al., 1996), the comprehensive analysis of complex chromosomal rearrangements present in tumor karyotypes was greatly improved (Fig. 4).

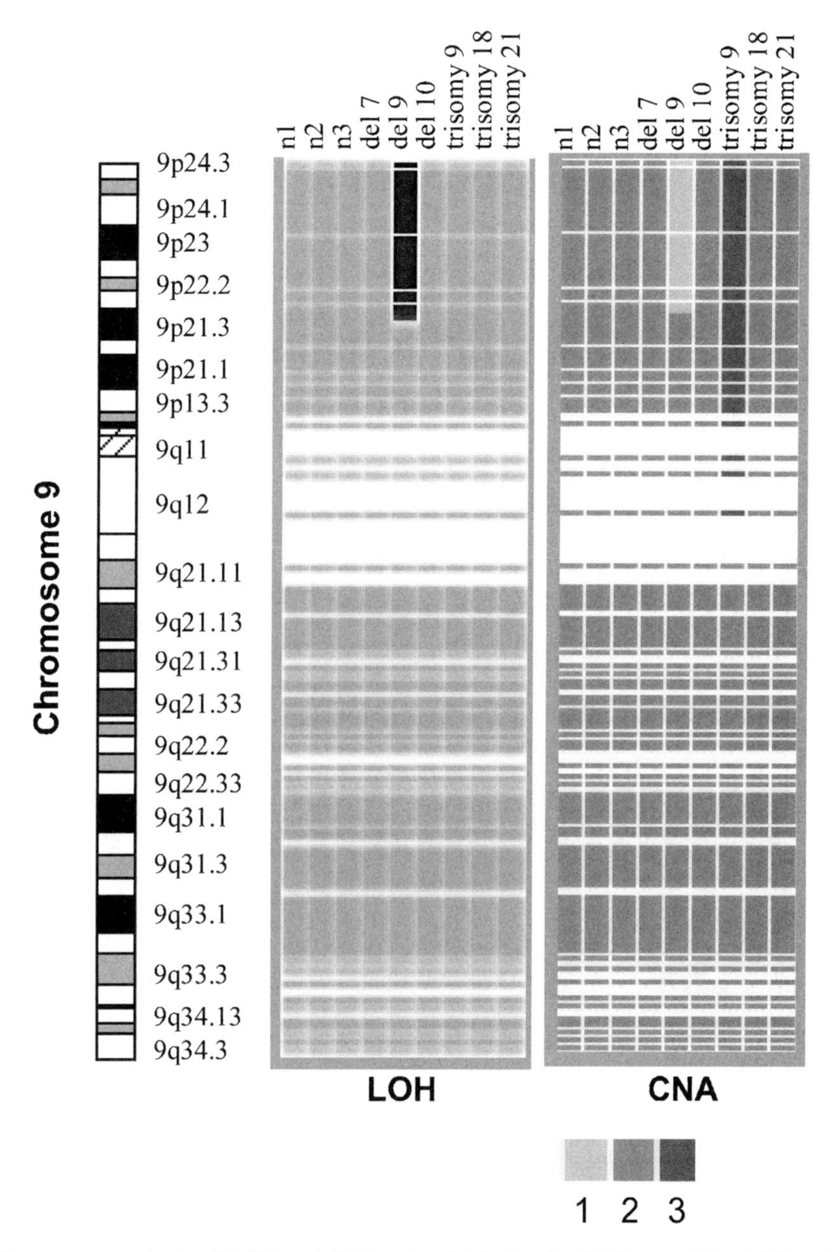

Fig. 3. Concurrent analysis of LOH and CNA using high-density SNP array. The LOH regions and the CNA regions were detected and demarcated as described previously (Lin et al., 2004; Zhao et al., 2004; Zhou et al., 2004c,d). The LOH and CNA patterns for chromosome 9 were shown for nine cell lines: normal cells (n1, n2, and n3), trisomy cells (trisomy 9, trisomy 18, and trisomy 21, with an extra copy of 9pter > q13, 18, 21, respectively) and deletion cells (del(7), del(9), and del(10), with deletions at 7pter > q34, 9pter > p21, 10qter > p11, respectively). Each column represents one sample, and each row represents a SNP marker. Color code for LOH profiling (*right panel*): *blue* = LOH; *light green* = retained; *gray* = uninformative; *white* = no call. Copy numbers (*left panel*) were represented by different intensity of *red colors* as indicated in the figure (adapted from Zhou et al., 2004d; with kind permission of Springer Science and Business Media) (*See Color Plates*)

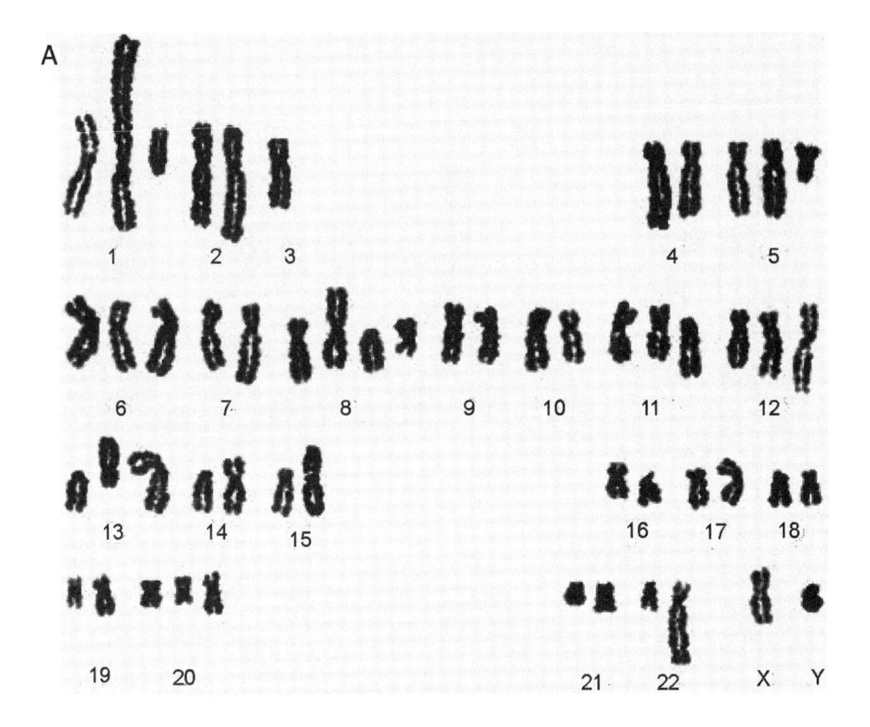

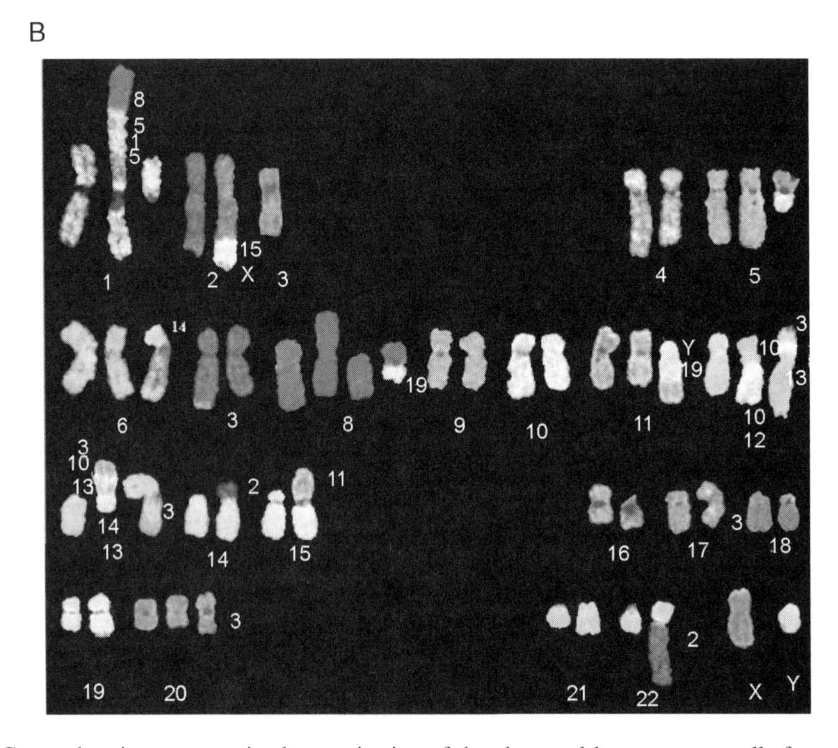

Fig. 4. Comprehensive cytogenetic characterization of the abnormal bone-marrow cells from an acute myeloid leukemia (AML) patient. (**A**) Standard trypsin Giemsa-banding technique-based karyotype analysis. (**B**) M-FISH was performed using Abbott-Vysis SpectraVysion assay. The hyperdiploid karyotype of this CML patient has numerous complex structural rearrangements. The simultaneous karyotype analysis and M-FISH analysis clearly depicted the complex nature of the genomic anomalies (*See Color Plates*)

2.4 Comprehensive Genomic Approaches

A central objective in cancer research is to comprehensively delineate the complex genomic aberrations that shape tumor cell behavior and clinical outcomes. One potential approach to this problem would be to combine molecular genetic technologies such as CGH or LOH with molecular cytogenetic analyses for comprehensive screening of genomic alterations. Each of these techniques has its own unique advantages, as well as individual limitations which motivate efforts to combine multiple approaches as shown in Fig. 5. In this example, the SNP array-based LOH and CGH analyses provide a high-resolution mapping of copy number abnormalities, but offer little information on chromosomal structure/spatial changes (e.g., translocations, the most common class of

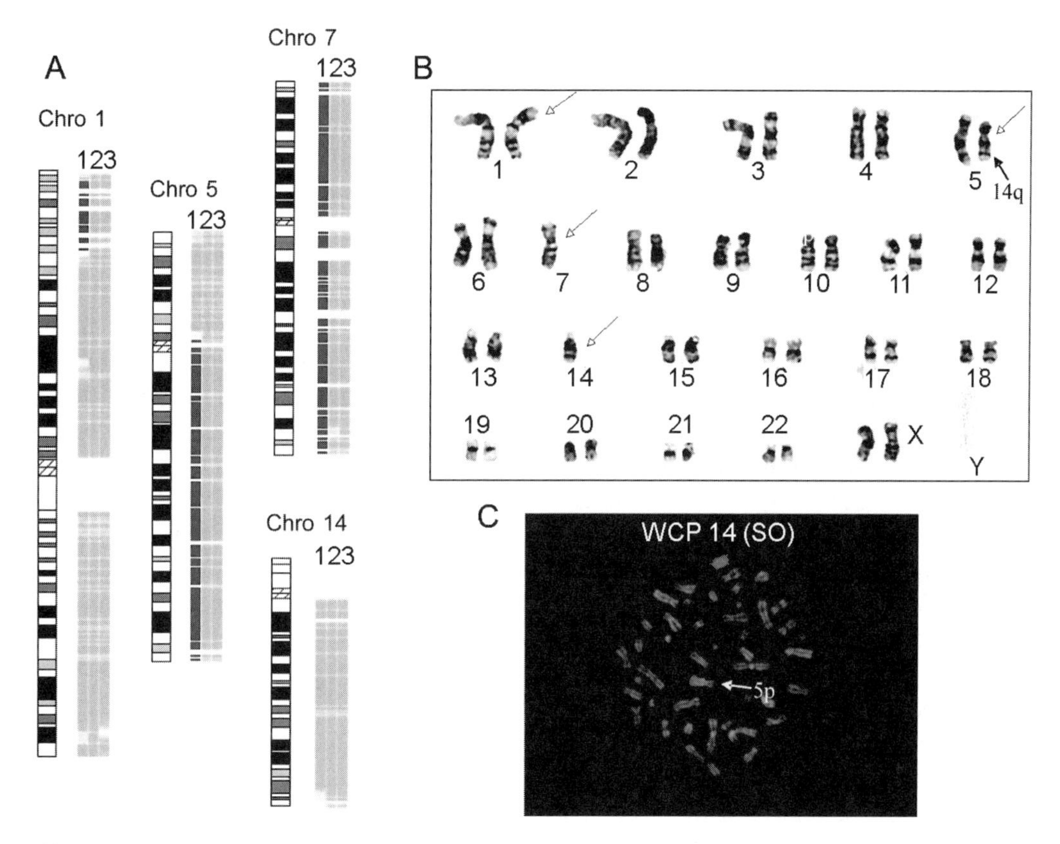

Fig. 5. Comprehensive genomic analyses of a myelodysplastic syndrome (MDS) using SNP array-based approach and complementary cytogenetic approaches. (**A**) Three MDS cases were analyzed with a 10K SNP mapping array. The LOH regions were detected and demarcated as described (Lin et al., 2004; Zhao et al., 2004; Zhou et al., 2004c,d). The LOH patterns for chromosome 1, 5, 7, 14 are shown. (**B**) The karyotype for Case 1 is presented. In Case 1, SNP array-based LOH demonstrated no loss of chromosome 14 material and a more extensive 5q deletion than interpreted by the karyotype. (**C**) Whole chromosome paint (WCP) of chromosome 14, identified two signals for chromosome 14 (*red*), one normal 14 and the other chromosome 14 translocated to 5q close to the pericentromeric region. Together with the results from (**A**) and (**B**), these data indicate a karyotype of 44,XX,del(1)(p32p36),der(5)t(5;14)(q13;q11.2),-7[20] for Case 1 (adapted from Zhou et al., 2006; with kind permission of Future Drugs Ltd) (*See Color Plates*)

somatic mutation registered in the cancer-gene census; Futreal et al., 2004). On the other hand, modern cytogenetic techniques provide a clear picture of the gross chromosomal structure/spatial alterations, but have limited resolution. This is illustrated in Fig. 5, where concurrent cytogenetic and SNP array-based LOH analysis were performed on three genomic samples from myelodysplastic syndrome (MDS). Two of the three cases exhibited concordant results between karyotyping and SNP array-based LOH. However for Case 1, as shown in Fig. 5a, the SNP array-based approach identified the loss of chromosome arm 5q, but failed to identify the translocation of chromosome 14 to chromosome 5 at the pericentromeric region. This translocation was identified by karyotyping and further confirmed by whole chromosome paint (Fig. 5b, c). These results illustrate the advantage of a multimodal approach to tumor genome analysis that combines the complementary strengths of array-based and cytogenetic approaches.

Recent technical advances in microarray-based gene expression analysis provide opportunities to significantly improve the diagnosis, treatment and prognostic staging of cancer. This continuing development of microarray-based expression analysis and the large public depositories of microarray data have motivated new efforts to extract additional biological information from these data in addition to the static RNA transcript levels. One such attempt involves inferring chromosomal structural changes from spatially linked alterations in microarray expression data. Several array CGH studies have shown a genome-wide correlation of gene expression with copy number alterations and have proved useful in individual amplicon refinement (Pollack et al., 2002; Wolf et al., 2004). For example, through tissue microarray FISH and RT-PCR, a minimally amplified region around ERBB2 (Her2) was identified in a large number of breast tumors; in addition, gene amplification was found to be correlated with increased gene expression in a subset of those samples (Kauraniemi et al., 2003). Recently, several groups have observed that chromosomal alterations can lead to regional gene expression biases in human tumors and tumor-derived cell lines (Phillips et al., 2001; Virtaneva et al., 2001; Crawley and Furge, 2002; Zhou et al., 2004a, 2005). A recent study also demonstrated the correlation between SNP array-based LOH profiles and expression profiles (Wang et al., 2004). These studies suggest that a fraction of gene expression values (15–25%) are regulated in concordance with chromosomal DNA content (Phillips et al., 2001; Virtaneva et al., 2001; Crawley and Furge, 2002; Zhou et al., 2004a, 2005). Several statistical methods have been developed and have shown promising results for detecting DNA copy number abnormalities based on differential gene expression (Crawley and Furge, 2002; Myers et al., 2004; Zhou et al., 2004a, 2005). As shown in Fig. 6, one recently developed statistical model successfully identified a cytogenetically confirmed 10p chromosomal deletion based on the microarray expression data, and substantially increased the precision with which the boundaries of that deletion were mapped to a region between 10p14 and 10p12 (Zhou et al., 2005). A key discovery using the high-resolution microarray-based analysis was that only a small interstitial region of 10p was deleted, with much of the 10pter-proximal region intact and only the segments between 10p12 and 10p14 showing marked abnormality. These results were further confirmed by subtelomere FISH, showing that it is feasible to use microarray differential expression data to identify significant DNA copy number abnormalities, and that RNA-based gene expression analyses are concordant with DNA-based measures of chromosomal structural alteration. The development of bioinformatics techniques for "reverse inference" of DNA alterations from RNA expression data offers a new approach for genomic profiling that can provide crossvalidation of functional genomic alterations at multiple biological levels when combined with DNA-based approaches such as CGH and LOH.

Additional functional genomic information can be derived from microarray gene expression data using bioinformatic analyses of upstream transcription factor dynamics. Several tools have recently been developed to identify aberrant transcription factor activity based on sequence similarities in the promoters of large groups of genes showing altered expression (Frith et al., 2004; Cole et al., 2005). Aberrant transcription factor activity plays a central role in many solid tumors, and reverse inference of such alterations from microarray gene expression data provides another mechanism for crossvalidating the results of structural genomic surveys suggesting that a particular transcription control pathway might be altered in a tumor.

3 RELEVANCE TO CANCER THERAPIES

Oncologists have long sought targeted therapies for cancer that focus on the specific genetic lesions present in an individual patient's tumor. The implementation of this concept requires precise characterization of the disease as well as knowledge of the patient's background (e.g., genetic and environmental characteristics). For tumors with a relatively narrow range of critical genetic defects (e.g., acute promyelocytic leukemia and chronic phase chronic myeloid leukemia), the development and deployment of targeted therapies has been more easily accomplished than in more complex and heterogeneous tumor types (e.g., breast cancer, and non-small-cell lung cancer — NSCLC). The emergence of rapid, high-resolution genome analysis tools provides an opportunity to tackle these more heterogeneous malignancies at both the levels of basic research (e.g., defining pathogenetic mechanisms) and clinical treatment selection (e.g., determining which patients are suitable for therapies targeting a particular growth-control pathway). The following are a few successful examples of targeted therapies for cancers and the continuing developments in the field.

A prime case of such a relationship between genetic features and patient care is the CLL. CLL is a clinically heterogeneous disease characterized by the accumulation of malignant CD5 + B cells. It is the most common type of leukemia in adults, accounting for up to 25% of all newly diagnosed leukemia. Some patients with CLL follow a nonaggressive course and often do not require treatment, while others exhibit a rapid progression. One major recent advance has been the identification of molecular and genetic prognostic factors including cytogenetic characteristics (11q deletion), mutational status (e.g., mutations in the immunoglobulin heavy-chain variable genes), gene expression markers (e.g., CD38 and ZAP-70) and some serum markers, that can be used in early-stage patients to identify those likely to progress rapidly (Stilgenbauer and Dohner, 2005). This affords the opportunity to tailor management for patients based on the predictable aggressiveness of their disease. Molecular and genetic findings are thus increasingly influencing management decisions in CLL and ultimately in drug discovery.

Another good example of using genetic characterizations to guide drug development and its deployment is the CML. Several efforts are being made to design drugs that target the BCR–ABL fusion protein that results from a specific chromosomal abnormality

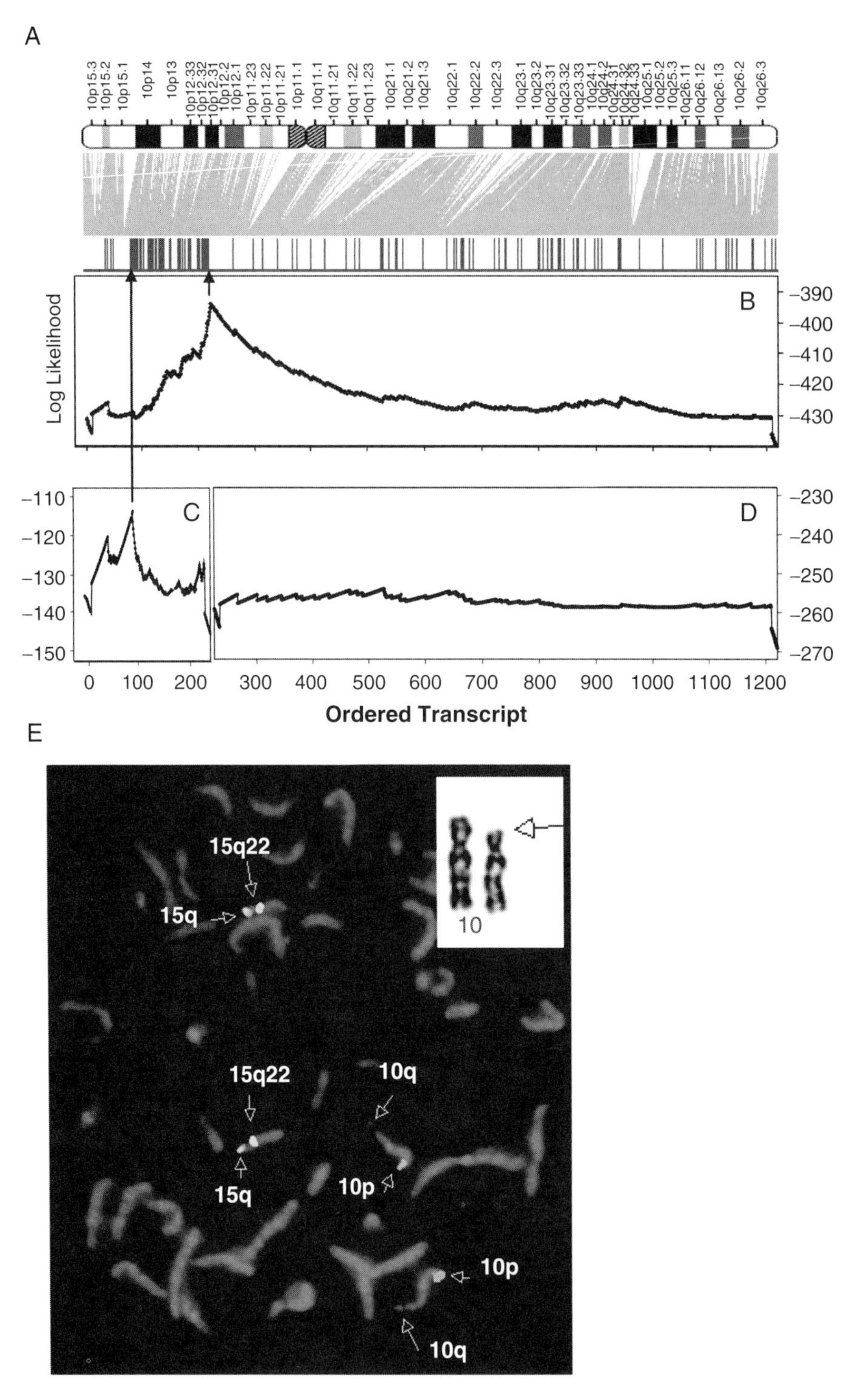

Fig. 6. Mapping the boundaries of chromosomal deletion by differential expression. (**A**) Affymetrix Human Genome U133 Plus 2.0 array was used to generate expression profiles for del(10) cells (GM03047) and matching control. Underexpressed transcripts in del(10) cells were identified using Affymetrix Microarray Suite 5.0, with decreased transcription declared when change *p* value was <0.002. The transcripts were ordered according to sequence on chromosome 10, with *red bars* indicating the transcription start site of genes identified as significantly underexpressed in del(10) cells relative to a tissue-matched normal control cell. (**B**) As detailed in our recent study (Zhou et al., 2004a, 2005),

found in CML patients. One such example is *Imatinib* (Norvatis Oncology) which has successfully controlled the progression of CML in most patients (Druker et al., 2001; O'Brien et al., 2003). However, it does not appear to cure the disease, as the relapse rate after 4.5½years is overall about 16% and higher in patients with advanced stages of the disease. Patients seem to develop resistance to this drug as the cancer cells mutate in such a way that the BCR—ABL protein is no longer recognized as a target by the drug. As such, a variety of CML clinical studies that are looking at combining drugs such as AMN107 (Bristol Myers Squibb) and imatinib are underway (Shah et al., 2004; Burgess et al., 2005). Other approaches are examining a combination of therapies with drugs that target other molecules in the pathway affected by BCR—ABL. Thus tests will be developed that can detect which BCR—ABL mutations develop so that treatment strategies can be tailored to each patient.

Increased knowledge of the mechanistic properties of malignant growth has facilitated the development of molecular-based therapies that can act on specific targets. Mutations in the epidermal growth factor receptor (EGFR) have been identified in NSCLC cells, and over expression of the EGFR and its ligands is a common feature of many cancers. As such, EGFR has become an attractive target for various antitumor drug development strategies. Anti-EGFR antibodies and EGFR tyrosine kinase inhibitors have shown efficacy in treating patients with advanced NSCLC, who have failed in previous therapy regimens (Arteaga, 2003; Lynch et al., 2004). These findings support the use of targeted agents for improving the clinical outcomes and quality of life in patients with advanced NSCLC. It is widely predicted that molecular targeted therapy for lung cancer will evolve away from the use of single agents to the use of Combinational therapy. Currently, researchers know of at least four genes that may be important in lung cancer when mutated: *EGFR*, *HER2*, *RAS*, and *BRAF* (Shigematsu and Gazdar, 2006). Identifying patients with mutations in these genes (as well as novel target genes) and using that knowledge to target all the abnormal proteins encoded by these mutated genes is a very promising approach.

Fig. 6. (continued) a single breakpoint model allowing differential density of underexpression was fit by maximum likelihood. The log likelihood associated with breakpoints at each ordinal position on chromosome 10 is plotted (*black line*) with the maximum likelihood value serving as the estimated origin of Copy Number Abnormality. *Gray lines* map ordinal positions of each assayed transcript to its chromosomal location. Significant change in the prevalence of underexpressed transcripts was identified at ordered transcript 224, 28.1 Mb from 10pter, agrees with previously defined the origin of deletion by cytogenetic analyses. (**C**) To determine whether deletion extended to the *p*-terminus, transcripts 1–223 were rescanned and a second significant change in the prevalence of underexpressed transcripts was identified at ordered transcript 85, 12.2 Mb from 10pter. (**D**) No significant change in the prevalence of underexpressed transcripts was identified in the region ranging from ordered transcript 224 to 10qter. Together with the results from panel (**B**), these data indicate a single partial deletion of chromosome 10p spanning the region between 10p14–10p12. (**E**) Subtelomere fluorescence in situ hybridization (FISH) verified results from the maximum likelihood expression-based analysis by confirming that the 10p deletion was interstitial with the intact subtelomere regions. Probes used are 10ptel006 (10pter probe, *green*); 10qtel24 (10qter probe, *red*); PML (15q22 probe, *aqua*); and AFMA224XHI (15qter probe, *yellow*). Two normal signals for both 10p and 10q subtelomeres were clearly identified. *Inset*: G-banded chromosome 10 of del(10) cell showing the deletion of *p* arm of the chromosome 10 (adapted from Zhou et al., 2005; with kind permission of BMJ Publishing Group) (*See Color Plates*)

In summary, advances in genome-wide profiling techniques will play a key role in the expansion of personalized therapy for cancer. The genome-wide assay technologies outlined above are part of the growing number of "top-down" approaches that provide comprehensive genomic profiles that can be correlated with biological/clinical status or functional aspects of tumors. In the near future, these analyses will help guide clinical treatment decisions, and in the long term, they may substantially advance our fundamental understanding of tumor progression. More traditional "bottom-up" studies (so-called basic studies), focused on individual genes or proteins, will continue to provide details that are not glimpsed by the top-down/global approaches. Conversely, focused studies may be misinterpreted due to the lack of global information. Thus, the integration of bottom-up and top-down information will play a critical role in defining the functional pathogenesis of cancer and shape its treatment. Integration of top-down and bottom-up analyses may also speed up the identification of interventions that might ameliorate malignant cell growth (e.g., selecting drugs that target growth aberrations downstream of those arising from defined sites of genetic damage).

REFERENCES

Albertson, D. G., C. Collins, F. McCormick and J. W. Gray (2003). Chromosome aberrations in solid tumors. *Nat Genet* **34**(4): 369–76

Arteaga, C. (2003). Targeting HER1/EGFR: a molecular approach to cancer therapy. *Semin Oncol* **30**(3 Suppl 7): 3–14

Avivi, I. and J. M. Rowe (2005). Prognostic factors in acute myeloid leukemia. *Curr Opin Hematol* **12**(1): 62–7

Bignell, G. R., J. Huang, J. Greshock, S. Watt, A. Butler, S. West, M. Grigorova, K. W. Jones, W. Wei, M. R. Stratton, et al. (2004). High-resolution analysis of DNA copy number using oligonucleotide microarrays. *Genome Res* **14**(2): 287–95

Brennan, C., Y. Zhang, C. Leo, B. Feng, C. Cauwels, A. J. Aguirre, M. Kim, A. Protopopov and L. Chin (2004). High-resolution global profiling of genomic alterations with long oligonucleotide microarray. *Cancer Res* **64**(14): 4744–8

Burgess, M. R., B. J. Skaggs, N. P. Shah, F. Y. Lee and C. L. Sawyers (2005). Comparative analysis of two clinically active BCR–ABL kinase inhibitors reveals the role of conformation-specific binding in resistance. *Proc Natl Acad Sci USA* **102**(9): 3395–400

Callagy, G., P. Pharoah, S. F. Chin, T. Sangan, Y. Daigo, L. Jackson and C. Caldas (2005). Identification and validation of prognostic markers in breast cancer with the complementary use of array-CGH and tissue microarrays. *J Pathol* **205**(3): 388–96

Caspersson, T., L. Zech, E. J. Modest, G. E. Foley, U. Wagh and E. Simonsson (1969a). Chemical differentiation with fluorescent alkylating agents in Vicia faba metaphase chromosomes. *Exp Cell Res* **58**(1): 128–40

Caspersson, T., L. Zech, E. J. Modest, G. E. Foley, U. Wagh and E. Simonsson (1969b). DNA-binding fluorochromes for the study of the organization of the metaphase nucleus. *Exp Cell Res* **58**(1): 141–52

Cole, S. W., W. Yan, Z. Galic, J. Arevalo and J. A. Zack (2005). Expression-based monitoring of transcription factor activity: the TELiS database. *Bioinformatics* **21**(6): 803–10

Crawley, J. J. and K. A. Furge (2002). Identification of frequent cytogenetic aberrations in hepatocellular carcinoma using gene-expression microarray data. *Genome Biol* **3**(12): RESEARCH0075

Davies, J. J., I. M. Wilson and W. L. Lam (2005). Array CGH technologies and their applications to cancer genomes. *Chromosome Res* **13**(3): 237–48

Druker, B. J., M. Talpaz, D. J. Resta, B. Peng, E. Buchdunger, J. M. Ford, N. B. Lydon, H. Kantarjian, R. Capdeville, S. Ohno-Jones, et al. (2001). Efficacy and safety of a specific inhibitor of the BCR–ABL tyrosine kinase in chronic myeloid leukemia. *N Engl J Med* **344**(14): 1031–7

Friend, S. H., R. Bernards, S. Rogelj, R. A. Weinberg, J. M. Rapaport, D. M. Albert and T. P. Dryja (1986). A human DNA segment with properties of the gene that predisposes to retinoblastoma and osteosarcoma. *Nature* **323**: 643

Frith, M. C., Y. Fu, L. Yu, J. F. Chen, U. Hansen and Z. Weng (2004). Detection of functional DNA motifs via statistical over-representation. *Nucleic Acids Res* **32**(4): 1372–81

Futreal, P. A., L. Coin, M. Marshall, T. Down, T. Hubbard, R. Wooster, N. Rahman and M. R. Stratton (2004). A census of human cancer genes. *Nat Rev Cancer* **4**(3): 177–83

Granville, C. A. and P. A. Dennis (2005). An overview of lung cancer genomics and proteomics. *Am J Respir Cell Mol Biol* **32**(3): 169–76

Henegariu, O., N. A. Heerema, P. Bray-Ward and D. C. Ward (1999). Colour-changing karyotyping: an alternative to M-FISH/SKY. *Nat Genet* **23**(3): 263–4

Hoque, M. O., C. C. Lee, P. Cairns, M. Schoenberg and D. Sidransky (2003). Genome-wide genetic characterization of bladder cancer: a comparison of high-density single-nucleotide polymorphism arrays and PCR-based microsatellite analysis. *Cancer Res* **63**(9): 2216–22

Huang, J., W. Wei, J. Zhang, G. Liu, G. R. Bignell, M. R. Stratton, P. A. Futreal, R. Wooster, K. W. Jones and M. H. Shapero (2004). Whole genome DNA copy number changes identified by high density oligonucleotide arrays. *Hum Genomics* **1**(4): 287–99

Imoto, H., S. Hirotsune, M. Muramatsu, K. Okuda, O. Sugimoto, V. M. Chapman and Y. Hayashizaki (1994). Direct determination of NotI cleavage sites in the genomic DNA of adult mouse kidney and human trophoblast using whole-range restriction landmark genomic scanning. *DNA Res* **1**(5): 239–43

Ishkanian, A. S., C. A. Malloff, S. K. Watson, R. J. DeLeeuw, B. Chi, B. P. Coe, A. Snijders, D. G. Albertson, D. Pinkel, M. A. Marra, et al. (2004). A tiling resolution DNA microarray with complete coverage of the human genome. *Nat Genet* **36**(3): 299–303

Janne, P. A., C. Li, X. Zhao, L. Girard, T. H. Chen, J. Minna, D. C. Christiani, B. E. Johnson and M. Meyerson (2004). High-resolution single-nucleotide polymorphism array and clustering analysis of loss of heterozygosity in human lung cancer cell lines. *Oncogene* **23**(15): 2716–26

Jeffrey, S. S., P. E. Lonning and B. E. Hillner (2005). Genomics-based prognosis and therapeutic prediction in breast cancer. *J Natl Compr Canc Netw* **3**(3): 291–300

Kallioniemi, A., O. P. Kallioniemi, D. Sudar, D. Rutovitz, J. W. Gray, F. Waldman and D. Pinkel (1992). Comparative genomic hybridization for molecular cytogenetic analysis of solid tumors. *Science* **258**(5083): 818–21

Kauraniemi, P., T. Kuukasjarvi, G. Sauter and A. Kallioniemi (2003). Amplification of a 280-kilobase core region at the ERBB2 locus leads to activation of two hypothetical proteins in breast cancer. *Am J Pathol* **163**(5): 1979–84

Knudson, A. G., Jr. (1971). Mutation and cancer: statistical study of retinoblastoma. *Proc Natl Acad Sci USA* **68**(4): 820–3

Knudson, A. G. (1996). Hereditary cancer: two hits revisited. *J Cancer Res Clin Oncol* **122**(3): 135–40

Lander, E. S., L. M. Linton, B. Birren, C. Nusbaum, M. C. Zody, J. Baldwin, K. Devon, K. Dewar, M. Doyle, W. FitzHugh, et al. (2001). Initial sequencing and analysis of the human genome. *Nature* **409**(6822): 860–921

Lieberfarb, M. E., M. Lin, M. Lechpammer, C. Li, D. M. Tanenbaum, P. G. Febbo, R. L. Wright, J. Shim, P. W. Kantoff, M. Loda, et al. (2003). Genome-wide loss of heterozygosity analysis from laser capture microdissected prostate cancer using single nucleotide polymorphic allele (SNP) arrays and a novel bioinformatics platform dChipSNP. *Cancer Res* **63**(16): 4781–5

Liehr, T., A. Heller, H. Starke and U. Claussen (2002). FISH banding methods: applications in research and diagnostics. *Expert Rev Mol Diagn* **2**(3): 217–25

Lin, M., L. J. Wei, W. R. Sellers, M. Lieberfarb, W. H. Wong and C. Li (2004). dChipSNP: significance curve and clustering of SNP-array-based loss-of-heterozygosity data. *Bioinformatics* **20**(8): 1233–40

Lindblad-Toh, K., D. M. Tanenbaum, M. J. Daly, E. Winchester, W. O. Lui, A. Villapakkam, S. E. Stanton, C. Larsson, T. J. Hudson, B. E. Johnson, et al. (2000). Loss-of-heterozygosity analysis of small-cell lung carcinomas using single-nucleotide polymorphism arrays. *Nat Biotechnol* **18**(9): 1001–5

Lisitsyn, N. and M. Wigler (1993). Cloning the differences between two complex genomes. *Science* **259**(5097): 946–51

Liyanage, M., A. Coleman, S. du Manoir, T. Veldman, S. McCormack, R. B. Dickson, C. Barlow, A. Wynshaw-Boris, S. Janz, J. Wienberg, et al. (1996). Multicolour spectral karyotyping of mouse chromosomes. *Nat Genet* **14**(3): 312–5

Lucito, R., J. Healy, J. Alexander, A. Reiner, D. Esposito, M. Chi, L. Rodgers, A. Brady, J. Sebat, J. Troge, et al. (2003). Representational oligonucleotide microarray analysis: a high-resolution method to detect genome copy number variation. *Genome Res* **13**(10): 2291–305

Lynch, T. J., D. W. Bell, R. Sordella, S. Gurubhagavatula, R. A. Okimoto, B. W. Brannigan, P. L. Harris, S. M. Haserlat, J. G. Supko, F. G. Haluska, et al. (2004). Activating mutations in the epidermal growth factor receptor underlying responsiveness of non-small-cell lung cancer to gefitinib. *N Engl J Med* **350**(21): 2129–39

Mantripragada, K. K., P. G. Buckley, T. D. de Stahl and J. P. Dumanski (2004). Genomic microarrays in the spotlight. *Trends Genet* **20**(2): 87–94

Martinez-Climent, J. A., A. A. Alizadeh, R. Segraves, D. Blesa, F. Rubio-Moscardo, D. G. Albertson, J. Garcia-Conde, M. J. Dyer, R. Levy, D. Pinkel, et al. (2003). Transformation of follicular lymphoma to diffuse large cell lymphoma is associated with a heterogeneous set of DNA copy number and gene expression alterations. *Blood* **101**(8): 3109–17

Mundle, S. D. and I. Sokolova (2004). Clinical implications of advanced molecular cytogenetics in cancer. *Expert Rev Mol Diagn* **4**(1): 71–81

Myers, C. L., M. J. Dunham, S. Y. Kung and O. G. Troyanskaya (2004). Accurate detection of aneuploidies in array CGH and gene expression microarray data. *Bioinformatics* **20**(18): 3533–43

Nowell, P. C. and D. A. Hungerford (1960). Chromosome studies on normal and leukemic human leukocytes. *J Natl Cancer Inst* **25**: 85–109

O'Brien, S. G., F. Guilhot, R. A. Larson, I. Gathmann, M. Baccarani, F. Cervantes, J. J. Cornelissen, T. Fischer, A. Hochhaus, T. Hughes, et al. (2003). Imatinib compared with interferon and low-dose cytarabine for newly diagnosed chronic-phase chronic myeloid leukemia. *N Engl J Med* **348**(11): 994–1004

Paris, P. L., A. Andaya, J. Fridlyand, A. N. Jain, V. Weinberg, D. Kowbel, J. H. Brebner, J. Simko, J. E. Watson, S. Volik, et al. (2004). Whole genome scanning identifies genotypes associated with recurrence and metastasis in prostate tumors. *Hum Mol Genet* **13**(13): 1303–13

Phillips, J. L., S. W. Hayward, Y. Wang, J. Vasselli, C. Pavlovich, H. Padilla-Nash, J. R. Pezullo, B. M. Ghadimi, G. D. Grossfeld, A. Rivera, et al. (2001). The consequences of chromosomal aneuploidy on gene expression profiles in a cell line model for prostate carcinogenesis. *Cancer Res* **61**(22): 8143–9

Pinkel, D. and D. G. Albertson (2005a). Array comparative genomic hybridization and its applications in cancer. *Nat Genet* **37**(Suppl): S11–7

Pinkel, D. and D. G. Albertson (2005b). Comparative genomic hybridization. *Annu Rev Genomics Hum Genet* **6**: 331–54

Pinkel, D., R. Segraves, D. Sudar, S. Clark, I. Poole, D. Kowbel, C. Collins, W. L. Kuo, C. Chen, Y. Zhai, et al. (1998). High resolution analysis of DNA copy number variation using comparative genomic hybridization to microarrays. *Nat Genet* **20**(2): 207–11

Pollack, J. R., C. M. Perou, A. A. Alizadeh, M. B. Eisen, A. Pergamenschikov, C. F. Williams, S. S. Jeffrey, D. Botstein and P. O. Brown (1999). Genome-wide analysis of DNA copy-number changes using cDNA microarrays. *Nat Genet* **23**(1): 41–6

Pollack, J. R., T. Sorlie, C. M. Perou, C. A. Rees, S. S. Jeffrey, P. E. Lonning, R. Tibshirani, D. Botstein, A. L. Borresen-Dale and P. O. Brown (2002). Microarray analysis reveals a major direct role of DNA copy number alteration in the transcriptional program of human breast tumors. *Proc Natl Acad Sci USA* **99**: 12963–8

Rosin, M. P., X. Cheng, C. Poh, W. L. Lam, Y. Huang, J. Lovas, K. Berean, J. B. Epstein, R. Priddy, N. D. Le, et al. (2000). Use of allelic loss to predict malignant risk for low-grade oral epithelial dysplasia. *Clin Cancer Res* **6**(2): 357–62

Rubio-Moscardo, F., J. Climent, R. Siebert, M. A. Piris, J. I. Martin-Subero, I. Nielander, J. Garcia-Conde, M. J. Dyer, M. J. Terol, D. Pinkel, et al. (2005). Mantle-cell lymphoma genotypes identified with CGH to BAC microarrays define a leukemic subgroup of disease and predict patient outcome. *Blood* **105**(11): 4445–54

Schrock, E., S. du Manoir, T. Veldman, B. Schoell, J. Wienberg, M. A. Ferguson-Smith, Y. Ning, D. H. Ledbetter, I. Bar-Am, D. Soenksen, et al. (1996). Multicolor spectral karyotyping of human chromosomes. *Science* **273**(5274): 494–7

Schwaenen, C., M. Nessling, S. Wessendorf, T. Salvi, G. Wrobel, B. Radlwimmer, H. A. Kestler, C. Haslinger, S. Stilgenbauer, H. Dohner, et al. (2004). Automated array-based genomic profiling in chronic lymphocytic leukemia: development of a clinical tool and discovery of recurrent genomic alterations. *Proc Natl Acad Sci USA* **101**(4): 1039–44

Shah, N. P., C. Tran, F. Y. Lee, P. Chen, D. Norris and C. L. Sawyers (2004). Overriding imatinib resistance with a novel ABL kinase inhibitor. *Science* **305**(5682): 399–401

Shigematsu, H. and A. F. Gazdar (2006). Somatic mutations of epidermal growth factor receptor signaling pathway in lung cancers. *Int J Cancer* **118**(2): 257–62

Speicher, M. R., S. Gwyn Ballard and D. C. Ward (1996). Karyotyping human chromosomes by combinatorial multi-fluor FISH. *Nat Genet* **12**(4): 368–75

Stilgenbauer, S. and H. Dohner (2005). Genotypic prognostic markers. *Curr Top Microbiol Immunol* **294**: 147–64

Venter, J. C., M. D. Adams, E. W. Myers, P. W. Li, R. J. Mural, G. G. Sutton, H. O. Smith, M. Yandell, C. A. Evans, R. A. Holt, et al. (2001). The sequence of the human genome. *Science* **291**(5507): 1304–51

Virtaneva, K., F. A. Wright, S. M. Tanner, B. Yuan, W. J. Lemon, M. A. Caligiuri, C. D. Bloomfield, A. de La Chapelle and R. Krahe (2001). Expression profiling reveals fundamental biological differences in acute myeloid leukemia with isolated trisomy 8 and normal cytogenetics. *Proc Natl Acad Sci USA* **98**(3): 1124–9

Vogelstein, B., E. R. Fearon, S. E. Kern, S. R. Hamilton, A. C. Preisinger, Y. Nakamura and R. White (1989). Alleotype of colorectal carcinomas. *Science* **244**: 201–211

Wang, D. G., J. B. Fan, C. J. Siao, A. Berno, P. Young, R. Sapolsky, G. Ghandour, N. Perkins, E. Winchester, J. Spencer, et al. (1998). Large-scale identification, mapping, and genotyping of single-nucleotide polymorphisms in the human genome. *Science* **280**(5366): 1077–82

Wang, T. L., C. Maierhofer, M. R. Speicher, C. Lengauer, B. Vogelstein, K. W. Kinzler and V. E. Velculescu (2002). Digital karyotyping. *Proc Natl Acad Sci USA* **99**(25): 16156–61

Wang, Z. C., M. Lin, L. J. Wei, C. Li, A. Miron, G. Lodeiro, L. Harris, S. Ramaswamy, D. M. Tanenbaum, M. Meyerson, et al. (2004). Loss of heterozygosity and its correlation with expression profiles in subclasses of invasive breast cancers. *Cancer Res* **64**(1): 64–71

Weir, B., X. Zhao and M. Meyerson (2004). Somatic alterations in the human cancer genome. *Cancer Cell* **6**(5): 433–8

Weiss, M. M., E. J. Kuipers, C. Postma, A. M. Snijders, D. Pinkel, S. G. Meuwissen, D. Albertson and G. A. Meijer (2004). Genomic alterations in primary gastric adenocarcinomas correlate with clinicopathological characteristics and survival. *Cell Oncol* **26**(5–6): 307–17

Wolf, M., S. Mousses, S. Hautaniemi, R. Karhu, P. Huusko, M. Allinen, A. Elkahloun, O. Monni, Y. Chen, A. Kallioniemi, et al. (2004). High-resolution analysis of gene copy number alterations in human prostate cancer using CGH on cDNA microarrays: impact of copy number on gene expression. *Neoplasia* **6**(3): 240–7

Zhao, X., C. Li, J. G. Paez, K. Chin, P. A. Janne, T. H. Chen, L. Girard, J. Minna, D. Christiani, C. Leo, et al. (2004). An integrated view of copy number and allelic alterations in the cancer genome using single nucleotide polymorphism arrays. *Cancer Res* **64**(9): 3060–71

Zhao, X., B. A. Weir, T. LaFramboise, M. Lin, R. Beroukhim, L. Garraway, J. Beheshti, J. C. Lee, K. Naoki, W. G. Richards, et al. (2005). Homozygous deletions and chromosome amplifications in human lung carcinomas revealed by single nucleotide polymorphism array analysis. *Cancer Res* **65**(13): 5561–70

Zhou, X., S. W. Cole, S. Hu and D. T. Wong (2004a). Detection of DNA copy number abnormality by microarray expression analysis. *Hum Genet* **114**(5): 464–7

Zhou, X., R. C. K. Jordan, S. Mok, M. J. Birrer and D. T. Wong (2004b). DNA copy number abnormality of oral squamous cell carcinoma detected by cDNA array-based CGH. *Cancer Genet Cytogenet* **151**(1): 90–92

Zhou, X., C. Li, S. C. Mok, Z. Chen and D. T. W. Wong (2004c). Whole genome loss of heterozygosity profiling on oral squamous cell carcinoma by high-density single nucleotide polymorphic allele (SNP) array. *Cancer Genet Cytogenet* **151**(1): 82–84

Zhou, X., S. C. Mok, Z. Chen, Y. Li and D. T. Wong (2004d). Concurrent analysis of loss of heterozygosity (LOH) and copy number abnormality (CNA) for oral premalignancy progression using the Affymetrix 10K SNP mapping array. *Hum Genet* **115**(4): 327–330

Zhou, X., S. W. Cole, Z. Chen, Y. Li and D. T. Wong (2005). Identification of discrete chromosomal deletion by binary recursive partitioning of microarray differential expression data. *J Med Genet* **42**(5): 416–9

Zhou, X., T. Yu, S. W. Cole and D. T. Wong (2006). Advancement in characterization of genomic alterations for improved diagnosis, treatment and prognostics in cancer. *Expert Rev Mol Diagn* **6**(1): 39–50

Color Plates

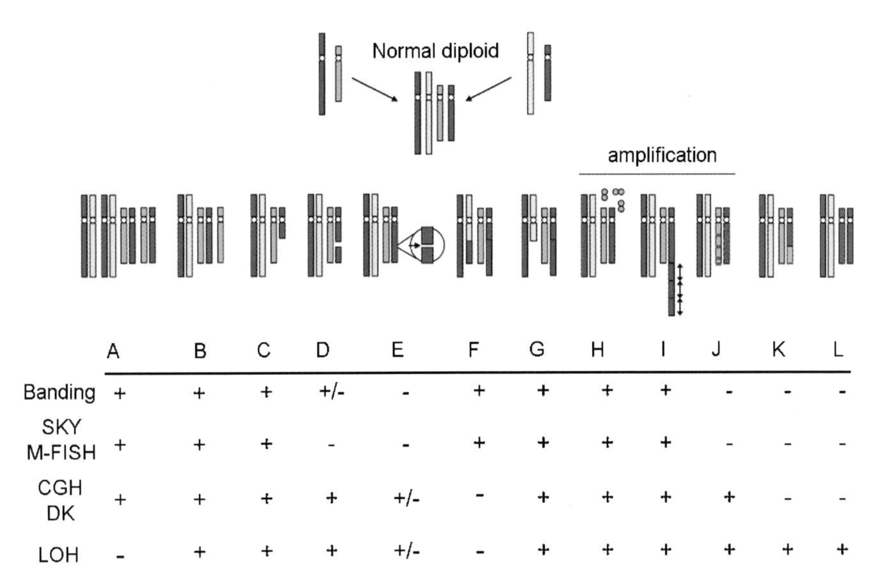

Chapter 7, Fig. 1. Detection and mapping of chromosomal abnormalities using different genomic and cytogenetic approaches. *A*, polyploid; *B*, aneuploid; *C*, gross deletion; *D*, interstitial deletion; *E*, microdeletion; *F*, reciprocal translocation; *G*, nonreciprocal translocation; *H*, double minutes; *I*, HSR; *J*, distributed insertion; *K*, somatic recombination; *L*, duplication and loss (adapted from Zhou et al., 2006; with kind permission of Future Drugs Ltd)

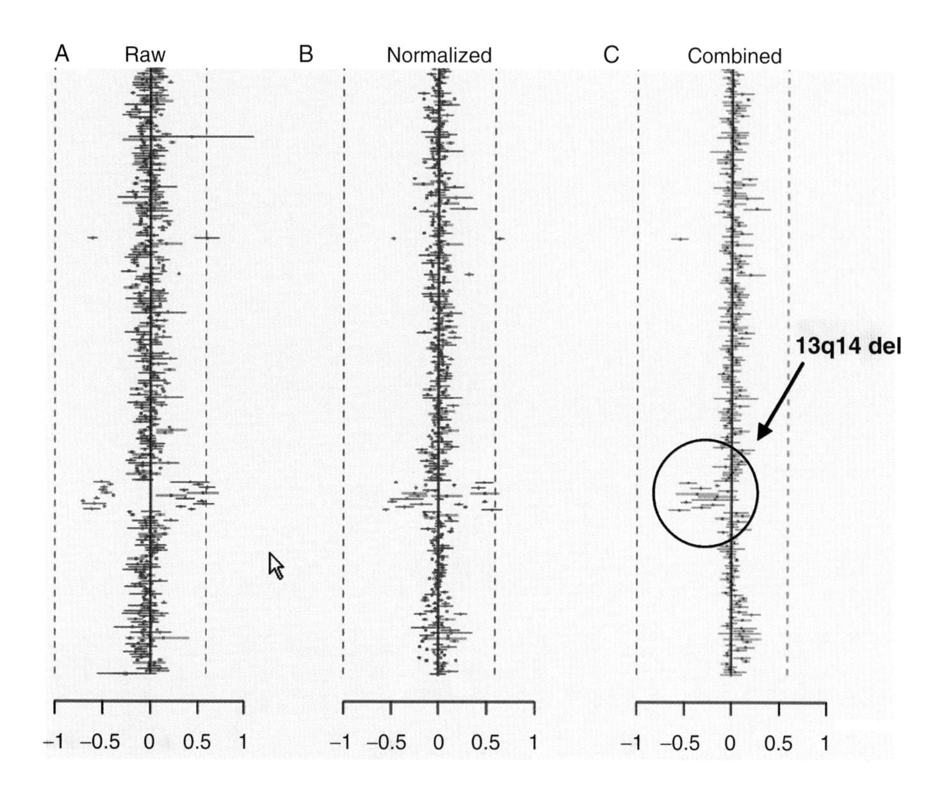

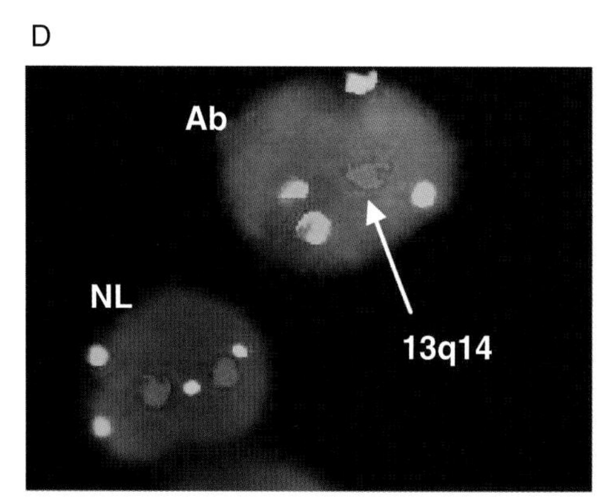

Chapter 7, Fig. 2. A typical BAC array-based CGH result on a CLL sample where a dye-swap experiment was performed. CGH was performed using array with BAC clones representing genomic regions that are commonly involved in CLL ((**a**) raw data, (**b**) normalized, and (**c**) combined data). A deletion of the 13q14 region was observed. (**d**) Targeted FISH analysis with DNA probes specific to 13q14 (*red*) confirmed this submicroscopic deletion (abnormal cell (Ab) with loss of one red signal when compared to the normal cell (NL))

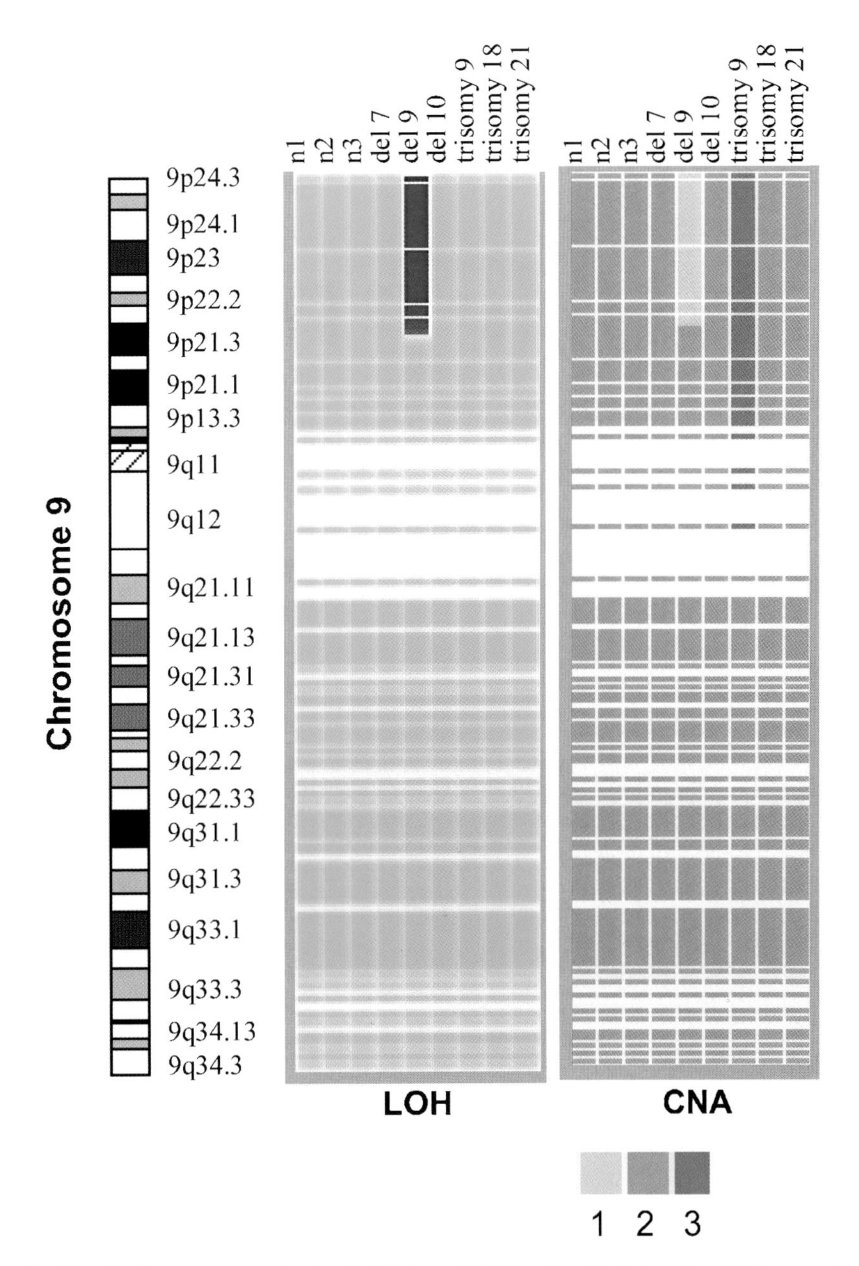

Chapter 7, Fig. 3. Concurrent analysis of LOH and CNA using high-density SNP array. The LOH regions and the CNA regions were detected and demarcated as described previously (Lin et al., 2004; Zhao et al., 2004; Zhou et al., 2004c,d). The LOH and CNA patterns for chromosome 9 were shown for nine cell lines: normal cells (n1, n2, and n3), trisomy cells (trisomy 9, trisomy 18, and trisomy 21, with an extra copy of 9pter > q13, 18, 21, respectively) and deletion cells (del(7), del(9), and del(10), with deletions at 7pter > q34, 9pter > p21, 10qter > p11, respectively). Each column represents one sample, and each row represents a SNP marker. Color code for LOH profiling (*right panel*): *blue* = LOH; *light green* = retained; *gray* = uninformative; *white* = no call. Copy numbers (*left panel*) were represented by different intensity of *red colors* as indicated in the figure (adapted from Zhou et al., 2004d; with kind permission of Springer Science and Business Media)

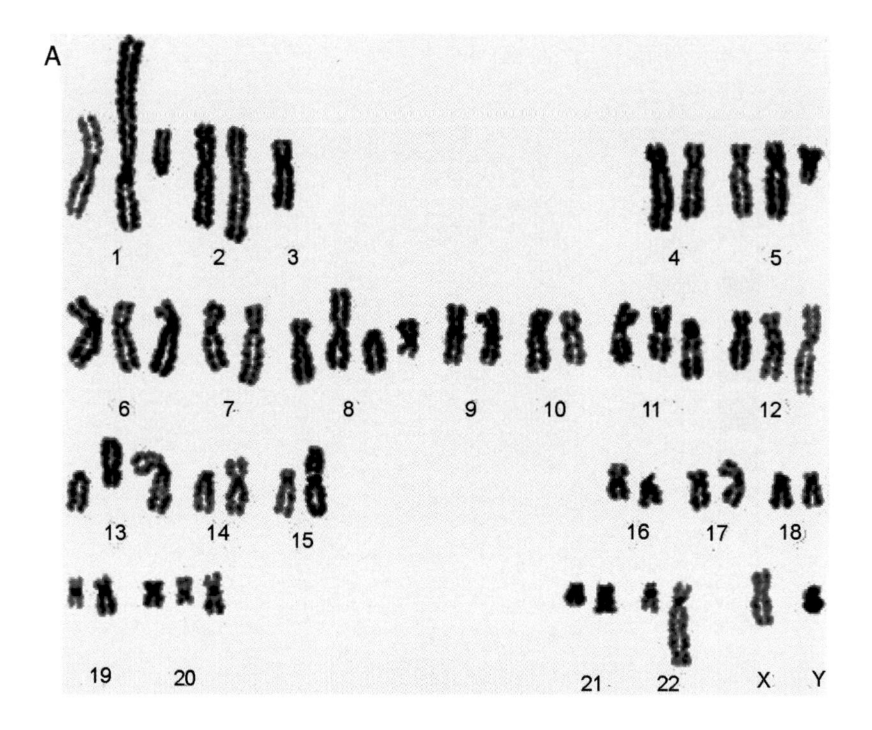

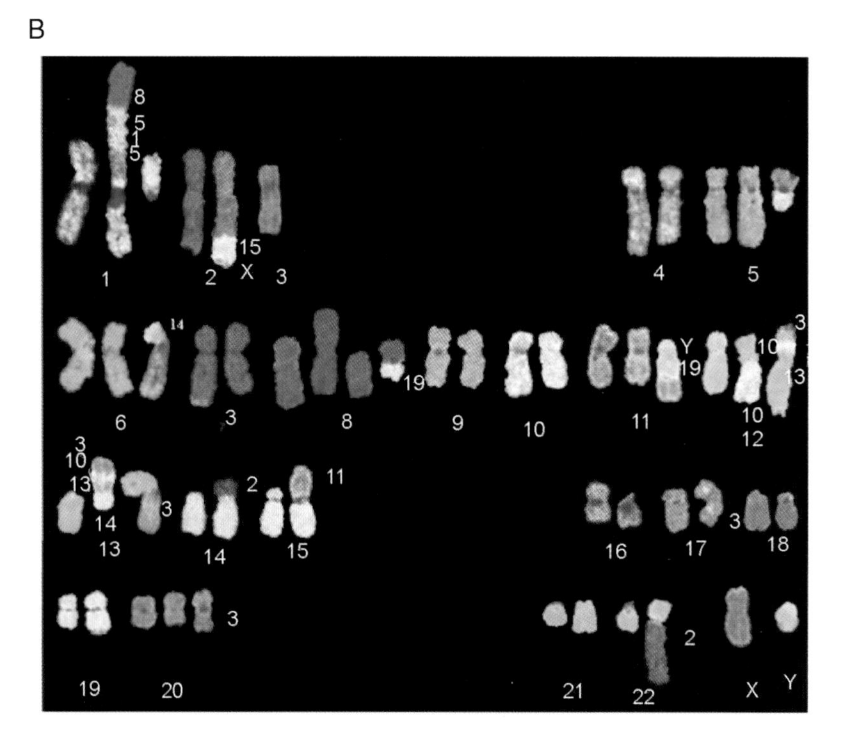

Chapter 7, Fig. 4. Comprehensive cytogenetic characterization of the abnormal bone-marrow cells from an acute myeloid leukemia (AML) patient. (**a**) Standard trypsin Giemsa-banding technique-based karyotype analysis. (**b**) M-FISH was performed using Abbott-Vysis SpectraVysion assay. The hyperdiploid karyotype of this CML patient has numerous complex structural rearrangements. The simultaneous karyotype analysis and M-FISH analysis clearly depicted the complex nature of the genomic anomalies

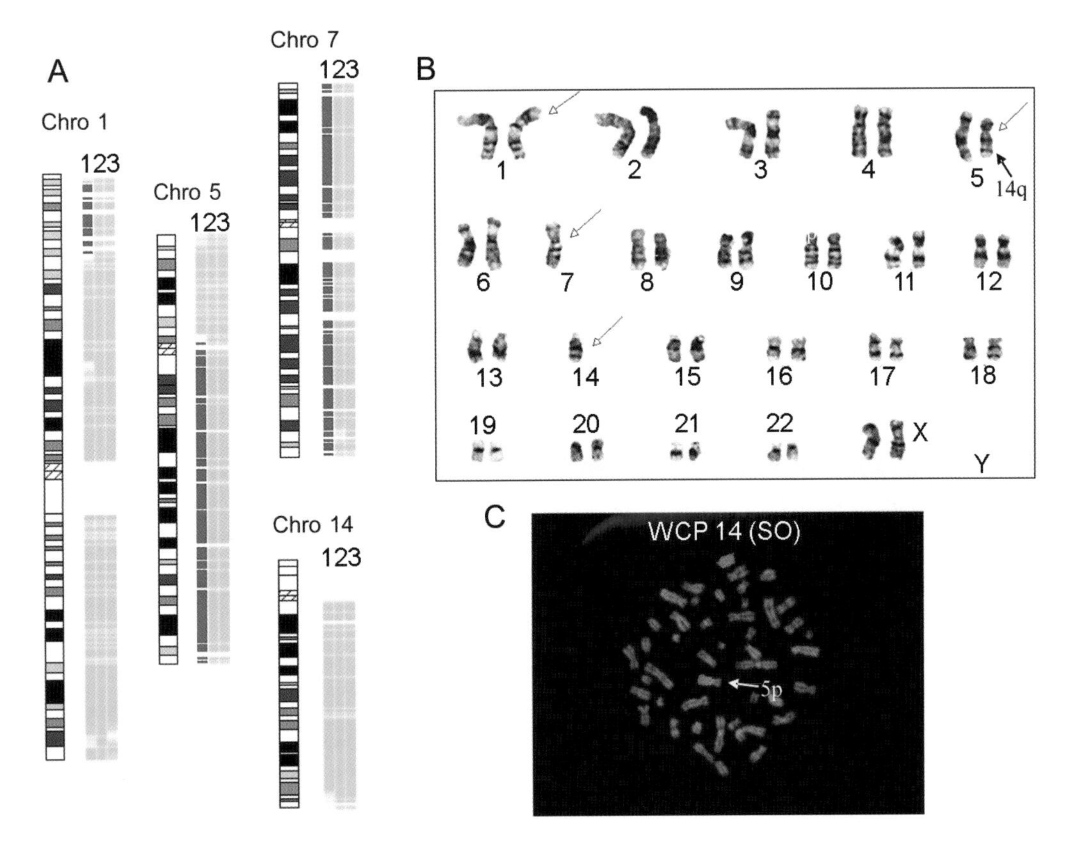

Chapter 7, Fig. 5. Comprehensive genomic analyses of a myelodysplastic syndrome (MDS) using SNP array-based approach and complementary cytogenetic approaches. (**a**) Three MDS cases were analyzed with a 10K SNP mapping array. The LOH regions were detected and demarcated as described (Lin et al., 2004; Zhao et al., 2004; Zhou et al., 2004c,d). The LOH patterns for chromosome 1, 5, 7, 14 are shown. (**b**) The karyotype for Case 1 is presented. In Case 1, SNP array-based LOH demonstrated no loss of chromosome 14 material and a more extensive 5q deletion than interpreted by the karyotype. (**c**) Whole chromosome paint (WCP) of chromosome 14, identified two signals for chromosome 14 (*red*), one normal 14 and the other chromosome 14 translocated to 5q close to the pericentromeric region. Together with the results from (**a**) and (**b**), these data indicate a karyotype of 44,XX,del(1)(p32p36),der(5)t(5;14)(q13;q11.2),-7[20] for Case 1 (adapted from Zhou et al., 2006; with kind permission of Future Drugs Ltd)

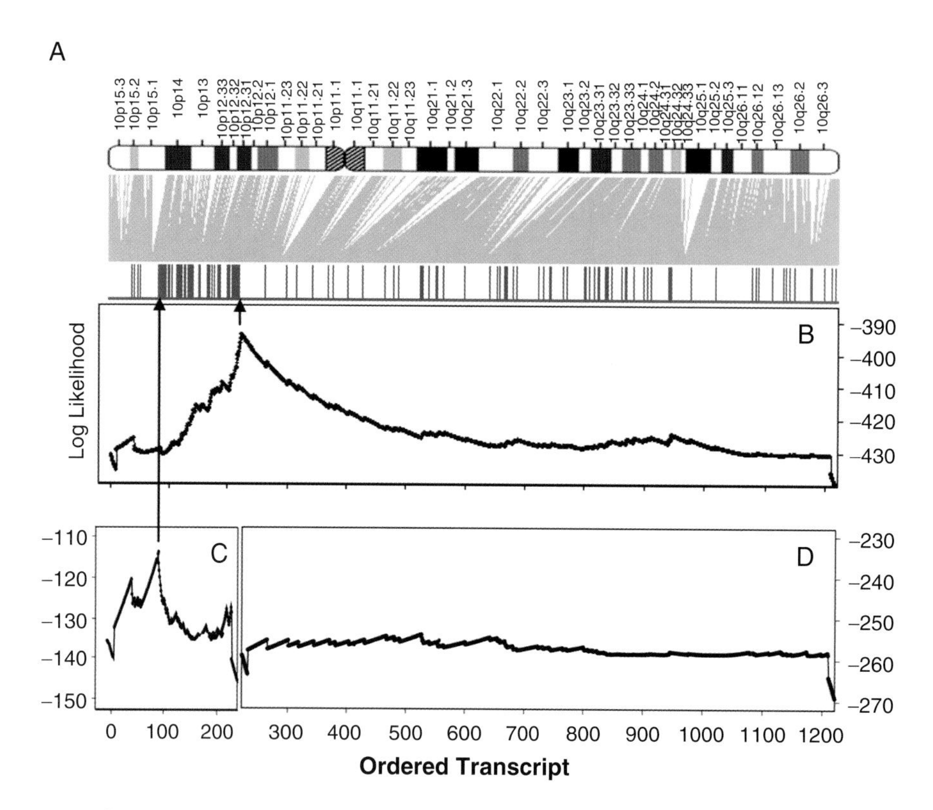

Chapter 7, Fig. 6. Mapping the boundaries of chromosomal deletion by differential expression. (**a**) Affymetrix Human Genome U133 Plus 2.0 array was used to generate expression profiles for del(10) cells (GM03047) and matching control. Underexpressed transcripts in del(10) cells were identified using Affymetrix Microarray Suite 5.0, with decreased transcription declared when change p value was <0.002. The transcripts were ordered according to sequence on chromosome 10, with *red bars* indicating the transcription start site of genes identified as significantly underexpressed in del(10) cells relative to a tissue-matched normal control cell. (**b**) As detailed in our recent study (Zhou et al., 2004a, 2005), a single breakpoint model allowing differential density of underexpression was fit by maximum likelihood. The log likelihood associated with breakpoints at each ordinal position on chromosome 10 is plotted (*black line*) with the maximum likelihood value serving as the estimated origin of Copy Number Abnormality. *Gray lines* map ordinal positions of each assayed transcript to its chromosomal location. Significant change in the prevalence of underexpressed transcripts was identified at ordered transcript 224, 28.1 Mb from 10pter, agrees with previously defined the origin of deletion by cytogenetic analyses. (**c**) To determine whether deletion extended to the p-terminus, transcripts 1–223 were rescanned and a second significant change in the prevalence of underexpressed transcripts was identified at ordered transcript 85, 12.2 Mb from 10pter. (**d**) No significant change in the prevalence of underexpressed transcripts was identified in the region ranging from ordered transcript 224 to 10qter. Together with the results from panel (**b**), these data indicate a single partial deletion of chromosome 10p spanning the region between 10p14–10p12

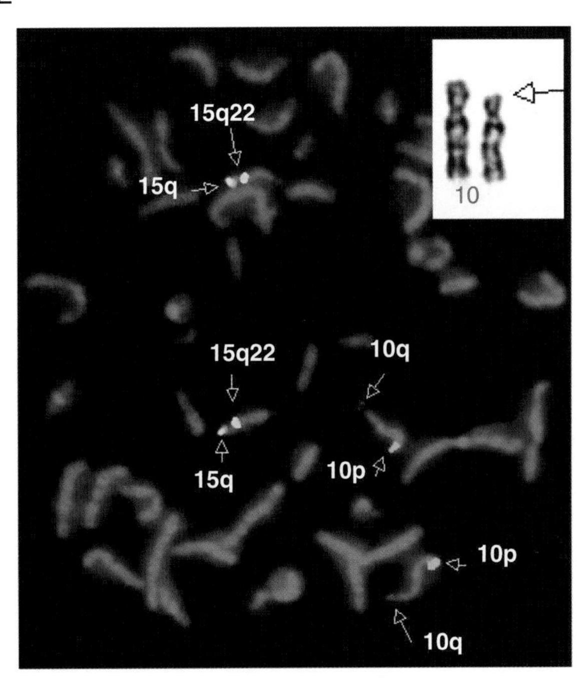

Chapter 7, Fig. 6. (continued) (**e**) Subtelomere fluorescence in situ hybridization (FISH) verified results from the maximum likelihood expression-based analysis by confirming that the 10p deletion was interstitial with the intact subtelomere regions. Probes used are 10ptel006 (10pter probe, *green*); 10qtel24 (10qter probe, *red*); PML (15q22 probe, *aqua*); and AFMA224XHI (15qter probe, *yellow*). Two normal signals for both 10p and 10q subtelomeres were clearly identified. *Inset*: G-banded chromosome 10 of del(10) cell showing the deletion of *p* arm of the chromosome 10 (adapted from Zhou et al., 2005; with kind permission of BMJ Publishing Group)

8 Gene Expression Profiling of the Leukemias: Oncogenesis, Drug Responsiveness, and Prediction of Clinical Outcome

Lars Bullinger, Hartmut Dohner, and Jonathan R. Pollack

ABSTRACT

Over the past decades, improvements in therapies have exposed a marked clinical heterogeneity among leukemias, and the identification of clonal karyotypic abnormalities has revealed morphologically similar leukemias as biologically heterogeneous hematopoietic disorders. In an attempt to define relevant entities biologically and clinically, current knowledge of morphologic, immunophenotypic, genetic and clinical features has been incorporated in a new WHO leukemia classification. Nevertheless, many of the newly-defined leukemia classes are still characterized by considerable heterogeneity, suggesting additional disease subtypes at the molecular level. Ultimately, an ideal classification system would be based on the underlying molecular genetic and epigenetic pathological mechanisms, which are reflected in aberrant gene expression. Recently, surveying leukemic gene expression patterns on a genomic scale has become feasible using DNA microarray technology. The combination of expression profiling methods with innovative bioinformatics analyses represents a powerful approach to gain novel insights into leukemogenesis, to investigate drug responsiveness, and to predict clinical outcome in leukemias. While many challenges remain ahead, we will outline how gene expression profiling might significantly contribute to a refined molecular taxonomy of leukemias. Furthermore, in combination with other whole genome approaches, we expect novel therapeutic targets as well as new therapies to be identified and improved prediction of drug response and clinical outcome to enable risk-adapted individualized treatment strategies for leukemia patients.

Key words: Leukemia, Microarray technology, Gene expression profiling, Class discovery, Class prediction, Drug response, Outcome prediction,

G.J. Gordon (ed.), *Cancer Drug Discovery and Development: Bioinformatics in Cancer and Cancer Therapy*,
DOI: 10.1007/978-1-59745-576-3_8,

1 INTRODUCTION

1.1 Leukemias: Historical Aspects

In 1845 Bennett and Virchow independently published two reports describing the first cases of leukemia (Bennett, 1845; Virchow, 1845), and Virchow was the first to call the disease "leukemia," or "white blood" (Virchow, 1845). Since then, advances in biochemistry, cytogenetics, cytochemistry, immunology, and subsequently molecular biology have led to the recognition of many different malignant hematopoietic diseases in the twentieth century (Hoffbrand and Fantini, 1999). During the past three decades, a dramatic expansion of our understanding of the hematopoietic system has revealed that leukemias demonstrate extraordinary morphologic, biologic, and clinical heterogeneity. Therefore, clinically-relevant classification systems that reflect the underlying tumor biology are needed.

1.2 Classification of Leukemias

Classification of leukemias is broadly related to the cell of origin (e.g., myeloid or lymphoid) as well as to the rapidity of the clinical course (e.g., acute or chronic), but over the past several years modern categorizations have identified specific leukemia subtypes on the basis of morphologic, immunophenotypic, genetic and clinical features. In an attempt to define biologically and clinically relevant leukemia entities, the current World Health Organization (WHO) incorporated this knowledge into a novel classification of tumors of hematopoietic and lymphoid tissues (Harris et al., 1999; Vardiman et al., 2002) (summarized in Table 1). The result is a more sophisticated classification that

Table 1

WHO classification of tumors of hematopoietic and lymphoid tissues

Acute leukemias	Chronic leukemias
Myeloid leukemias	
AML with recurrent genetic abnormalities	***Myeloproliferative diseases***
AML with t(8;21)(q22;q22), (*AML1/ETO*)	Chronic myelomonocytic leukemia
AML with inv(16)(p13q22), (*CBFβ/MYH11*)	Atypical chronic myeloid leukemia
AML with t(15;17)(q22;q12), (*PML/RARα*)	Juvenile myelomonocytic leukemia
AML with 11q23 (*MLL*) abnormalities	
AML with multilineage dysplasia	***Chronic myeloproliferative diseases***
	CML with t(9;22)(q34;q11), (*BCR/ABL*)
AML, therapy related	Chronic neutrophilic leukemia
	Chronic eosinophilic leukemia
AML not otherwise categorized	
Lymphoid leukemias	
Precursor B-cell neoplasms	***Mature B-cell neoplasms***
Precursor B-ALL	CLL
Burkitt leukemia	B-cell prolymphocytic leukemia
	Hairy cell leukemia
Precursor T-cell neoplasms	***Mature T-cell neoplasms***
Precursor T-ALL	T-cell prolymphocytic leukemia
	T-cell granular lymphocytic leukemia

for example divides acute myeloid leukemias (AML) into four large subclasses (AML with recurrent cytogenetic abnormalities, AML with multilineage dysplasia, therapy related AML, and AML not otherwise categorized), which are themselves further subdivided into several distinct AML subtypes. However, for many myeloid leukemia subtypes there is no specific genetic or pathogenic event discovered as yet. And even within well-defined AML subgroups, like cases with t(8;21)(q22;q22) or inv(16)((p13q22), considerable clinical heterogeneity is observed (Schlenk et al., 2004). Therefore, several molecularly distinct subtypes may exist within the same cytogenetic category, and to further refine the current classification of leukemias a deeper molecular understanding of these diseases is needed.

1.3 Cytogenetics and Molecular Genetics

The invention of chromosomal banding techniques heralded a new era for investigating leukemia biology. Chronic myelogenous leukemia (CML) was the first malignancy in which a recurring chromosomal abnormality, the Philadelphia chromosome (Nowell and Hungerford, 1960), was found to result from a translocation of genetic material from one chromosome to another [t(9;22)] (Rowley, 1973). A decade later the fusion gene resulting from this translocation, *BCR-ABL* (de Klein et al., 1982), was shown to be responsible for the myeloproliferation observed in CML (Konopka et al., 1985; Shtivelman et al., 1985). Since then, the discovery of well over 100 chromosome translocations, and the identification of a number of recurring chromosomal gains and losses in leukemic cells, have transformed our understanding of the genetic mechanisms involved in leukemogenesis (Rowley, 1999).

Recent advances in molecular genetics have shown that the expression of a single fusion gene, e.g., *RUNX1-CBFA2T1* or *CBFB-MYH11*, resulting from t(8;21) or inv(16), respectively, can block myeloid differentiation, but does not by itself cause leukemia (Okuda et al., 1998; Castilla et al., 1999). However, constitutively activated signaling molecules, such as FLT3 or RAS family members, can induce a myeloproliferative phenotype (Kelly et al., 2002; Chan et al., 2004). Today, there are further lines of evidence arguing for a multi-step pathogenesis of leukemia, and many pathogenetically relevant mutations have been identified both in myeloid (Frohling et al., 2005; Licht and Sternberg, 2005) and lymphoid (Armstrong and Look, 2005; Pui and Evans, 2006) malignancies.

1.4 Treatment Options: Novel Drugs

The development of combination therapies of cytotoxic drugs with or without stem-cell transplantation has increased the cure rate of leukemia (Brenner and Pinkel, 1999), especially in childhood acute lymphoblastic leukemia (ALL) with an overall cure rate of over 80% (Pui and Evans, 2006). However, despite the optimal use of cytotoxic drugs in combination with a stringent application of current prognostic factors for risk-directed therapy, cure rates for adults remain much lower, signaling a need for new targeted therapies.

Acute promyelocytic leukemia (APL) was the first human leukemia to be successfully treated with a molecularly targeted agent, all-trans retinoic acid (ATRA), that specifically targets the transforming potential of the fusion gene product, PML-RARA, resulting from the t(15;17) (Huang et al., 1988). Likewise, CML was the first disorder in which a small molecule inhibitor had been designed to specifically target the molecular defect caused by *BCR-ABL* (Buchdunger et al., 1996; Druker et al., 1996). Since then, many new drugs have been shown to be active against leukemias, including additional tyrosine kinase inhibitors (e.g., FLT3 inhibitors like PKC412), farnesyltransferase inhibitors (e.g., tipifarnib), demethylating agents (e.g., decitabine), histone deacetylase inhibitors (e.g., valproic acid), and monoclonal antibodies (e.g., the anti-CD33 antibody gemtuzumab ozogamicin) (Tallman, 2005), which are currently being investigated in clinical treatment trials. While novel treatment approaches are currently being developed, improved risk stratifications which reflect the biological and clinical heterogeneity of the disease are also needed to guide efficient patient management.

1.5 Prognostic Factors

Cytogenetics represents one of the most powerful prognostic factors in leukemias. In AML, patients are assigned into risk-groups based on the underlying leukemia karyotype (Grimwade et al., 1998; Slovak et al., 2000; Byrd et al., 2002). Recently, the identification of novel molecular markers has permitted the further dissection of existing prognostic subclasses, like for example the large group of AML patients presenting with normal karyotype disease (Frohling et al., 2005; Licht and Sternberg, 2005). In this AML group internal tandem duplications (ITD) of the *FLT3* gene (Frohling et al., 2002), partial tandem duplications (PTD) of the *MLL* gene (Dohner et al., 2002), as well as mutations of *CEBPA* (Frohling et al., 2004) and *NPM1* (Dohner et al., 2005) are of prognostic relevance, as are the expression levels of *EVI1* (Barjesteh van Waalwijk van Doorn-Khosrovani et al., 2003) and *BAALC* (Baldus et al., 2003). Similarly, in chronic lymphoid leukemias (CLL) cytogenetic aberrations confer important markers (Dohner et al., 2000; Stilgenbauer et al., 2002) that provide prognostic information in addition to the immunoglobulin variable heavy chain gene (*VH*) mutational status (Krober et al., 2002; Dighiero, 2005). However, despite such recent progress, for many leukemia subtypes there is still no commonly accepted risk stratification as leukemogenic mechanisms are not yet fully understood.

2 ONCOGENESIS: INSIGHTS INTO LEUKEMIA BIOLOGY

2.1 Class Prediction in Leukemias

2.1.1 Gene Expression Profiling in Leukemias

Today, genomics offer a range of experimental approaches to capture the molecular variation underlying the biological and clinical heterogeneity of leukemias. Of these novel technologies, genome-wide gene expression profiling (GEP) based on DNA microarrays represents one of the most powerful tools (Liotta and Petricoin, 2000; Ramaswamy and Golub, 2002). The utility and promise of DNA microarrays for the study of human malignancies was first demonstrated in an analysis of AML and ALL samples (Golub et al., 1999). In this seminal study by Golub et al., leukemia became an early test case to demonstrate the potential of gene-expression profiling based classification of tumors. An unsupervised class discovery method (self-organizing maps) recognized the distinction between AML and ALL without previous knowledge of these classes, and a supervised class predictor, derived from neighborhood analysis and applied using a weighted voting procedure, assigned the class of new leukemia cases with high accuracy and predictive strength (Golub et al., 1999). Unexpectedly, many of the discriminatory genes were not markers of hematopoietic lineage, but encoded critical genes related to cancer pathogenesis, thereby suggesting that genes useful for cancer classification may also provide insight into cancer biology.

2.1.2 Gene Expression Patterns Associated with Recurrent Cytogenetic Aberrations

Since the publication of this groundbreaking work, DNA microarray technology has contributed significantly to the field of leukemia research (Staudt, 2003; Ebert and Golub, 2004). First, it was shown that AML cases with trisomy 8 differ from cases with normal cytogenetics based on a gene dosage effect for genes located on chromosome 8 (Virtaneva et al., 2001). Similar gene dosage effects have since been found for other chromosomal gains and losses in AML (Schoch et al., 2005; Rucker et al., 2006), as well as in other leukemias, e.g., CLL (Haslinger et al., 2004).

Using supervised analytical approaches, Yeoh et al. were the first to describe that in childhood ALL, distinct expression profiles identify prognostically important leukemia subtypes, including T-ALL, *E2A-PBX1*, *BCR-ABL*, *TEL-AML1*, *MLL* rearrangement, and ALL with a hyperdiploid karyotype (>50 chromosomes) (Yeoh et al., 2002). Confirmed by subsequent studies in both childhood and adult ALL (Armstrong et al., 2002; Ross et al., 2003; Fine et al., 2004), gene expression based discrimination could also be demonstrated for cytogenetically defined AML subgroups, including cases with inv(16), t(8;21), t(15;17) and t(11q23)/*MLL* (Schoch et al., 2002; Debernardi et al., 2003; Kohlmann et al., 2003; Bullinger et al., 2004; Ross et al., 2004; Valk et al., 2004) (Fig. 1). Importantly, these distinct biologically meaningful gene signatures can also be used to accurately predict the respective cytogenetically defined leukemia subgroups (Ross et al., 2004; Haferlach et al., 2005), and the classifiers generated from pediatric leukemia samples can be used to accurately classify adult leukemia cases exhibiting the same genetic aberrations (Kohlmann et al., 2004; Ross et al., 2004). Based on the robustness of these signatures with regard to technical aspects of specimen sampling and target preparation (Mitchell et al., 2004; Kohlmann et al., 2005), GEP may provide a useful alternative approach for the diagnosis of known leukemia subgroups with high accuracy (Haferlach et al., 2005; Kern et al., 2005).

2.1.3 Gene Expression Patterns Associated with Molecular Genetic Aberrations

While the first GEP studies of CLL demonstrated a homogeneous phenotype irrespective of the *VH* mutational status (Klein et al., 2001; Rosenwald et al., 2001), supervised analyses revealed in both studies a *VH* mutation-associated expression pattern leading to the identification of a clinically important surrogate marker for the *VH* mutational status, ZAP-70 (see also Sect. 4.1.2) (Rosenwald et al., 2001).

Since then, molecular genetic aberration-associated gene expression patterns have also been described in AML for *FLT3* ITDs (Bullinger et al., 2004; Lacayo et al., 2004; Valk et al., 2004; Neben et al., 2005), *CEBPA* (Valk et al., 2004), and *NPM1* mutations (Alcalay et al., 2005; Verhaak et al., 2005). Supervised analyses have also identified

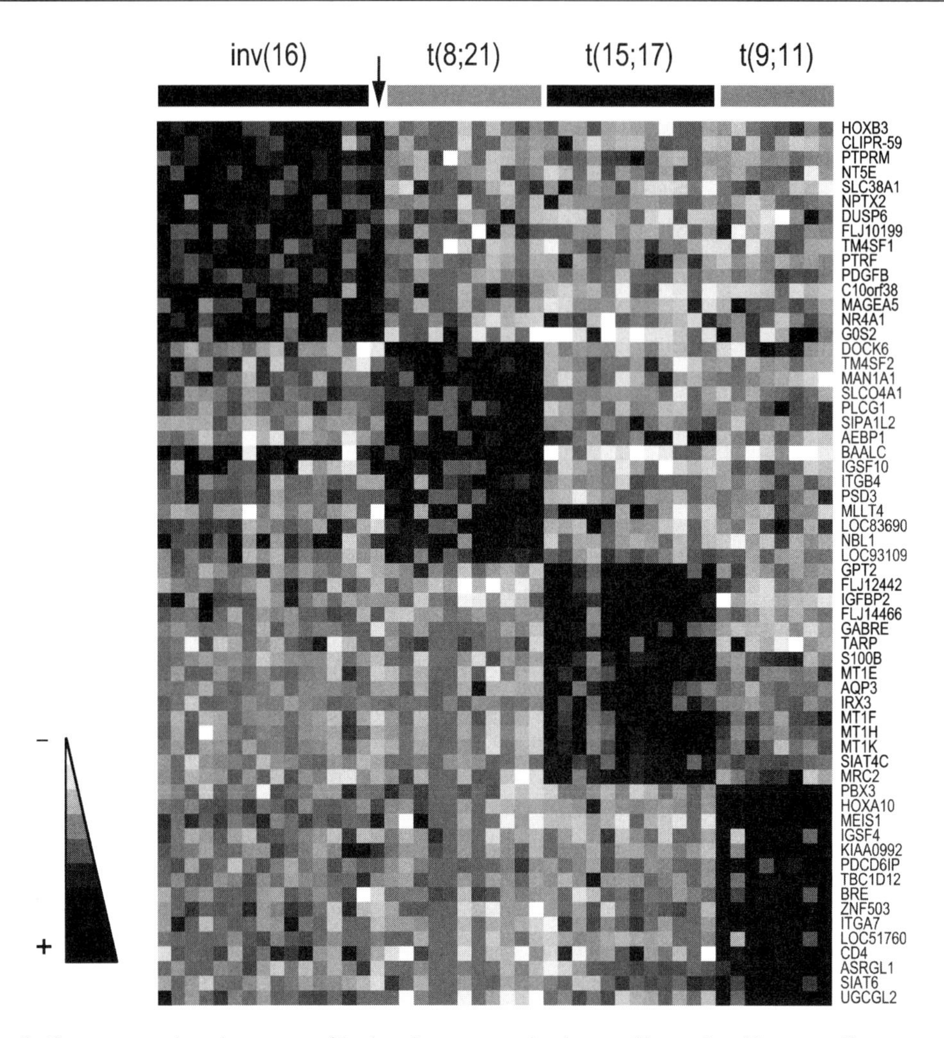

Fig. 1. Gene-expression signatures of leukemia cytogenetic classes. Shown is a "heat map" representation of selected genes identified by supervised analysis whose expression is significantly correlated with specific AML cytogenetic aberrations, here including inv(16), t(8;21), t(15;17) and t(9;11). Gene-expression levels are depicted in gray-scale, where *darker shades* indicate higher expression levels. The analysis shown was performed using significance analysis of microarrays (SAM) (Tusher et al., 2001) using publicly available microarray data (Bullinger et al., 2004). Gene-expression signatures such as these can be used to classify the cytogenetic group of new specimens with high accuracy. The *arrow* indicates a specimen characterized as "normal karyotype" by cytogenetics, but for which RT-PCR subsequently identified the diagnostic *CBFB-MYH11* fusion transcript characteristic of inv(16) cases

significant patterns correlating with centrosome aberrations (Neben et al., 2004). However, in contrast to translocations involving the *MLL* gene, which exhibit a distinct gene expression signature (Armstrong et al., 2002), no characteristic pattern has been identified for cases with *MLL* PTD in large studies (Bullinger et al., 2004; Ross et al., 2004). This suggests that cases with *MLL* PTD might be more heterogeneous at a molecular level and quite distinct from AML with t(11q23). Likewise, no signature has yet been associated with AML cases with *NRAS* mutations (Neben et al., 2005), highlighting that not all genetic alterations in leukemia result in definably altered gene expression patterns. Possible explanations might be that the respective mutations affect signaling pathways at a post-transcriptional level (i.e., without altering mRNA levels), or that the associated expression patterns are subtle and/or masked by additional underlying pathogenic mechanisms.

2.2 Class Discovery in Leukemias

By applying unsupervised analytical approaches to gene expression data, novel clinically-significant subtypes of cancer can be identified. The first gene-expression based discovery of new clinically-relevant tumor subclasses, diffuse large B-cell lymphoma (DLBCL), was again reported for a hematologic malignancy, (Alizadeh et al., 2000). Using hierarchical cluster analysis, Alizadeh et al. identified two molecularly distinct DLBCL subtypes characterized by patterns indicative of different stages of B-cell differentiation. These patterns have since been identified in independent studies, and their clinical importance has been validated as well (Rosenwald et al., 2002; Shipp et al., 2002).

Besides the pivotal work by Golub et al. (1999), one of the first studies, demonstrating the potential of GEP based class discovery in leukemia, examined therapy-related AML (t-AML) cases (Qian et al., 2002). The authors identified t-AML subgroups displaying a common pattern typical of arrested differentiation in early progenitor cells, but each t-AML subgroup also exhibited a characteristic gene expression signature providing new insights in the biology of t-AML (Qian et al., 2002).

In a recent study of 285 AML patients, Valk et al. identified 16 clusters (subgroups) of AML specimens by unsupervised cluster analysis of gene-expression patterns (Valk et al., 2004). Specimens with favorable cytogenetics generally exhibited homogeneous clustering, while novel clusters were often characterized by high frequencies of specific molecular alterations, like for example two clusters (their clusters #1 and #16) which both harbored mainly cases with t(11q23)/*MLL* abnormalities. However, Valk et al. also observed molecular variation within "homogenously grouped" classes, like cases with an inv(16) or a t(8;21), when they included more than the original 2,856 probe sets in their unsupervised analysis (Valk et al., 2004).

Accordingly, in our own studies of AML we have detected considerable molecular heterogeneity within the cytogenetically well-characterized t(8;21) and inv(16) groups, with each class being separated into mainly two groups based on unsupervised clustering using 6,283 genes (Bullinger et al., 2004). Since the primary translocation/inversion events themselves are not sufficient for leukemogenesis (Okuda et al., 1998; Castilla et al., 1999), distinct patterns of gene expression within each of these t(8;21) and inv(16) subgroups may suggest alternative cooperating mutations/deregulated pathways leading to transformation.

Interestingly, cases with normal karyotype in our study also segregated mainly into two distinct groups (Fig. 2), each of which included a small number of cases from other cytogenetic classes (Bullinger et al., 2004). *FLT3* aberrations were more prevalent in one subgroup, while FAB (French American British) M4/M5 morphologic subtypes were more highly represented in the other subgroup. Notably, Kaplan-Meier analysis identified a statistically significant difference in overall survival between the two subclasses (Bullinger et al., 2004). In agreement with these results, Valk et al. identified normal karyotype-predominated clusters associated with *FLT3* ITD, as well as a cluster including mainly specimens from patients with AML of subtype M4 or M5 (Valk et al., 2004).

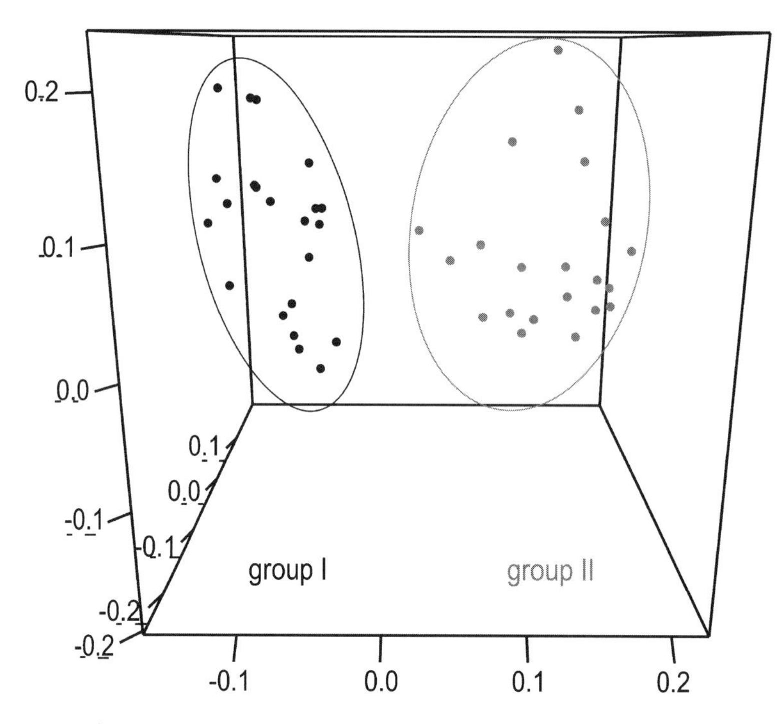

Fig. 2. Discovery of new molecular subtypes of leukemia. Shown are the results of an unsupervised principal components analysis (PCA) displaying a projection of the three principal components of variable gene expression for AML specimens with normal karyotype. PCA identifies two novel subgroups of normal karyotype AML cases (indicated by *black and gray dots*, respectively) based on distinct patterns of genes expression. Group II cases exhibit more frequent myelomonocytic differentiation, while group I cases more often harbor *FLT3* mutations and are associated with shorter overall survival. Data are from (Bullinger et al., 2004)

In addition, the Valk et al. study (Valk et al., 2004) revealed a distinctive cluster associated with increased *EVI1* expression and poor treatment outcome, and another cluster also associated with shorter survival times included cases with a variety of known adverse cytogenetic markers, such as monosomies 7 and 5, and the translocation t(9;22) (Valk et al., 2004). The leukemic cells of these patients displayed a signature similar to CD34 + cell samples, suggesting that these leukemic cells might be resistant to therapy, like CD34 + hematopoietic progenitors. These initial studies clearly demonstrate the power of GEP for the discovery of novel leukemia subtypes (Bullinger and Valk, 2005).

3 DRUG RESPONSIVENESS

3.1 Gene Expression Based Analysis of Drug Effects in Leukemia

3.1.1 Molecular Signatures of Retinoic Acid Treatment

Monitoring the effects of treatment with all-trans retinoic acid (ATRA) in APL derived cell lines was one of the first applications of DNA microarray technology in leukemias (Tamayo et al., 1999). In NB4 cells, ATRA was found to induce *UBE1* (ubiquitin-activating enzyme E1-like) over-expression, triggering the apoptosis of APL cells and the degradation of the *PML-RARA* fusion gene product (Tamayo et al., 1999; Kitareewan et al., 2002). Other ATRA-regulated genes in NB4 included members of the tumor necrosis (TNF) pathway, suggesting that this pathway might intersect with ATRA signaling (Park et al., 2003; Witcher et al., 2003). ATRA and TNF together led to increased NF-κB activity followed by a synergistic induction of NF-κB target genes (Witcher et al., 2003). This supports the idea that ATRA primes cells are more susceptible to the differentiation effects of other pathways. Interestingly, there is an enrichment of NF-kB binding sites in the promoters of ATRA target genes, further supporting a role for this pathway in regulating cell survival in response to ATRA (Meani et al., 2005).

3.1.2 Sensitivity of CML to Imatinib Mesylate

Genome-wide DNA microarray technology has also been used successfully to evaluate the sensitivity of CML to imatinib mesylate (Kaneta et al., 2002; Ohno and Nakamura, 2003; Tipping et al., 2003), which targets the ABL kinase activity of the BCR-ABL fusion. Using a cell line model, differentially-expressed genes correlated with imatinib mesylate resistance could be identified, suggesting that alternative pathways might maintain viability and promote growth independently of BCR-ABL (Tipping et al., 2003). Furthermore, in an analysis of 18 specimens from CML patients, responders could be clearly separated from non-responders based on the expression patterns of only a few genes (Kaneta et al., 2002; Ohno and Nakamura, 2003). Though examination of larger sample sets will be required to validate these findings, gene expression patterns associated with imatinib resistance might help to guide treatment decisions in CML.

3.1.3 Fludarabine Response Signature in CLL

The purine analog fludarabine is a component of current standard treatment regimens for B cell CLL. While fludarabine can induce apoptosis in CLL cells in vitro, a number of molecular mechanisms might contribute to its observed cytotoxicity. Using GEP, Rosenwald et al. have recently investigated the molecular consequences of fludarabine treatment of CLL patient samples (Rosenwald et al., 2004). Both in vitro and in vivo fludarabine exposure resulted in a consistent "response signature" characterized by p53 target genes and genes involved in DNA repair. Functional analyses in isogenic p53 wild-type and null lymphoblastoid cell lines provided further evidence that many of the fludarabine response signature genes were also p53 target genes. These findings provide a molecular explanation for the drug resistance and the aggressive clinical course often seen in p53 mutated CLL patients, and further suggest the importance of only treating patients that warrant therapy, as fludarabine treatment might have the potential to select p53 mutant CLL cells (Rosenwald et al., 2004).

3.1.4 In Vitro Response to L-Asparaginase

L-asparaginase is an important component of most treatment regimens for ALL, and both in vitro and in vivo resistance to L-asparaginase has been associated with poor long-term outcome (Pui and Evans, 2006). A better molecular understanding of the resistance mechanisms may provide for improved management of ALL patients. In vitro exposure to L-asparaginase in cell lines and pediatric ALL samples followed by GEP, revealed the increased expression of tRNA synthetases, solute transporters, activating transcription factors, and CCAAT/enhancer binding protein family members (Fine et al., 2005). These changes appear to reflect a consistent coordinated response to asparagine starvation in both cell lines and clinical samples, which is independent of asparagine synthetase base-line expression levels. Therefore, targeting particular genes involved in the response to amino acid starvation in ALL cells may provide a novel way to overcome L-asparaginase resistance (Fine et al., 2005).

3.2 Prediction of Drug Responsiveness in Leukemia

3.2.1 Gene Expression Based Chemosensitivity Prediction

The first genomics-based approach to the prediction of drug response was the development of an algorithm for classification of cell line chemosensitivity based on gene expression profiles (Staunton et al., 2001). Using expression patterns of 60 human cancer cell lines (the NCI-60 panel), for which the chemosensitivity profiles of thousands of chemical compounds had been determined, the authors were able to generate gene expression-based classifiers to predict sensitivity or resistance for 232 compounds. Evaluation on independent data sets demonstrated that a substantial number of the expression-based classifiers performed accurately, indicating that genomic approaches to chemosensitivity prediction are feasible (Staunton et al., 2001).

3.2.2 Chemosensitivity Prediction in Leukemia

To illuminate the distinct nature of cellular responses provoked by chemotherapeutic agents, Cheok et al. profiled gene expression in childhood ALL cells before and after in vivo treatment with methotrexate and mercaptopurine given either alone or in combination (Cheok et al., 2003). A signature consisting of 124 genes accurately discriminated among the randomly assigned treatments in 60 patients. The signature, which included genes involved in apoptosis, mismatch repair, cell cycle control and stress response, exhibited differences in cellular response to drug combinations versus single agents, and indicated that different ALL subtypes share common pathways of genomic response to the same treatment.

In a subsequent study testing for in vitro sensitivity to prednisolone, vincristine, asparaginase, and daunorubicin, William Evans' group identified differentially expressed genes in drug-sensitive and drug-resistant ALL leukemia cells from 173 children (Holleman et al., 2004). Based on the gene-expression patterns that differed according to sensitivity or resistance to the four drugs, the authors created a combined gene-expression score of resistance, which was shown to be of prognostic relevance in childhood ALL (see also Sect. 4.1.3).

3.3 Drug Discovery

3.3.1 Applications of GEP in Drug Discovery

DNA microarray technology plays an integral part in the drug target discovery process, and provides a valuable tool for the optimization and clinical validation of novel compounds. GEP aids in the initial identification and prioritization of potential therapeutic targets such as those with relevant patterns of differential expression. Subsequently expression profiling assists in drug discovery and toxicology, where various bioinformatics approaches can be used to deduce the mechanism of action of new drugs as well as off-target effects from expression profiles (Gerhold et al., 2002). While it is important to consider the limitations of interpreting drug responses through measurements of mRNA abundance alone, expression profiling nonetheless provides a useful tool in pharmacogenomics, promising a better characterization of patient populations and a better prediction of prognosis and drug response (Walgren et al., 2005).

3.3.2 Anti-leukemic Drug Discovery Based on Responsiveness Signatures

Recently, a gene expression-based high-throughput screening approach was described using microarrays to screen for chemical compounds with differentiation-inducing activity in AML (Stegmaier et al., 2004). Following definition and validation of a five-gene differentiation signature derived from microarray data, a high throughput screening method using multiplexed RT-PCR, single base extension reaction and MALDI-TOF (matrix-assisted laser desorption/ionization time-of flight) mass spectrometry, was designed to detect the five-gene pattern. Treatment of HL-60 cells with 1,739 different compounds revealed eight chemicals that reliably induced the differentiation signature (Stegmaier et al., 2004). Among these, DAPH1 (4,5-dianilinophthalimide) has been reported to inhibit epidermal growth factor receptor (EGFR) kinase activity (Buchdunger et al., 1994).

In a subsequent study the authors showed that the Food and Drug Administration (FDA)—approved EGFR inhibitor gefitinib similarly promoted the differentiation of AML cell lines and primary patient—derived AML blasts in vitro (Stegmaier et al., 2005). However, the analyzed AML cells did not express EGFR suggesting an EGFR-independent mechanism of gefitinib induced differentiation. Thus, this high-throughput procedure proved successful for the systematic identification of clinically useful compounds for which the key targets are not yet known.

4 PREDICTION OF CLINICAL OUTCOME

4.1 Identification of Surrogate Markers for Known Prognostic Factors

4.1.1 Signatures Predictive of "Favorable" and "Unfavorable" Cytogenetics

As mentioned above, GEP allows the identification and prediction of specific signatures correlated with "favorable-risk" cytogenetic aberrations like for example AML cases with inv(16), t(8;21), or t(15;17) (Bullinger et al., 2004; Valk et al., 2004; Haferlach et al., 2005), or B cell precursor ALL cases with t(12;21), or hyperdiploidy (more than 50 chromosomes per leukemia cell) (Yeoh et al., 2002; Ross et al., 2003). Similarly, "unfavorable-risk" cytogenetics subgroups, as well as prognostically relevant molecular genetic aberrations can be predicted (Bullinger and Valk, 2005).

4.1.2 ZAP-70, Clinical Implementation of a Surrogate Marker Identified by GEP

Among the surrogate markers identified by GEP ZAP-70 represents the first to be clinically implemented. Coding for a tyrosine kinase essential for T cell signaling but not expressed in normal B cells, *ZAP-70* exhibited fivefold higher expression levels in patients with unmutated *VH* genes compared to those with mutated genes, and was the best discriminator between the two groups (Rosenwald et al., 2001). Recently, several studies have validated the clinical usefulness of this novel marker (Orchard et al., 2004;

Rassenti et al., 2004), especially as the expression of ZAP-70 can readily be measured at the protein level by flow cytometry, which is commonly available in large clinical centers specialized for leukemia treatment.

4.1.3 Prediction of Treatment Resistance

Recently, the expression of differentially-expressed genes in leukemias sensitive or resistant to chemotherapeutic agents has been shown to be significantly and independently associated with treatment outcome in ALL, and results have been confirmed in an independent set of patients treated with the same therapies (Holleman et al., 2004; Lugthart et al., 2005). Similarly, Heuser et al. were able to identify a characteristic gene-expression pattern associated with resistance to chemotherapy in AML, which was an independent prognostic factor and could be used to predict outcome in an independent test set of AML patients, and in multivariate analysis (Heuser et al., 2005). Comparison of refractory patients and those who responded to induction chemotherapy also led to the identification of prognostic genes in adult T cell ALL (Chiaretti et al., 2004). Furthermore, expression profiling has been used to predict minimal residual disease (MRD) in leukemias with high accuracy. Hypothesizing that treatment resistance reflects an intrinsic feature of ALL cells, which can be predicted before treatment, Cario et al. profiled expression of ALL samples with high MRD load (Cario et al., 2005). Compared with MRD negative samples, a prognostic signature distinguishing resistant from sensitive ALL samples could be defined.

4.2 Identification of Prognostic Signatures

4.2.1 Supervised Approaches

The pivotal study of Golub et al. was also the first to apply supervised analysis to explore genes predictive of response to chemotherapy (Golub et al., 1999). Although not statistically significant, candidates with potential biological significance in AML patients with treatment failure included over-expressed *HOXA9*, a gene known to be expressed in a subset of AML cases (Dorsam et al., 2004), and recently associated with *NPM1* mutations (Alcalay et al., 2005). Using a supervised approach in childhood AML, a prognostic signature was generated by comparing patients with a good and poor outcome (Yagi et al., 2003); however, in an independent data set the respective gene expression pattern did not allow a significant risk stratification (Ross et al., 2004).

In CML, comparing CD34 + cells from patients with an "aggressive" and an "indolent" clinical course led to the identification of novel markers, *CD7* and *PR-3* (proteinase 3), as predictors of longer survival in CML patients (Yong et al., 2006). While these findings await validation by independent studies, it is important to note that these results were derived from samples from the "pre-imatinib" era; therefore the reported signature will also need to be evaluated in CML patients treated with imatinib.

4.2.2 Semi-supervised Methods to Predict Patient Outcome

A limitation of strictly supervised approaches to outcome prediction is that, survival and survival time, because they are impacted by many things other then the tumor cells themselves, are likely to be very noisy surrogates for the underlying prognostically relevant tumor subclasses. On the opposite side, strictly unsupervised analyses of leukemias

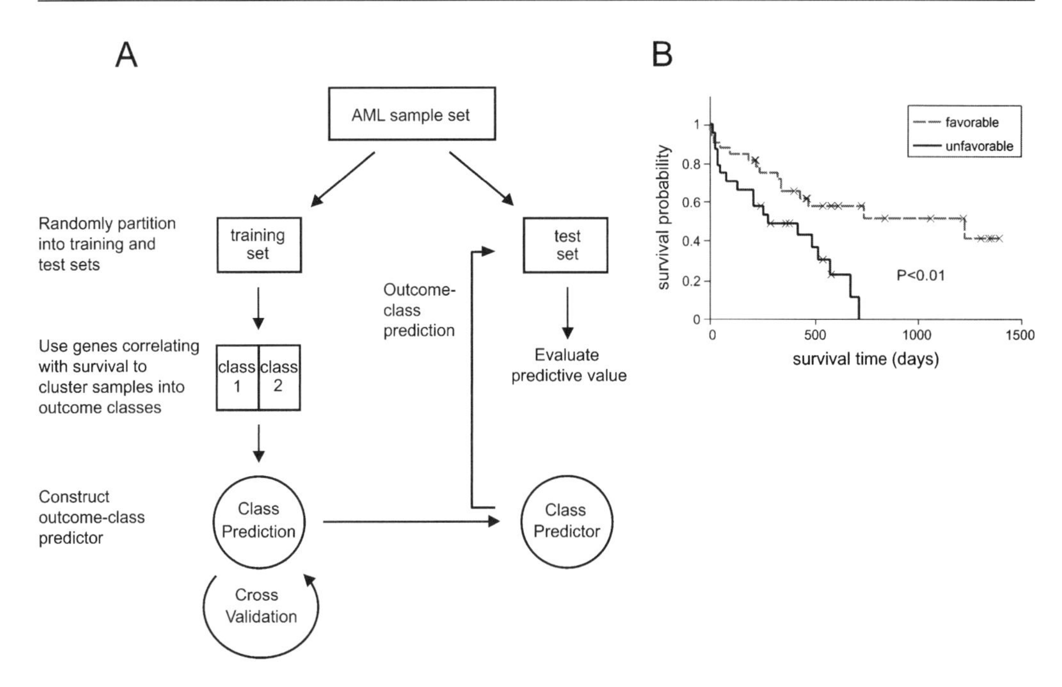

Fig. 3. Semi-supervised approach for leukemia outcome prediction. (**A**) Schematic overview of a supervised clustering strategy. AML specimens are randomized into separate training and test sets. In the training set, genes whose expression correlates with survival are used to cluster samples into favorable and unfavorable outcome classes. An optimal gene-expression predictor is constructed for these outcome classes, and is then validated by predicting outcomes in the independent test set. (**B**) Evaluation of outcome predictor. Kaplan-Meier survival analysis of the independent test set validates the gene-expression classifier (here comprising 133 genes) as a significant predictor of overall survival. Data are from (Bullinger et al., 2004)

are driven largely by cytogenetic groupings of already known prognostic value. Therefore, to discover gene-expression signatures with additional prognostic value, we applied a novel semi-supervised strategy which combines the strengths of supervised and unsupervised approaches (Bair and Tibshirani, 2004). The idea was to use the subset of genes correlating with survival time in a supervised clustering of specimens, to reveal the underlying prognostically relevant tumor subtypes, and then to build a predictor for these subtypes (Fig. 3). Applying this approach to a cohort of AML patients, we defined and validated a 133 gene signature as a significant independent outcome predictor, both across all cytogenetic classes and within the large subset of clinically-important AML cases with normal karyotype (Bullinger et al., 2004). This signature was also recently validated in an independent set of 68 AML cases with normal karyotype using a different microarray platform (Marcucci et al., 2006).

4.2.3 Integrative Approaches to Outcome Prediction

Recently, Glinsky et al. (2005) integrated expression signatures from murine models of prostate cancer progression and of BMI1-driven stem cell self-renewal to define an 11-gene signature predictive of clinical outcome in human prostate cancer, as well as ten other human cancer types including AML. This study highlights the likelihood that many different gene sets are likely to emerge, which both capture important underlying biology and provide prognostic information. This study also underscores the importance of the public-availability of microarray data sets with associated clinical annotation, which will be instrumental in evaluating the various proposed prognostic signatures.

While this and the other above-mentioned findings are definitely encouraging, further validation of results in larger cohorts and in independent studies are clearly required before clinical implementation becomes feasible in leukemias.

5 CONCLUSIONS

5.1 GEP: An Important New Facet to Study Leukemias

In the past several years, hematologic malignancies have been an attractive area of study using genomic approaches like DNA microarray technology (Ramaswamy and Golub, 2002; Staudt, 2003; Ebert and Golub, 2004), and GEP has contributed an important new facet to the exploration of leukemias (Bullinger et al., 2005; Chiaretti et al., 2005; Kern et al., 2005). The existing knowledge in hematopoiesis and leukemia genetics has also guided gene expression data interpretation, thereby facilitating the generation of biologically meaningful hypotheses.

In the future, this technology will further contribute to a comprehensive molecular leukemia classification. Characteristic expression patterns will support individualization of cancer treatment and help to individually discern cases with high probability of resistance to therapy or a high relapse risk. Improved risk-adapted management of leukemia patients will also be supplemented by novel therapeutics developed with the aid of DNA microarray technology. Ultimately, a single microarray assay might be sufficient to adequately diagnose leukemias, predict their course, and enable individualized treatment strategies.

5.2 Validation of Microarray Data

Relating gene expression patterns to patient outcome provides increased challenges for translating initial research findings into robust diagnostics validated to be of clinical benefit. However, to prevent unsuccessful validation attempts the initial study should already be based on sophisticated algorithms including for example, a multiple random validation strategy, as results might otherwise be overoptimistic (Michiels et al., 2005). Furthermore, there remain unresolved data analysis issues that merit further research but which if appropriately addressed may facilitate validation of findings, e.g., the examination of intersections between sets of findings (Allison et al., 2006).

Although reported classifiers often show a seemingly impressive accuracy for predicting outcome, besides internal split-sample validation and cross validation, there are many reasons to demand external validation based on truly independent data, and the objective of this external validation should be to determine whether using a prognostic classifier results in patient benefit (Simon, 2005).

5.3 Future Challenges: Integration of Whole Genome Approaches

While primary analyses have begun to decipher the molecular heterogeneity of leukemia, an outstanding challenge for the future will be the integration of DNA microarray technology and other whole genome approaches to validate the numerous biologic hypotheses suggested by gene expression data. Integrative analyses that evaluate the leukemia transcriptome in the context of other data sources, such as SNP (single nucleotide polymorphism) arrays, array comparative genome hybridization (array CGH), tiling arrays, promoter arrays, and proteomics, will allow the extraction of additional biological insights from the data. Novel integrative computational and analytical approaches include meta-analysis, functional enrichment analysis, transcriptional network analysis and integrative model system analysis (Rhodes and Chinnaiyan, 2005). However, a prerequisite for a successful integration will be to define a common language for communicating genomic profiles across diverse experimental systems, as well as the development of integrative bioinformatics solutions for sharing and analyzing such profiles.

REFERENCES

Alcalay, M., E. Tiacci, R. Bergomas, et al. (2005). Acute myeloid leukemia bearing cytoplasmic nucleophosmin (NPMc + AML) shows a distinct gene expression profile characterized by up-regulation of genes involved in stem-cell maintenance. Blood 106(3): 899–902.

Alizadeh, A. A., M. B. Eisen, R. E. Davis, et al. (2000). Distinct types of diffuse large B-cell lymphoma identified by gene expression profiling. Nature 403(6769): 503–511.

Allison, D. B., X. Cui, G. P. Page, et al. (2006). Microarray data analysis: from disarray to consolidation and consensus. Nat Rev Genet 7(1): 55–65.

Armstrong, S. A., A. T. Look (2005). Molecular genetics of acute lymphoblastic leukemia. J Clin Oncol 23(26): 6306–6315.

Armstrong, S. A., J. E. Staunton, L. B. Silverman, et al. (2002). MLL translocations specify a distinct gene expression profile that distinguishes a unique leukemia. Nat Genet 30(1): 41–47.

Bair, E., R. Tibshirani (2004). Semi-supervised methods to predict patient survival from gene expression data. PLoS Biol 2(4): E108.

Baldus, C. D., S. M. Tanner, A. S. Ruppert, et al. (2003). BAALC expression predicts clinical outcome of de novo acute myeloid leukemia patients with normal cytogenetics: a Cancer and Leukemia Group B Study. Blood 102(5): 1613–1618.

Barjesteh van Waalwijk van Doorn-Khosrovani, S., C. Erpelinck, W. L. van Putten, et al. (2003). High EVI1 expression predicts poor survival in acute myeloid leukemia: a study of 319 de novo AML patients. Blood 101(3): 837–845.

Bennett, J. H. (1845). Case of hypertrophy of the spleen and liver, in which death took place from suppuration of the blood. Edinburgh Med Surg J 64: 413–423.

Brenner, M. K., D. Pinkel (1999). Cure of leukemia. Semin Hematol 36(4 Suppl 7): 73–83.

Buchdunger, E., U. Trinks, H. Mett, et al. (1994). 4,5-Dianilinophthalimide: a protein-tyrosine kinase inhibitor with selectivity for the epidermal growth factor receptor signal transduction pathway and potent in vivo antitumor activity. Proc Natl Acad Sci USA 91(6): 2334–2338.

Buchdunger, E., J. Zimmermann, H. Mett, et al. (1996). Inhibition of the Abl protein-tyrosine kinase in vitro and in vivo by a 2-phenylaminopyrimidine derivative. Cancer Res 56(1): 100–104.

Bullinger, L., P. J. Valk (2005). Gene expression profiling in acute myeloid leukemia. J Clin Oncol 23(26): 6296–6305.

Bullinger, L., K. Dohner, E. Bair, et al. (2004). Use of gene-expression profiling to identify prognostic subclasses in adult acute myeloid leukemia. N Engl J Med 350(16): 1605–1616.

Bullinger, L., H. Dohner, J. R. Pollack (2005). Genomics in myeloid leukemias: an array of possibilities. Rev Clin Exp Hematol 9(1): E2.

Byrd, J. C., K. Mrozek, R. K. Dodge, et al. (2002). Pretreatment cytogenetic abnormalities are predictive of induction success, cumulative incidence of relapse, and overall survival in adult patients with de novo acute myeloid leukemia: results from Cancer and Leukemia Group B (CALGB 8461). Blood 100(13): 4325–4336.

Cario, G., M. Stanulla, B. M. Fine, et al. (2005). Distinct gene expression profiles determine molecular treatment response in childhood acute lymphoblastic leukemia. Blood 105(2): 821–826.

Castilla, L. H., L. Garrett, N. Adya, et al. (1999). The fusion gene Cbfb-MYH11 blocks myeloid differentiation and predisposes mice to acute myelomonocytic leukaemia. Nat Genet 23(2): 144–146.

Chan, I. T., J. L. Kutok, I. R. Williams, et al. (2004). Conditional expression of oncogenic K-ras from its endogenous promoter induces a myeloproliferative disease. J Clin Invest 113(4): 528–538.

Cheok, M. H., W. Yang, C. H. Pui, et al. (2003). Treatment-specific changes in gene expression discriminate in vivo drug response in human leukemia cells. Nat Genet 34(1): 85–90.

Chiaretti, S., J. Ritz, R. Foa (2005). Genomic analysis in lymphoid leukemias. Rev Clin Exp Hematol 9(1): E3.

Chiaretti, S., X. Li, R. Gentleman, et al. (2004). Gene expression profile of adult T-cell acute lymphocytic leukemia identifies distinct subsets of patients with different response to therapy and survival. Blood 103(7): 2771–2778.

Debernardi, S., D. M. Lillington, T. Chaplin, et al. (2003). Genome-wide analysis of acute myeloid leukemia with normal karyotype reveals a unique pattern of homeobox gene expression distinct from those with translocation-mediated fusion events. Genes Chromosomes Cancer 37(2): 149–158.

Dighiero, G. (2005). CLL biology and prognosis. Hematology (Am Soc Hematol Educ Program): 278–284.

Dohner, H., S. Stilgenbauer, A. Benner, et al. (2000). Genomic aberrations and survival in chronic lymphocytic leukemia. N Engl J Med 343(26): 1910–1916.

Dohner, K., K. Tobis, R. Ulrich, et al. (2002). Prognostic significance of partial tandem duplications of the MLL gene in adult patients 16 to 60 years old with acute myeloid leukemia and normal cytogenetics: a study of the Acute Myeloid Leukemia Study Group Ulm. J Clin Oncol 20(15): 3254–3261.

Dohner, K., R. F. Schlenk, M. Habdank, et al. (2005). Mutant nucleophosmin (NPM1) predicts favorable prognosis in younger adults with acute myeloid leukemia and normal cytogenetics: interaction with other gene mutations. Blood 106(12): 3740–3746.

Dorsam, S. T., C. M. Ferrell, G. P. Dorsam, et al. (2004). The transcriptome of the leukemogenic homeoprotein HOXA9 in human hematopoietic cells. Blood 103(5): 1676–1684.

Druker, B. J., S. Tamura, E. Buchdunger, et al. (1996). Effects of a selective inhibitor of the Abl tyrosine kinase on the growth of Bcr-Abl positive cells. Nat Med 2(5): 561–566.

Ebert, B. L., T. R. Golub (2004). Genomic approaches to hematologic malignancies. Blood 104(4): 923–932.

Fine, B. M., M. Stanulla, M. Schrappe, et al. (2004). Gene expression patterns associated with recurrent chromosomal translocations in acute lymphoblastic leukemia. Blood 103(3): 1043–1049.

Fine, B. M., G. J. Kaspers, M. Ho, et al. (2005). A genome-wide view of the in vitro response to l-asparaginase in acute lymphoblastic leukemia. Cancer Res 65(1): 291–299.

Frohling, S., R. F. Schlenk, J. Breitruck, et al. (2002). Prognostic significance of activating FLT3 mutations in younger adults (16 to 60 years) with acute myeloid leukemia and normal cytogenetics: a study of the AML Study Group Ulm. Blood 100(13): 4372–4380.

Frohling, S., R. F. Schlenk, I. Stolze, et al. (2004). CEBPA mutations in younger adults with acute myeloid leukemia and normal cytogenetics: prognostic relevance and analysis of cooperating mutations. J Clin Oncol 22(4): 624–633.

Frohling, S., C. Scholl, D. G. Gilliland, et al. (2005). Genetics of myeloid malignancies: pathogenetic and clinical implications. J Clin Oncol 23(26): 6285–6295.

Gerhold, D. L., R. V. Jensen, S. R. Gullans (2002). Better therapeutics through microarrays. Nat Genet 32(Suppl): 547–551.

Glinsky, G. V., O. Berezovska, A. B. Glinskii (2005). Microarray analysis identifies a death-from-cancer signature predicting therapy failure in patients with multiple types of cancer. J Clin Invest 115(6): 1503–1521.

Golub, T. R., D. K. Slonim, P. Tamayo, et al. (1999). Molecular classification of cancer: class discovery and class prediction by gene expression monitoring. Science 286(5439): 531–537.

Grimwade, D., H. Walker, F. Oliver, et al. (1998). The importance of diagnostic cytogenetics on outcome in AML: analysis of 1,612 patients entered into the MRC AML 10 trial. The Medical Research Council Adult and Children's Leukaemia Working Parties. Blood 92(7): 2322–2333.

Haferlach, T., A. Kohlmann, S. Schnittger, et al. (2005). Global approach to the diagnosis of leukemia using gene expression profiling. Blood 106(4): 1189–1198.

Harris, N. L., E. S. Jaffe, J. Diebold, et al. (1999). World Health Organization classification of neoplastic diseases of the hematopoietic and lymphoid tissues: report of the Clinical Advisory Committee meeting-Airlie House, Virginia, November 1997. J Clin Oncol 17(12): 3835–3849.

Haslinger, C., N. Schweifer, S. Stilgenbauer, et al. (2004). Microarray gene expression profiling of B-cell chronic lymphocytic leukemia subgroups defined by genomic aberrations and VH mutation status. J Clin Oncol 22(19): 3937–3949.

Heuser, M., L. U. Wingen, D. Steinemann, et al. (2005). Gene-expression profiles and their association with drug resistance in adult acute myeloid leukemia. Haematologica 90(11): 1484–1492.

Hoffbrand, A. V., B. Fantini (1999). Achievements in hematology in the twentieth century: an introduction. Semin Hematol 36(4 Suppl 7): 1–4.

Holleman, A., M. H. Cheok, M. L. den Boer, et al. (2004). Gene-expression patterns in drug-resistant acute lymphoblastic leukemia cells and response to treatment. N Engl J Med 351(6): 533–542.

Huang, M. E., Y. C. Ye, S. R. Chen, et al. (1988). Use of all-trans retinoic acid in the treatment of acute promyelocytic leukemia. Blood 72(2): 567–572.

Kaneta, Y., Y. Kagami, T. Katagiri, et al. (2002). Prediction of sensitivity to STI571 among chronic myeloid leukemia patients by genome-wide cDNA microarray analysis. Jpn J Cancer Res 93(8): 849–856.

Kelly, L. M., Q. Liu, J. L. Kutok, et al. (2002). FLT3 internal tandem duplication mutations associated with human acute myeloid leukemias induce myeloproliferative disease in a murine bone marrow transplant model. Blood 99(1): 310–318.

Kern, W., A. Kohlmann, S. Schnittger, et al. (2005). Role of gene expression profiling for diagnosing acute leukemias. Rev Clin Exp Hematol 9(1): E1.

Kitareewan, S., I. Pitha-Rowe, D. Sekula, et al. (2002). UBE1L is a retinoid target that triggers PML/RARalpha degradation and apoptosis in acute promyelocytic leukemia. Proc Natl Acad Sci USA 99(6): 3806–3811.

de Klein, A., A. G. van Kessel, G. Grosveld, et al. (1982). A cellular oncogene is translocated to the Philadelphia chromosome in chronic myelocytic leukaemia. Nature 300(5894): 765–767.

Klein, U., Y. Tu, G. A. Stolovitzky, et al. (2001). Gene expression profiling of B cell chronic lymphocytic leukemia reveals a homogeneous phenotype related to memory B cells. J Exp Med 194(11): 1625–1638.

Kohlmann, A., C. Schoch, S. Schnittger, et al. (2003). Molecular characterization of acute leukemias by use of microarray technology. Genes Chromosomes Cancer 37(4): 396–405.

Kohlmann, A., C. Schoch, S. Schnittger, et al. (2004). Pediatric acute lymphoblastic leukemia (ALL) gene expression signatures classify an independent cohort of adult ALL patients. Leukemia 18(1): 63–71.

Kohlmann, A., C. Schoch, M. Dugas, et al. (2005). Pattern robustness of diagnostic gene expression signatures in leukemia. Genes Chromosomes Cancer 42(3): 299–307.

Konopka, J. B., S. M. Watanabe, J. W. Singer, et al. (1985). Cell lines and clinical isolates derived from Ph1-positive chronic myelogenous leukemia patients express c-abl proteins with a common structural alteration. Proc Natl Acad Sci USA 82(6): 1810–1814.

Krober, A., T. Seiler, A. Benner, et al. (2002). V(H) mutation status, CD38 expression level, genomic aberrations, and survival in chronic lymphocytic leukemia. Blood 100(4): 1410–1416.

Lacayo, N. J., S. Meshinchi, P. Kinnunen, et al. (2004). Gene expression profiles at diagnosis in de novo childhood AML patients identify FLT3 mutations with good clinical outcomes. Blood 104(9): 2646–2654.

Licht, J. D., D. W. Sternberg (2005). The molecular pathology of acute myeloid leukemia. Hematology (Am Soc Hematol Educ Program): 137–142.

Liotta, L., E. Petricoin (2000). Molecular profiling of human cancer. Nat Rev Genet 1(1): 48–56.

Lugthart, S., M. H. Cheok, M. L. den Boer, et al. (2005). Identification of genes associated with chemotherapy crossresistance and treatment response in childhood acute lymphoblastic leukemia. Cancer Cell 7(4): 375–386.

Marcucci, G., M. D. Radmacher, A. S. Ruppert, et al. (2006). Independent validation of prognostic relevance of a previously reported gene-expression signature in acute myeloid leukemia (AML) with normal cytogenetics (NC): A Cancer and Leukemia Group B (CALGB) Study. Blood, ASH Annual Meeting Abstracts 106: 755.

Meani, N., S. Minardi, S. Licciulli, et al. (2005). Molecular signature of retinoic acid treatment in acute promyelocytic leukemia. Oncogene 24(20): 3358–3368.

Michiels, S., S. Koscielny, C. Hill (2005). Prediction of cancer outcome with microarrays: a multiple random validation strategy. Lancet 365(9458): 488–492.

Mitchell, S. A., K. M. Brown, M. M. Henry, et al. (2004). Inter-platform comparability of microarrays in acute lymphoblastic leukemia. BMC Genomics 5(1): 71.

Neben, K., B. Tews, G. Wrobel, et al. (2004). Gene expression patterns in acute myeloid leukemia correlate with centrosome aberrations and numerical chromosome changes. Oncogene 23(13): 2379–2384.

Neben, K., S. Schnittger, B. Brors, et al. (2005). Distinct gene expression patterns associated with FLT3- and NRAS-activating mutations in acute myeloid leukemia with normal karyotype. Oncogene 24(9): 1580–1588.

Nowell, P. C., D. A. Hungerford (1960). A minute chromosome in human chronic granulocytic leukemia. Science 132: 1497.

Ohno, R., Y. Nakamura (2003). Prediction of response to imatinib by cDNA microarray analysis. Semin Hematol 40(2 Suppl 2): 42–49.

Okuda, T., Z. Cai, S. Yang, et al. (1998). Expression of a knocked-in AML1-ETO leukemia gene inhibits the establishment of normal definitive hematopoiesis and directly generates dysplastic hematopoietic progenitors. Blood 91(9): 3134–3143.

Orchard, J. A., R. E. Ibbotson, Z. Davis, et al. (2004). ZAP-70 expression and prognosis in chronic lymphocytic leukaemia. Lancet 363(9403): 105–111.

Park, D. J., P. T. Vuong, S. de Vos, et al. (2003). Comparative analysis of genes regulated by PML/RAR alpha and PLZF/RAR alpha in response to retinoic acid using oligonucleotide arrays. Blood 102(10): 3727–3736.

Pui, C. H., W. E. Evans (2006). Treatment of acute lymphoblastic leukemia. N Engl J Med 354(2): 166–178.

Qian, Z., A. A. Fernald, L. A. Godley, et al. (2002). Expression profiling of CD34 + hematopoietic stem/progenitor cells reveals distinct subtypes of therapy-related acute myeloid leukemia. Proc Natl Acad Sci USA 99(23): 14925–14930.

Ramaswamy, S., T. R. Golub (2002). DNA microarrays in clinical oncology. J Clin Oncol 20(7): 1932–1941.

Rassenti, L. Z., L. Huynh, T. L. Toy, et al. (2004). ZAP-70 compared with immunoglobulin heavy-chain gene mutation status as a predictor of disease progression in chronic lymphocytic leukemia. N Engl J Med 351(9): 893–901.

Rhodes, D. R., A. M. Chinnaiyan (2005). Integrative analysis of the cancer transcriptome. Nat Genet 37(Suppl): S31–S37.

Rosenwald, A., A. A. Alizadeh, G. Widhopf, et al. (2001). Relation of gene expression phenotype to immunoglobulin mutation genotype in B cell chronic lymphocytic leukemia. J Exp Med 194(11): 1639–1647.

Rosenwald, A., G. Wright, W. C. Chan, et al. (2002). The use of molecular profiling to predict survival after chemotherapy for diffuse large-B-cell lymphoma. N Engl J Med 346(25): 1937–1947.

Rosenwald, A., E. Y. Chuang, R. E. Davis, et al. (2004). Fludarabine treatment of patients with chronic lymphocytic leukemia induces a p53-dependent gene expression response. Blood 104(5): 1428–1434.

Ross, M. E., X. Zhou, G. Song, et al. (2003). Classification of pediatric acute lymphoblastic leukemia by gene expression profiling. Blood 102(8): 2951–2959.

Ross, M. E., R. Mahfouz, M. Onciu, et al. (2004). Gene expression profiling of pediatric acute myelogenous leukemia. Blood 104(12): 3679–3687.

Rowley, J. D. (1973). Identificaton of a translocation with quinacrine fluorescence in a patient with acute leukemia. Ann Genet 16(2): 109–112.

Rowley, J. D. (1999). The role of chromosome translocations in leukemogenesis. Semin Hematol 36(4 Suppl 7): 59–72.

Rucker, F. G., L. Bullinger, S. Schwaenen, et al. (2006). Disclosure of candidate genes in acute myeloid leukemia with complex karyotypes using microarray-based molecular characterization. J Clin Oncol 24: 3887–3894.

Schlenk, R. F., A. Benner, J. Krauter, et al. (2004). Individual patient data-based meta-analysis of patients aged 16 to 60 years with core binding factor acute myeloid leukemia: a survey of the German Acute Myeloid Leukemia Intergroup. J Clin Oncol 22(18): 3741–3750.

Schoch, C., A. Kohlmann, S. Schnittger, et al. (2002). Acute myeloid leukemias with reciprocal rearrangements can be distinguished by specific gene expression profiles. Proc Natl Acad Sci USA 99(15): 10008–10013.

Schoch, C., A. Kohlmann, M. Dugas, et al. (2005). Genomic gains and losses influence expression levels of genes located within the affected regions: a study on acute myeloid leukemias with trisomy 8, 11, or 13, monosomy 7, or deletion 5q. Leukemia 19(7): 1224–1228.

Shipp, M. A., K. N. Ross, P. Tamayo, et al. (2002). Diffuse large B-cell lymphoma outcome prediction by gene-expression profiling and supervised machine learning. Nat Med 8(1): 68–74.

Shtivelman, E., B. Lifshitz, R. P. Gale, et al. (1985). Fused transcript of abl and bcr genes in chronic myelogenous leukaemia. Nature 315(6020): 550–554.

Simon, R. (2005). Roadmap for developing and validating therapeutically relevant genomic classifiers. J Clin Oncol 23(29): 7332–7341.

Slovak, M. L., K. J. Kopecky, P. A. Cassileth, et al. (2000). Karyotypic analysis predicts outcome of preremission and postremission therapy in adult acute myeloid leukemia: a Southwest Oncology Group/ Eastern Cooperative Oncology Group Study. Blood 96(13): 4075–4083.

Staudt, L. M. (2003). Molecular diagnosis of the hematologic cancers. N Engl J Med 348(18): 1777–1785.

Staunton, J. E., D. K. Slonim, H. A. Coller, et al. (2001). Chemosensitivity prediction by transcriptional profiling. Proc Natl Acad Sci USA 98(19): 10787–10792.

Stegmaier, K., K. N. Ross, S. A. Colavito, et al. (2004). Gene expression-based high-throughput screening (GE-HTS) and application to leukemia differentiation. Nat Genet 36(3): 257–263.

Stegmaier, K., S. M. Corsello, K. N. Ross, et al. (2005). Gefitinib induces myeloid differentiation of acute myeloid leukemia. Blood 106(8): 2841–2848.

Stilgenbauer, S., L. Bullinger, P. Lichter, et al. (2002). Genetics of chronic lymphocytic leukemia: genomic aberrations and V(H) gene mutation status in pathogenesis and clinical course. Leukemia 16(6): 993–1007.

Tallman, M. S. (2005). New strategies for the treatment of acute myeloid leukemia including antibodies and other novel agents. Hematology (Am Soc Hematol Educ Program): 143–150.

Tamayo, P., D. Slonim, J. Mesirov, et al. (1999). Interpreting patterns of gene expression with self-organizing maps: methods and application to hematopoietic differentiation. Proc Natl Acad Sci USA 96(6): 2907–2912.

Tipping, A. J., M. W. Deininger, J. M. Goldman, et al. (2003). Comparative gene expression profile of chronic myeloid leukemia cells innately resistant to imatinib mesylate. Exp Hematol 31(11): 1073–1080.

Tusher, V. G., R. Tibshirani, G. Chu. (2001). Significance analysis of microarrays applied to the ionizing radiation response. Proc Natl Acad Sci USA 98(9): 5116–5121.

Valk, P. J., R. G. Verhaak, M. A. Beijen, et al. (2004). Prognostically useful gene-expression profiles in acute myeloid leukemia. N Engl J Med 350(16): 1617–1628.

Vardiman, J. W., N. L. Harris, R. D. Brunning (2002). The World Health Organization (WHO) classification of the myeloid neoplasms. Blood 100(7): 2292–2302.

Verhaak, R. G., C. S. Goudswaard, W. van Putten, et al. (2005). Mutations in nucleophosmin (NPM1) in acute myeloid leukemia (AML): association with other gene abnormalities and previously established gene expression signatures and their favorable prognostic significance. Blood 106(12): 3747–3754.

Virchow, R. (1845). Weisses Blut. N Notiz Geb Natur Heilk 36: 151–156.

Virtaneva, K., F. A. Wright, S. M. Tanner, et al. (2001). Expression profiling reveals fundamental biological differences in acute myeloid leukemia with isolated trisomy 8 and normal cytogenetics. Proc Natl Acad Sci USA 98(3): 1124–1129.

Walgren, R. A., M. A. Meucci, H. L. McLeod (2005). Pharmacogenomic discovery approaches: will the real genes please stand up? J Clin Oncol 23(29): 7342–7349.

Witcher, M., D. T. Ross, C. Rousseau, et al. (2003). Synergy between all-trans retinoic acid and tumor necrosis factor pathways in acute leukemia cells. Blood 102(1): 237–245.

Yagi, T., A. Morimoto, M. Eguchi, et al. (2003). Identification of a gene expression signature associated with pediatric AML prognosis. Blood 102(5): 1849–1856.

Yeoh, E. J., M. E. Ross, S. A. Shurtleff, et al. (2002). Classification, subtype discovery, and prediction of outcome in pediatric acute lymphoblastic leukemia by gene expression profiling. Cancer Cell 1(2): 133–143.

Yong, A. S., R. M. Szydlo, J. M. Goldman, et al. (2006). Molecular profiling of CD34 + cells identifies low expression of CD7, along with high expression of proteinase 3 or elastase, as predictors of longer survival in patients with CML. Blood 107(1): 205–212.

9 Personalized Medicine in the Clinical Management of Colorectal Cancer

Anthony El-Khoueiry and Heinz Josef Lenz

Key Words: Colorectal cancer, Colon cancer, Prognosis, Chemotherapy, Personalized treatment,

1 INTRODUCTION

The practice of personalized medicine has been an objective of physicians for centuries. Attempts at tailoring therapy to the individual patient can be found in the day-to-day approach to the treatment of many diseases. For example, the choice of antibiotics to treat bacterial pneumonia depends on the age of the patient, the comorbidities, and the most likely infectious agent based on their exposure history. Similarly, the choice of the most appropriate antihypertensive agent depends on the patient's age, race, renal function, and other comorbidities. In the setting of treating cancer with chemotherapy, the aim of personalized medicine would be to maximize efficacy and minimize toxicity. In other words, personalized medicine would provide the oncologist with the necessary tools to decide when to treat a patient and with what agent(s). Such an approach has been brought closer to practical application by the science of pharmacogenomics (Rioux, 2000; Lenz, 2003).

Rather than trying to individualize therapy based on broad clinical and environmental characteristics as described above, pharmacogenomics uses novel genomic technologies to elucidate the influence of a patient's genetic make-up on the variability in individual response to drugs (McLeod and Yu, 2003). Various technologies have been developed to identify genetic variability between patients and elucidate a genetic signature or profile that can be used to tailor therapy and move away from the "one size fits all" approach. In this chapter, we will discuss the role of pharmacogenomics in the clinical management of colorectal cancer to illustrate the concept of personalized medicine. As we review the current data on molecular markers and their predictive or prognostic role, we will also highlight some of the pitfalls and limitations that need to be overcome as we progress in our quest for personalized medicine.

G.J. Gordon (ed.), *Cancer Drug Discovery and Development: Bioinformatics in Cancer and Cancer Therapy*,
DOI: 10.1007/978-1-59745-576-3_9, © Humana Press, a part of Springer Science + Business Media, LLC 2009

2 PERSONALIZED MEDICINE IN THE TREATMENT OF METASTATIC COLORECTAL CANCER

2.1 The Current Reality in the Clinical Management of Metastatic Colorectal Cancer

The last decade has witnessed a significant evolution in the treatment of metastatic colorectal cancer (mCRC). The incorporation of novel cytotoxic chemotherapeutic agents such as oxaliplatin and irinotecan, as well as novel targeted agents such as bevacizumab (BV) and cetuximab (CB), has resulted in a significant survival benefit for patients with mCRC (Tournigand et al., 2004; Hurwitz et al., 2004; Cunningham et al., 2003). The median survival has increased from 12 months with 5-fluorouracil (5-FU) and leucovorin (LV) to over 24 months with the more recent combination therapies. The standard approach to the treatment of mCRC involves the sequencing of several active regimens that include both cytotoxic and targeted agents. The combinations of 5-FU, LV, oxaliplatin and BV (FOLFOX/BV) or 5-FU, LV, irinotecan and BV (FOLFIRI/BV) are both effective frontline choices for the majority of patients with response rates surpassing 50% (Venook et al., 2006; Hochster et al., 2006). If a patient is known to have progression of disease on frontline therapy, several second and third line therapeutic options exist. The choice of second and third line therapies depends on the aim of treatment, the type of first line therapy, and the toxicity profile of the different chemotherapeutic options. While it is beyond the scope of this chapter to discuss the clinical algorithm of the treatment of mCRC, we will highlight the potential role of pharmacogenomics in influencing the choice of therapy. Specifically, we will review the current data about the predictive molecular markers of efficacy and toxicity and how they may be incorporated in the treatment approach to mCRC.

2.2 Predictive Markers of Efficacy

2.2.1 5-FU (Fig. 1)

5-FU remains an essential component of most combination chemotherapy regimens used in the treatment of mCRC. However, new combinations that do not include 5-FU have become available and may offer a superior therapeutic option if a patient has genetic markers of resistance to 5-FU. Such combinations include irinotecan and oxaliplatin (IROX) or irinotecan and cetuximab, with or without bevacizumab (Goldberg et al., 2004; Saltz et al., 2005). The enzymes involved in 5-FU metabolism have been evaluated for their role in affecting the probability of response to 5-FU based therapy (Rich et al., 2004). Thymidine phosphorylase (TP) and thymidine kinase (TK) produce fluorodeoxyuridine monophosphate (FdUMP) which forms a stable ternary complex with thymidilate synthase (TS) and folic acid, resulting in inhibition of DNA synthesis. The inhibition of TS deprives the cell from its sole de novo source of thymidine. Dihydropyrimidine dehydrogenase (DPD) is involved in the catabolism of 5-FU and its elimination.

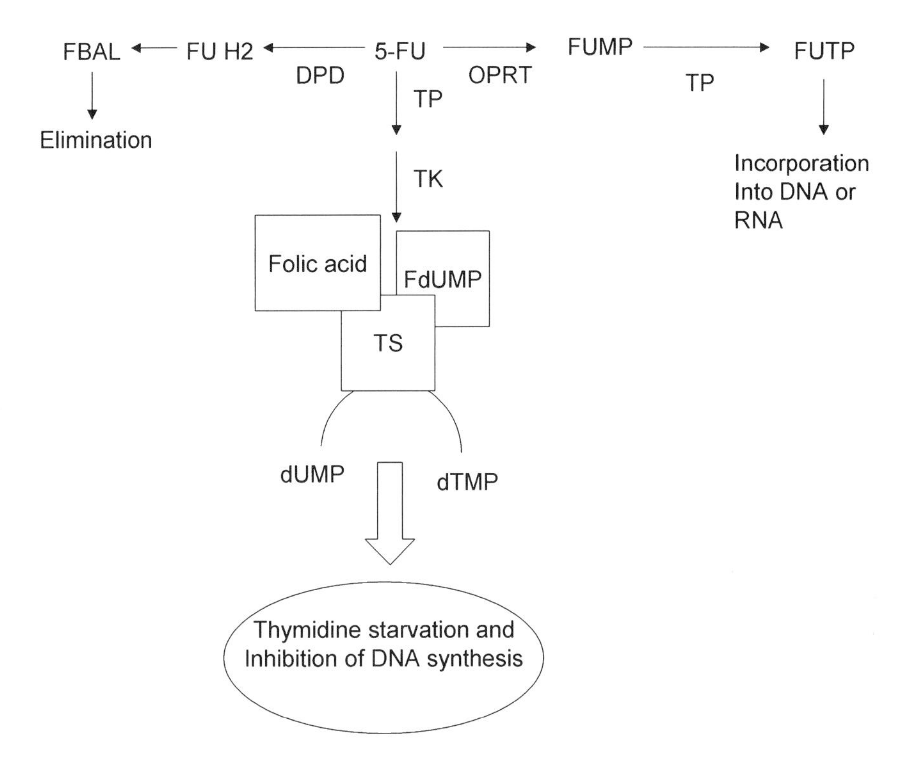

Fig. 1. 5-Fluorouracil or its prodrugs, such as capecitabine, continue to play a central role in the treatment of GI malignancies. The enzymes involved in 5-FU metabolism have been evaluated for their role in affecting the probability of response to 5-FU based therapy. Thymidine phosphorylase (TP) and thymidine kinase (TK) produce FdUMP which forms a stable ternary complex with TS and folic acid resulting in inhibition of DNA synthesis. OPRT is responsible for the conversion of 5-FU to FUMP which is subsequently phosphorylated to the active metabolite FUTP. FUTP incorporates into RNA, disrupting normal RNA processing and function. Dihydropyrimidine dehydrogenase (DPD) is responsible for the catabolism of 5-FU and its elimination. *FUMP* fluorouridine monophosphate, *FUTP* fluorouridine triphosphate, *FUH2* dihydrofluorouracil, *FBAL* fluoro-β-alanine, *FdUMP* fluorodeoxyuridine monophosphate, *dUMP* deoxyuridine monophosphate, *dTMP* deoxythymidine monophosphate, *OPRT* orotate phosphoribosyltransferase

2.2.2 TS

Several molecular mechanisms that affect the expression of the TS gene have been found to influence the probability of response to 5-FU based therapy. These molecular mechanisms include increased expression at the RNA or protein level, the number 28-base pair (bp) repeats in the 5′ promoter region, deletions in the 3′ tail region, and single nucleotide polymorphisms (SNPs) within the 28-bp repeats in the 5′ region.

2.2.3 TS Gene Expression

Leichman et al. (1997) demonstrated a significant inverse relationship between intratumoral TS gene expression and response to infusional 5-FU in a retrospective study of 46 patients with mCRC. The median TS mRNA level of responders was 1.9 $\times 10^{-3}$ vs. the mRNA level of 5.6 $\times 10^{-3}$ in nonresponders. The median survival for patients with intratumoral TS mRNA level $\leq 3.5 \times 10^{-3}$ was 13.6 months as compared to 8.2 months in patients with a level $<3.5 \times 10^{-3}$ ($p = 0.02$). Similarly, increased TS expression at the protein level as evaluated by immunohistochemical staining (IHC), has been correlated with increased resistance to infusional 5-FU based therapy in 108 patients with mCRC. Patients with positive IHC for TS had an objective response rate (RR) of 15 vs. 30% in patients with negative IHC for TS (HR 2.0; 95% CI: 3.5–1.2; $p < 0.04$) (Paradiso et al., 2000).

2.2.4 TS Gene Polymorphisms

The evaluation of gene expression at the RNA level depends on a special technology that utilizes laser capture microdissection to separate tumor cells from normal mucosa, followed by RNA isolation, cDNA preparation and subsequent quantitation of the relevant genes using a fluorescence-based real-time detection (Bonner et al., 1997). The use of IHC to evaluate gene expression at the protein level is limited by the quality of the antibodies used and the level of experience of the laboratory, and is labor intensive. Genomic polymorphisms which can be evaluated using peripheral blood and PCR technology may offer a potentially cheaper and more efficient alternative to gene expression analysis. Polymorphisms that affect the activity or function of an enzyme critical for the efficacy of a drug may be associated with clinical outcome and serve as predictors of response (McLeod and Yu, 2003). This concept is highlighted well with the example of the 28-bp repeat sequence in the 5′ terminal regulatory region of the TS gene. In a retrospective study of 52 patients with metastatic CRC, TS gene expression levels were 3.6 times higher in patients who were homozygous for the triple repeat (3R/3R) 28-bp sequence compared to those homozygous for the double repeat variant (2R/2R) (p = 0.004). Patients homozygous for the double repeats (2R/2R) showed a significantly better response rate to 5-FU compared to those homozygous for the triple repeats (3R/3R) (50 vs. 9%; $p = 0.04$) (Pullarkat et al., 2001). Kawakami et al. (2001) reported higher TS protein expression levels with the 3R/3R genotype, which suggested improved translational efficiency as a potential mechanistic explanation for the genotype-dependent difference in expression. A prospective study of 102 patients with metastatic CRC treated with 5-FU revealed that the 2R/2R genotype conferred the longest survival (median survival of 19 months for 2R/2R, 10 months for 2R/3R and 14 months for 3R/3R; $p = 0.025$). All three groups had similar objective response rates (Etienne et al., 2002). The other polymorphisms in the TS gene such as a 6-bp deletion in the 3′ region and a G to C SNP have been identified but their clinical significance awaits further evaluation (Mandola et al., 2003, 2004)

2.2.5 Is the Response to 5-FU Dependent on TS Only?

As highlighted previously, the metabolism and mechanism of action of 5-FU are influenced by several genes, including TS, DPD, and TP. In a study of 38 patients with metastatic CRC treated with 5-FU and leucovorin, the range of pretreatment intratumoral expression levels of TP was much narrower among the responding (16-fold) vs. the nonresponding patients (205-fold). None of the patients with a TP level over 18 responded ($p = 0.037$) (Mandola et al., 2003). These results appear to be inconsistent with in vitro data that showed a correlation between increased TP levels and increased sensitivity to 5-FU. This apparent inconsistency may be due to the fact that TP is an angiogenic factor, also known as platelet-derived endothelial cell growth factor. The angiogenic role of TP is likely to be more relevant in vivo where higher TP levels may be markers of more aggressive tumors with higher invasive and malignant potential (Allen and Johnston, 2005). In another retrospective study of 33 patients with mCRC treated with 5-FU based therapy, the range of DPD expression was narrower among responding patients (0.60×10^{-3} to 2.5×10^{-3}, 4.2-fold) compared with that of the non-responders (0.2×10^{-3} to 16×10^{-3}, 80-fold). In the same study, there was no correlation between the expression of TS and that of DPD, suggesting that the two genes are independently regulated. Several patients with DPD levels below the cutoff value of 2.5×10^{-3} were nonresponders and had elevated TS levels. The patients who had low levels of both TS and DPD had a RR of 92%. When the expression level of TP was included in the analysis, all the responders could be identified by low expression of all three genes. Nonresponders had at least one gene with high expression values (Salonga et al., 2000). These data suggest that the predictive power of TS expression for clinical outcome can be improved by the incorporation of other independent response determinants involved in the metabolism of 5-FU.

2.2.6 How Far Are We from Being Able to Predict Response to 5-FU in the Clinic?

Being able to determine which patients should receive 5-FU as part of their therapy for mCRC based on the current molecular determinants of response would constitute a critical step towards the application of the concept of personalized medicine. However, the incorporation of the molecular data discussed above into clinical practice requires validation in a prospective fashion as well as a better understanding of the complexity of TS gene expression control and its interaction with other genes in the metabolism of 5-FU. One of the criticisms of the data about the predictive value of TS expression is that it is derived in a retrospective manner and in small groups of patients. Recently, the Medical Research Council (MRC) investigators attempted to address this issue by conducting an exploratory analysis of 13 molecular markers in relation to the clinical outcome of 846 patients with mCRC treated with 5-FU, 5-FU and irinotecan, or 5-FU and oxaliplatin in a prospective clinical trial (FOCUS). In the test set of 846 patients, high TS level expression by IHC and tumor grade were significantly associated with failure-free survival (FFS). However, in the validation set of 449 patients, there was no statistically significant association between TS level and FFS. When an exploratory analysis was preformed on both the test and validation sets combined, high TS expression by IHC remained significantly associated with reduced FFS (Richman et al., 2006). The lack of correlation between TS expression and FFS in the validation set may be related to the choice of FFS as a surrogate for clinical outcome when evaluating a predictive marker such as TS. FFS is influenced by the number of patients who fail treatment secondary to toxicity and not only progression. Furthermore, the source of tissue utilized to perform the IHC evaluation may impact the degree of TS expression and its clinical outcome. Several investigators have demonstrated that TS expression by IHC in the primary tumor site does not fully correlate with TS expression in other metastatic sites (Aschele et al., 2000; Corsi et al., 2002). Simultaneously, clinical outcome correlated with TS expression at the metastatic site and not at the primary site. Other methodological pitfalls include the method used to evaluate TS gene expression. While IHC does not require complex technology and is cheap, the use of RT-PCR to quantify gene expression at the RNA level is quantitative and less subject to variation based on technique and the reader's interpretation. As we move into the age of prospective validation of molecular predictors, it is important that the methods used to assess TS gene expression be standardized, the cutoff points for RNA expression be uniform, and the statistical design be appropriate to capture the significance of an association without a high degree of false negativity or positivity. Furthermore, in the case of TS, the evaluation of gene expression levels in conjunction with the polymorphisms in the untranslated regions may result in a better assessment of the predictive value of the gene.

2.2.7 Molecular Predictors of Response to Oxaliplatin and Irinotecan

The use of combination chemotherapy is standard in the approach to treatment of mCRC. As a consequence, the ability to predict the response to one combination vs. another requires that we identify predictors of response to oxaliplatin and irinotecan which are combined with 5-FU. Being able to predict response to combinations like FOLFOX and FOLFIRI could allow us to improve the probability of response to front-line chemotherapy. Choosing the "right" treatment for a patient can have several clinical advantages. For example, a patient who is symptomatic from widely mCRC could achieve more immediate palliation prior to worsening of the performance status which could lead to missing the window of opportunity for treatment. In patients who have limited metastatic disease to the liver or lung, being able to choose the most effective regimen could increase the chances of cure by down-staging the disease and allowing for subsequent resection of the metastases, which could translate into a cure.

2.2.8 Oxaliplatin

Oxaliplatin (Eloxatin, Sanofi-Aventis) is a platinum analog that inhibits DNA synthesis through the formation of intrastrand DNA adducts (Simpson et al., 2003). The diaminocyclohexane moiety, which distinguishes oxaliplatin from other platinum compounds, is thought to enhance its cytotoxicity through the formation of bulkier adducts. Several intracellular mechanisms are involved in the inherited or acquired resistance to oxaliplatin. In addition to the balance of cellular uptake vs. efflux, increased detoxification and DNA repair have been found to influence the sensitivity to oxaliplatin. The excision repair crosscomplementation group 1 enzyme (ERCC1), XRCC1, and the xeroderma pigmentosum group (XPD) are all members of the nucleotide excision repair (NER) pathway involved in the repair of DNA damage induced by the formation of platinum-DNA adducts. An increased ability to repair DNA damage through the NER pathway is thought to lead to increased resistance to platinum drugs (Kweekel et al., 2005). Elevated mRNA levels of ERCC1 were found to be associated with a 4.2-fold increased risk of death (95% CI: 1.4, 13.3; $p = 0.008$) in a group of patients with mCRC treated with FOLFOX as second line therapy (Shirota et al., 2001). Genomic polymorphisms of ERCC1, XPD and XRCC1 have been found to be associated with the probability of response to second line chemotherapy with FOLFOX in patients with mCRC. A polymorphism in exon 10 of XRCC1 causing an arginine to glycine (Gln) substitution is thought to result in decreased ability for repair of DNA damage by DNA adducts. The presence of at least one Gln allele has been associated with fivefold increased risk to fail 5-FU/oxaliplatin therapy (Stoehlmacher et al., 2001). An A to C substitution in exon 23 of the XPD gene leads to a lysine (Lys) to glutamine (Gln) substitution. In patients treated with FOLFOX, the presence of one or more Gln alleles resulted in a lower RR (10 vs. 24% in patients with Lys/Lys) (Park et al., 2001). Glutathione-S-transferase P1 (GSTP1) belongs to a family of enzymes that catalyze the conjugation of reduced glutathione, thereby protecting cellular macromolecules from damage caused by carcinogenic and chemotherapeutic agents (Srivastava et al., 1999). A polymorphism leading to an isoleucine to valine substitution at position 313 leads to decreased enzymatic activity of the GSTP1–105Val variant. Among 107 metastatic CRC patients treated with FOLFOX, the homozygous Val/Val genotype appeared to be correlated with improved survival as compared to the Val/Ile or Ile/Ile genotypes (25 vs. 13 vs. 9.6 months; $p < 0.001$) (Stoehlmacher et al., 2002). However, these results are in contrast to the study by McLeod et al. (2003) who could not confirm an association between this GSTP1 polymorphism and oxaliplatin-based chemotherapy. The most recent genetic polymorphism to be associated with clinical outcome of 5-FU and oxaliplatin is the SCN1A polymorphism of a neuronal voltage-gated sodium channel (VGSC), which was evaluated in a group of 173 patients with mCRC treated with 5-FU and oxaliplatin after failure of 5-FU or 5-FU and irinotecan. Patients with SCN1A T1067A SNP T/T genotype showed a significantly better response rate (21.9% [23/105] vs. 11.3% [5/44]; $p = 0.02$), time to progression (4.6 vs. 3.4 months; $p = 0.02$) and OS (12.3 vs. 8 months; $p = 0.002$) compared to patients with the T/A genotype (Nagashima et al., 2006).

At the clinical application level, it is unlikely that response to a combination chemotherapeutic regimen would depend on one gene or one pathway only. Therefore, it is imperative to identify more comprehensive genetic signatures or profiles that would account for the multiple genes involved and that would separate responders from nonresponders. In this context, Stoehlmacher et al. (2004) were able to separate patients with mCRC who were treated with FOLFOX into three groups using a combination analysis of favorable genotypes from polymorphisms in XPD-751, ERCC1–118, GSTP1–105 and TS-3′ untranslated region. Patients possessing two or more favorable genotypes survived a median of 17.4 months compared with 5.4 months in patients with no favorable genotype. Patients with one favorable genotype had an intermediate survival of 10.2 months. Daud et al. (2006) are performing a phase II study that randomizes patients with metastatic CRC to capecitabine, irinotecan and bevacizumab vs. capecitabine, oxaliplatin and bevacizumab. Patient responses in the two groups will be correlated with molecular analysis of liver biopsies, and a U133A gene chip will be used to develop a microarray classifier for treatment responders and nonresponders.

2.2.9 Irinotecan

Irinotecan is a prodrug that is activated by carboxylesterase (CES) into SN38 which achieves its antitumor effect by inhibiting DNA topoisomerase I, an enzyme involved in the relaxation of supercoiled DNA. SN38 is metabolized in the human liver by UGT1A1 to an inactive compound SN38G (Mathijssen et al., 2001). UGT1A1 polymorphisms are thought to predict toxicity to irinotecan, a subject which will be discussed in more detail subsequently. Other members of the UGT1A family are involved in the metabolism of irinotecan including hepatic UGT1A6 and UGT1A9 and extrahepatic UGT1A7. UGT1A7 is expressed in the gastrointestinal tract and may influence the disposition of SN38 within the gut. In a phase II study of 67 patients with mCRC treated with capecitabine and irinotecan, low enzyme activity UGT1A7 genotypes, UGT1A7*2/*2 (six patients) and UGT1A7*3/*3 (seven patients), were significantly associated with improved response probability (85 vs. 44% in patients with other genotypes; $p = 0.013$). A similar finding was noted with UGT1A9–118 $(dT)_{9/9}$ genotype (RR of 74% with UGT1A9–118 $(dT)_{9/9}$ vs. 43% in patients with other genotypes). The UGT1A9–118 $(dT)_{10}$ allele has been associated with 2.6-fold greater transcriptional activity than the –118 $(dT)_9$ allele in vitro. These data would suggest that the UGT1A9–118 $(dT)_{9/9}$ genotype results in lower UGT1A9 activity and therefore higher levels of the active metabolite SN-38 and lower levels of SN-38G (Carlini et al., 2005).

Preliminary data from a retrospective study with 33 metastatic CRC patients suggest that gene expression levels of ERCC1, GSTP1, epidermal growth factor receptor (EGFR) and multidrug resistance (MDR1) may be associated with response to irinotecan-based therapy (Yang et al., 2005). These data need to be validated in larger cohorts.

At this point, FOLFOX and FOLFIRI are thought to have equivalent efficacy in the frontline treatment of mCRC (Tournigand et al., 2004). The available data on molecular predictors of efficacy are limited by the retrospective nature of the studies, the small numbers of patients, and the need for a better understanding of the functional significance of several of the polymorphisms involved. Several ongoing studies are evaluating some of these markers as well as others in a prospective manner. At the same time, the continuously changing landscape of the treatment of mCRC presents a challenge to our ability to be able to validate molecular markers. This challenge is due to the evolution of the treatment combinations and the incorporation of targeted agents such as bevacizumab and cetuximab into the treatment algorithm. Efforts are ongoing to identify molecular predictors of response to bevacizumab and cetuximab.

2.3 Molecular Predictors of Toxicity

When treating patients with mCRC, the ability to predict toxicity to specific drugs may influence the choice of therapy with the hope of minimizing side effects and preserving quality of life. The prolongation in overall survival afforded by the new therapeutic combinations is accompanied by an increased risk of toxicity, which is partially related to the longer exposure to the drugs in question. Sixty two percent of patients with mCRC treated with FOLFOX on clinical trial N9741 had to discontinue treatment for reasons other than progression. Among those patients, 53% discontinued treatment due to toxicity, including neurotoxicity, myelosuppression and hypersensitivity. As a result, the time to treatment failure (TTF) which accounts for treatment failure for reasons other than disease progression was shorter than the time to tumor progression (TTP) (5.8 vs. 9.2 months) (Green et al., 2005). This example highlights the importance of tailoring the treatment to the individual patient in a way that minimizes toxicity while not compromising efficacy. We discuss examples of potential predictive markers of toxicity to 5-FU, irinotecan and oxaliplatin later.

2.3.1 5-FU

The number of repeats (2R vs. 3R) of a 28-bp sequence in the 5′UTR of TS has been found to correlate with TS expression as discussed above. The 3R allele is associated with increased TS expression, which is the main target of 5-FU. In a group of 52 patients with mCRC treated with bolus 5-FU and LV, grade 2 or more adverse events were reported in all patients with the 2R/2R genotype and in 14 out of 41 patients with a 3R allele ($p < 0.0001$, simple logistic regression analysis). In the same study, a SNP (G→A) in exon 3 of orotate phosphoribosyltransferase (OPRT) was found to be associated with the risk of toxicity. OPRT is responsible for the conversion of 5-FU to fluorouridine monophosphate (FUMP), which is subsequently phosphorylated to the active metabolite fluorouridine triphosphate (FUTP). The G to A mutation is thought to increase the activity of OPRT, which leads to higher levels of the active metabolite FUTP. All 20 patients with G213A allele experienced adverse events of more than grade 2 compared to 5 out of 32 without G213A allele ($p < 0.0001$, simple logistic regression analysis) (Ichikawa et al., 2003).

DPD is an enzyme that accounts for about 80% of the catabolism of 5-FU. A G to A mutation in the invariant GT splice donor site flanking exon 14 (IVS14+1G>A) is thought to lead to decreased DPD activity and consequently severe, potentially lethal toxicity (Van Kuilenburg et al., 2001).

2.3.2 Irinotecan

As noted previously, UGT1A1 is involved in the glucoronidation of SN38, the active metabolite of irinotecan, to SN38G, the inactive metabolite. Polymorphisms that decrease the activity of UGT1A1 would subsequently be expected to increase the exposure to SN38, and potentially worsen the risk of toxicity. A 2-bp (TA) insertion in the TATA box in the promoter region of UGT1A1 leads to 7 TA repeats instead of 6 and results in the variant allele UGT1A1*28. UGT1A1*28 leads to decreased protein expression of UGT1A1 and increased risk of neutropenia with irinotecan in patients homozygous for the 7 TA repeats (Iyer et al., 2002; Innocenti et al., 2004). These data resulted in a drug label change for irinotecan that incorporated the UGT1A1 molecular assay information. However, the same polymorphism did not predict the risk of toxicity in patients who received bolus 5-FU/LV and irinotecan (IFL) in the clinical trial N9741 which randomized patients with mCRC to FOLFOX vs. IFL vs. the combination of irinotecan and oxaliplatin (IROX). Statistically significant associations were found between the 7/7 genotype and the risk of grade 4 neutropenia in the patients treated with IROX ($p = 0.004$) and in all patients ($p = 0.007$) (McLeod et al., 2006). Part of the explanation for the failure of the 7/7 genotype to predict toxicity in patients treated with IFL is the low frequency of the 7/7 genotype. Another reason may be the fact that several other enzymes play a role in the transport and metabolism of irinotecan and SN-38 and should be incorporated into the toxicity prediction algorithm. Furthermore, the clinical impact of identifying patients with the UGT1A1 7/7 genotype is limited as it was found to reduce the risk of grade 4 neutropenia from 18 to 17% only in the same study. Multiple efforts are ongoing to develop a more comprehensive assessment of the risk of toxicity with irinotecan. For example, Innocenti et al. (2005) examined the role of several transporters involved in the metabolism of irinotecan along with UGT1A1. The superfamily of ATP-binding cassette (ABC) transporters is responsible for the elimination of irinotecan and its metabolites through biliary and intestinal excretion. ABCC2 has been found to have at least six SNPs. The organic anion transporter polypeptide-1B1 (OATP-1B1), also known as SLCO1B1, is found at the basolateral membrane of hepatocytes and transports SN-38 from blood into liver. Twelve haplotypes of ABCC2 were identified and haplotype 4 was correlated with SN-38G/SN-38 AUC ratio ($p < 0.0001$). SLCO1B1*5 (521T>C) CT and CC genotypes had a higher irinotecan AUC compared to TT genotype ($p = 0.0001$). This difference is attributed to reduced transport function of the SLCO1B1*5 variant. Using a multivariate model, SLCO1B1, ABCC2 and UGT1A1 gene variants along with total bilirubin were associated with the risk of neutropenia in this retrospective study of 65 patients. Once again, such data highlight the importance of the interplay among several genetic variations in determining toxicity to a drug.

2.3.3 Oxaliplatin

Chronic peripheral sensory neuropathy is a dose-limiting toxicity of oxaliplatin, with grade 3 neuropathy, being reported in as many as 50% of patients who reached doses over 1,000 mg/m^{-2} (Cersosimo, 2005). Recently, GSTP1 I105V polymorphism was found to be associated with early onset of oxaliplatin-induced neurotoxicity. In a retrospective evaluation of 299 patients who received FOLFOX on Intergroup study N9741, four genetic variants in genes involved in detoxification and DNA repair were evaluated for their potential role as predictors of susceptibility to oxaliplatin mediated neuropathy. Patients with the GSTP1 C/C polymorphism (equivalent to the Val/Val genotype noted above) were more likely to discontinue FOLFOX due to neuropathy than patients with T/T (23.7 vs. 9.2%) or with C/T (10%; $p = 0.039$). The cumulative dose to onset of neuropathy was lower for patients with the C/C polymorphism as compared to the T/T or C/T patients ($p = 0.05$) (Grothey et al., 2005). More recently, polymorphisms in the neuronal VGSCs were assessed for their role in predicting oxaliplatin toxicity. In a prospective study of 152 patients with mCRC treated with infusional 5-FU and oxaliplatin, the SCN1A T1067A SNP T/T genotype showed a significant association with a lower risk of grade 3/4 toxicity ($p = 0.002$) (Nagashima et al., 2006). These preliminary data offer hope of the possibility of being able to determine which patients are at increased risk for oxaliplatin-related peripheral neuropathy but need to be validated in other prospective studies. With regard to clinical applicability, being able to determine if a patient is at increased risk for peripheral neuropathy would influence the treating physician to choose an irinotecan-based combination as the frontline regimen or to minimize the total dose of oxaliplatin administered by using a "stop-and-go" approach to therapy as illustrated in the OPTIMOX 1 and 2 studies (Maindrault-Goebel et al., 2006).

3 PERSONALIZED MEDICINE IN THE ADJUVANT TREATMENT OF COLORECTAL CANCER

The objective of adjuvant chemotherapy is to reduce the risk of recurrence of a cancer after curative therapy, most commonly surgery, via the inhibition of micrometastases. FOLFOX has become the standard adjuvant treatment for patients with stage III colon cancer (Wolmark et al., 2005; De Gramont et al., 2005). However, there is controversy about the use of adjuvant chemotherapy in all patients with stage II colon cancer. The concern is about exposing patients with a low risk of recurrence to chemotherapy and potential toxicity unnecessarily. Prognostic molecular markers that help predict tumor behavior as well as host/tumor interactions could be helpful in risk stratification and in restriction of treatment to patients who would derive the most benefit. It is beyond the scope of this chapter to review the entire body of data related to prognostic markers for colorectal cancer. However, we will use the example of TS and microsatellite instability to illustrate the possibility of individualizing adjuvant chemotherapy for colon cancer patients. TS expression has been shown to have prognostic value with low intratumoral levels predicting longer survival (Johnston et al., 1994; Lenz et al., 1998). More recently, low TS expression assessed by immunohistochemistry in tumors from 1,326 patients with stage II and III CRC was found to be a statistically valid independent prognostic factor. While the prognostic value of TS has been established, its role in predicting benefit from adjuvant 5-FU based chemotherapy has been somewhat controversial, possibly because of the small patient numbers and the difference in the methodology used to assess the expression level (IHC vs. RT-PCR) (Johnston et al., 2005; Popat et al., 2004). Several recent and ongoing studies attempt to address this issue. For example, in a series of 121 consecutive patients with II and III colon cancer in Korea treated with oral 5-FU therapy (doxifluridine), patients with 3R/3R TS genotype had shorter survival than patients with 2R/2R genotype (53 vs. 80%; $p = 0.0481$) (Suh et al., 2005). Even though these data are consistent with the studies that show a higher level of TS expression with the 3R/3R genotype, other publications highlight the complexity of the prognostic role of TS polymorphisms. Dotor et al. (2006) examined the prognostic value of 3 TS gene polymorphisms in a prospective cohort of 129 patients with CRC. The patients received adjuvant bolus 5-FU plus levamisol for 1 year or bolus 5-FU and leucovorin for 6 months as adjuvant therapy depending on the date of enrollment. The TS enhancer region was evaluated for the presence of 2R or 3R 28-bp repeats and for the functional G>C SNP present in the second repeat of 3R alleles. In addition, the TS1494del6 genotype was evaluated to determine the presence or absence of a 6-bp deletion polymorphism which is associated with decreased intratumoral TS levels. Haplotype and mutational analyses of the TS gene were also performed. The TSER 3R/3R genotype was associated with better prognosis ($p = 0.02$). SNP genotyping of the 3R alleles did not manifest any significant association with clinical outcome. The presence of the −6 bp allele in the tumor was associated with a reduced risk of death. Haplotype analysis revealed that patients with the 3R/−6 bp haplotype showed a significant OS benefit compared with patients with the 2R/+6 bp haplotype. The apparent contradiction in the prognostic value of the 28-bp repeats between the two studies is likely to be multifactorial. Potential explanations include the fact that the relationship between the number of 28-bp repeats and TS expression has to be better elucidated. Furthermore, the choice of the predictive marker has to be consistent with the mechanism of action of the drug. In the study by Dotor et al., patients were treated with bolus 5-FU, while TS has a role as a potential predictive marker in patients treated with infusional 5-FU. The different modes of administration affect the mechanism of action of 5-FU, with the infusional mode resulting in TS inhibition whereas the bolus administration is thought to act by incorporation into RNA (Sobrero et al., 1997; Matsusaka et al., 2003). Therefore, TS expression may not be the most appropriate marker for patients treated with bolus 5-FU.

MSI is present in about 15% of CRCs and reflects the inactivation of mismatch repair (MMR) genes. Analysis of pooled data from published studies by Popat et al. reveals that CRCs with high MSI have a better prognosis (HR 0.65, 95% CI: 0.59–0.71). MSI tumors appeared to derive no benefit from adjuvant 5-FU but the data are limited with HR of 1.24 and a 95% CI of 0.72–2.14 (Popat et al., 2005). While there is general agreement about the positive prognostic impact of deficient MMR or high MSI, it remains unclear whether patients with deficient MMR benefit from chemotherapy. Two recent studies suggest that the benefit of adjuvant 5-FU based chemotherapy is restricted to MMR competent tumors (Jover et al., 2006; Lanza et al., 2006). The role of MMR or MSI status in predicting benefit from chemotherapy needs to be evaluated further with the current standard adjuvant chemotherapy combination of 5-FU/LV and oxaliplatin. The irinotecan-based combination, FOLFIRI, failed to show superiority over 5-FU/LV in the adjuvant treatment of patients with CRC in several clinical trials. However, patients with high MSI tumors may benefit from adjuvant chemotherapy with 5-FU/LV and irinotecan as shown in early results from CALGB 89803 (Bertagnolli et al., 2006). If these findings are validated, they would provide strong support for the concept of personalized medicine and will result in the incorporation of irinotecan in the adjuvant treatment of a subset of patients with CRC with tumors expressing high MSI.

Several other genes have been evaluated for their prognostic role in stage II and III colon cancer. It is likely that the most accurate prognostic assessment will depend on more than one gene in one pathway. To address this issue, there are multiple attempts to develop a molecular signature that can be used for molecular staging of patients with stage II and III colon cancer. The goal is to achieve better risk stratification to avoid over or under treating of patients. Eschrich et al. (2005) used a 32,000 cDNA microarray to evaluate 78 human colon cancer specimens and correlated the results with survival. The authors identified a set of 43 core genes which was 90% accurate in predicting a 36-month overall survival, and was better than Duke's staging ($p = 0.03878$). Johnston et al. (2006) used another method to determine a prognostic genetic signature. They developed a disease-specific microarray for colorectal cancer using a high-throughput transcriptome-based approach that encodes over 52,500 transcripts expressed in normal and diseased colorectal tissue. Using this approach, a gene signature containing 48 genes demonstrated 100% accuracy in the prediction of recurrence in 32 patients with stage II CRC. These studies and others offer both great promise and significant challenge. Many technical issues need to be simplified and standardized. For example, microarrays still require microdissection, a process that is not widely available or easy to perform. At the same time, while arrays result in a molecular profile, they do not reveal the functional significance of the genes or pathways involved. On the other hand, a candidate gene approach may lead to important genes being overlooked or missed. As the scientific community attempts to address these issues and many others, prospective validation of these techniques and signatures is imperative prior to their application in the clinic.

4 HOW DO WE BRING THE CONCEPT OF PERSONALIZED MEDICINE TO THE DAILY PRACTICE OF ONCOLOGY? (FIG. 2)

In a discussion of several abstracts during the tumor biology oral presentation session at the annual meeting of the American Society of Clinical Oncology, Dr. Carmen Allegra's outline suggested developmental milestones for clinical markers. Phase 1A studies represent the first clinical investigation of a marker using retrospective

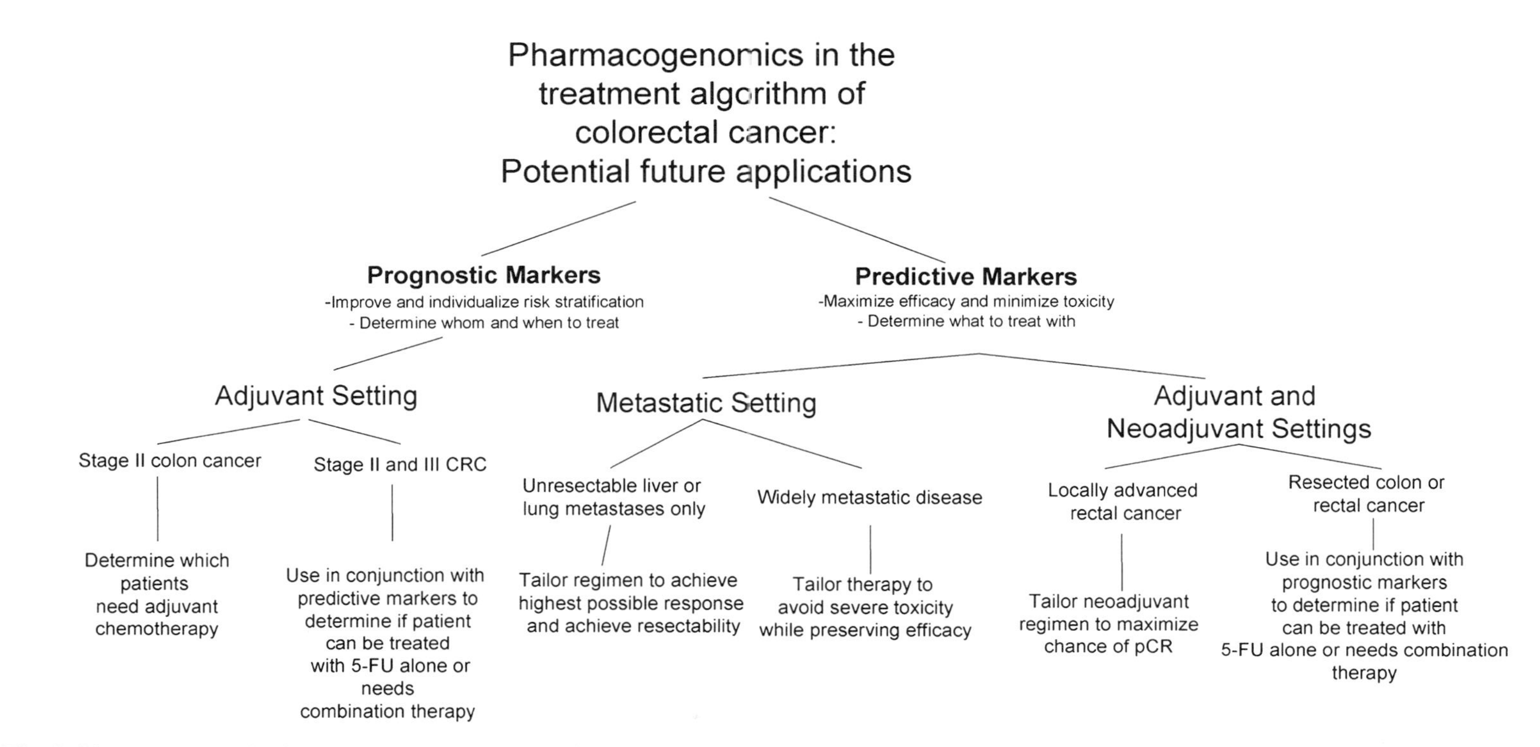

Fig. 2. Pharmacogenomics in the treatment algorithm of colorectal cancer: potential future application

data and tissues and retrospectively determined "cut-points." Phase 1B studies are retrospective ones that refute or support initial studies. Phase II studies represent a prospective validation of findings from phase 1A and 1B studies; they may still use retrospective tissue but must have predefined marker cut-points/gene sets and have adequate power. In other terms, this is consistent with the concept of a test set. Phase III studies consist of a prospective demonstration of clinical benefit through the use of the marker for therapeutic decision making. Multiple investigators are now involved in studies that are designed to prospectively demonstrate the superiority of molecularly chosen therapy over the standard approach. A current study by McLeod et al. (2005) assigns patients with T3 and T4 rectal cancer to 5-FU vs. irinotecan-based chemotherapy based on the likelihood of response to 5-FU. Patients homozygous for the 3R repeats in the 5′UTR region of TS are considered "bad risk" patients and are assigned to receive irinotecan and radiation, whereas the 2R/3R and 2R/2R patients receive 5-FU and radiation. Preliminary results in the 13 "bad risk" patients with 3R/3R genotype who received irinotecan and radiation reveal 85% down-staging with 62% pathologically complete response rate. Such provocative data need to be validated in a randomized setting, but they still serve to illustrate the potential advantage of tailored therapy. One of the favored clinical trials designs to determine the relevance of a marker is illustrated in a study that assigned patients with nonsmall cell lung cancer to treatment on a control arm (docetaxel and cisplatin) vs. a genotypic arm. In the latter arm, patients with high ERCC1 levels who are expected to be platinum resistant received gemcitabine and docetaxel, whereas patients with low ERCC1 expression received docetaxel and cisplatin. The final results of this study are pending, but preliminary ones reveal a significant role for ERCC1 in predicting response and therefore influencing the choice of therapy (Rosell et al., 2005).

As we move into the future, the benefits of the tools used to individualize therapy should extend beyond the predictive and prognostic markers. For instance, if a molecular marker identifies the risk of extrahepatic metastases in patients with CRC, a clinical trial could be designed with the aim of determining whether patients with elevated expression of the marker should undergo resection of the liver metastases for cure. The implication is that molecular markers could lead to a change in the endpoints of clinical trials. Drug development is another field that should benefit from the field of pharmacogenomics. If a genetic marker of resistance to a drug is identified, the gene involved could represent the ideal target for drug development to overcome the resistance. Similarly, a better understanding of the mechanism of resistance to specific drugs in patients could provide support for more rational therapeutic combinations.

5 CONCLUSION

Personalized medicine in the treatment of colorectal cancer will be of great benefit to patients as it allows the treating physician to tailor the treatment in a manner that maximizes benefit or response and minimizes toxicity. It will also allow patients to avoid being exposed to treatment that is not needed in the adjuvant setting based on better risk stratification. However, applying these concepts in the clinic requires a large coordinated effort involving molecular biology, informatics and well designed clinical trials. Ongoing and planned studies promise to bring us closer to this reality.

REFERENCES

Allen W, Johnston P. Have we made progress in pharmacogenomics? The implementation of molecular markers in colon cancer. Pharmacogenomics 6:1–9, 2005

Aschele C, Debernardis D, Tunesi G, et al. Thymidylate synthase protein expression in primary colorectal cancer compared with the corresponding distant metastases and relationship with the clinical response to 5-fluorouracil. Clin Cancer Res 6:4797–4802, 2000

Bertagnolli MM, Compton CC, Niedzwiecki D, et al. Microsatellite instability predicts improved response to adjuvant therapy with irinotecan, 5-fluorouracil and leucovorin in stage III colon cancer. J Clin Oncol, ASCO Annual Meeting Proceedings. Part I. Vol. 24, No. 18S (June 20 Supplement), 10003, 2006

Bonner RF, Emmert-Buck M, Cole K, et al. Laser capture microdissection: molecular analysis of tissue. Science 278:1481–1483, 1997

Carlini L, Meropol N, Beve J, et al. UGT1A7 and UGT1A9 polymorphisms predict response and toxicity in colorectal cancer patients treated with capecitabine/irinotecan. Clin Cancer Res 11:1226–1236, 2005

Cersosimo R. Oxaliplatin-associated neuropathy: a review. Ann Pharmacol 39:128–134, 2005

Corsi DC, Ciaparrone M, Zannoni G, et al. Predictive value of thymidylate synthase expression in resected metastases of colorectal cancer. Eur J Cancer 38:527–534, 2002

Cunningham D, Humblet Y, Siena S, et al. Cetuximab (C225) alone or in combination with irinotecan (CPT-11) in patients with epidermal growth factor (EGFR) positive, irinotecan refractory metastatic colorectal cancer (MCRC). Presented at 39th Annual Meeting of the American Society of Clinical Oncology, Chicago, IL, May 31–June 3, 2003. Abstract 1012

Daud A, Chodkiewicz C, Garrett C, et al. Microarray analysis of colon cancer outcome: preliminary results of a randomized, phase II study. Am Soc Clin Oncol GI Symp 2006. Abstract 288

De Gramont A, Boni C, Navarro M, et al. Oxaliplatin/5FU/LV in the adjuvant treatment of stage II and stage III colon cancer: efficacy results with a median follow-up of 4 years. Am Soc Clin Oncol Annu Meet 2005. Abstract 3501

Dotor E, Cuatrecases M, Martinez-Iniesta M, et al. Tumor thymidylate synthase 1494del6 genotype as a prognostic factor in colorectal cancer patients receiving fluorouracil-based adjuvant treatment. J Clin Oncol 24(10):1603–1611, 2006

Eschrich S, Yang I, Bloom G, et al. Molecular staging for survival prediction of colorectal cancer patients. J Clin Oncol 23:3526–3535, 2005

Etienne MC, Chazal M, Laurent-Puig P, et al. Prognostic value of tumoral thymidilate synthase and p53 in metastatic colorectal cancer patients receiving fluorouracil-based chemotherapy: phenotypic and genotypic analyses. J Clin Oncol 20:2832–2843, 2002

Goldberg R, Sargent D, Morton R, et al. A randomized controlled trial of fluorouracil plus leucovorin, irinotecan, and oxaliplatin combinations in patients with previously untreated metastatic colorectal cancer. J Clin Oncol 22:2084–2091, 2004

Green E, Sargent DJ, Goldberg RM, et al. Detailed analysis of oxaliplatin-associated neurotoxicity in Intergroup trial N9741. Presented at the ASCO GI Cancer Symposium, 2005. Abstract 182

Grothey A, McLeod HL, Gree EM. Glutathione S-transferase P1 I105V (GSTPa I105V) polymorphism is associated with early onset of oxaliplatin-induced neurotoxicity. Am Soc Clin Oncol Annu Meet 2005. Abstract 3509

Hochster HS, Hart LL, Ramanathan RK, et al. Safety and efficacy of oxaliplatin/fluoropyrimidine regimens with or without bevacizumab as first-line treatment of metastatic colorectal cancer (mCRC): final analysis of the TREE-study. J Clin Oncol, ASCO Annual Meeting Proceedings. Part I. Vol. 24, No. 18S (June 20 Supplement), 3510, 2006

Hurwitz H, Fehrenbacher L, Novotny W, et al. Bevacizumab plus irinotecan, fluorouracil, and leucovorin for metastatic colorectal cancer. N Engl J Med 350:2335–2342, 2004

Ichikawa W, Takahashi T, Nihei Z, et al. Polymorphisms of orotate phosphoribosyl transferase (OPRT) gene and thymidylate synthase tandem repeat (TSTR) predict adverse events (AE) in colorectal cancer (CRC) patients treated with 5-fluorouracil (FU) plus leucovorin (LV). Proc Am Soc Clin Oncol 22:265, 2003. Abstract 1063

Innocenti F, Undevia SD, Iyer L, et al. Genetic variants in the *UDP-glucuronosyltransferase 1A1* gene predict the risk of severe neutropenia of irinotecan. J Clin Oncol 22:1382–1388, 2004

Innocenti F, Undevia SD, Rosner GL, et al. Irinotecan (CPT-11) pharmacokinetics (PK) and neutropenia: interaction among UGT1A1 and transporter genes. Am Soc Clin Oncol Annu Meet 2005. Abstract 2006

Iyer L, Das S, Janisch L, et al. UGT1A1*28 polymorphism as a determinant of irinotecan disposition and toxicity. Pharmacogenomics J 2:43–47, 2002

Johnston PG, Fisher ER, Rockette HE, et al. The role of thymidilate synthase expression in prognosis and outcome of adjuvant chemotherapy in patients with rectal cancer. J Clin Oncol 12:2640–2647, 1994

Johnston PG, Benson A, Catalano P, et al. The clinical significance of thymidilate synthase (TS) expression in primary colorectal cancer: an intergroup combined analysis. Am Soc Clin Oncol Annu Meet 2005. Abstract 3510

Johnston PG, Mulligan K, Kay E, et al. A genetic signature in stage II colorectal cancer derived from formalin fixed paraffin embedded tissue (FFPE) using a unique disease specific colorectal array. J Clin Oncol, ASCO Annual Meeting Proceedings. Vol. 24, No. 18S, 2006. Abstract 3519

Jover R, Zapater P, Castells A, et al. Mismatch repair status in the prediction of benefit from adjuvant fluorouracil chemotherapy in colorectal cancer. Gut 55(6):848–855, 2006

Kawakami K, Salonga D, Park JM, et al. Different lengths of a polymorphic repeat sequence in the thymidylate synthase gene affect translational efficiency but not its gene expression. Clin Cancer Res 7:4096–4101, 2001

Kweekel DM, Gelderblom H, Guchelaar HJ. Pharmacology of oxaliplatin and the use of pharmacogenomics to individualize therapy. Cancer Treat Rev 31:90–105, 2005

Lanza G, Gafa R, Santini A, et al. Immunohistochemical test for MLH1 and MSH2 expression predicts clinical outcome in stage II and III colorectal cancer patients. J Clin Oncol 24(15):2359–2367, 2006

Leichman CG, Lenz HJ, Leichman L, et al. Quantitation of intratumoral thymidilate synthase expression predicts for disseminated colorectal cancer response and resistance to protracted infusion fluorouracil and leucovorin. J Clin Oncol 15:3223–3229, 1997

Lenz HJ. Pharmacogenomics in colorectal cancer. Semin Oncol 30:47–53, 2003

Lenz HJ, Danenberg KD, Leichman CG, et al. P53 and thymidilate synthase expression in untreated stage II colon cancer: associations with recurrence, survival, and site. Clin Cancer Res 4:1227–1234, 1998

Maindrault-Goebel F, Lledo G, Chibaudel B, et al. OPTIMOX2, a large randomized phase II study of maintenance therapy or chemotherapy-free intervals (CFI) after FOLFOX in patients with metastatic colorectal cancer (MRC). A GERCOR study. J Clin Oncol, ASCO Annual Meeting Proceedings. Part I. Vol. 24, No. 18S (June 20 Supplement), 3504, 2006

Mandola M, Stoehlamcher J, Muller-Weeks S, et al. A novel single nucleotide polymorphism within the 5 tandem repeat polymorphism of the thymidylate synthase gene abolishes USF-1 binding and alters transcriptional activity. Cancer Res 63:2898–2904, 2003

Mandola MV, Stoehlamcher J, Zhang W, et al. A 6 bp polymorphism in the thymidylate synthase gene causes message instability and is associated with decreased intratumoral TS mRNA levels. Pharmacogenetics 15:319–327, 2004

Mathijssen RH, van Alphen RJ, Verweij J, et al. Clinical pharmacokinetics and metabolism of irinotecan (CPT-11). Clin Cancer Res 7:2182–2194, 2001

Matsusaka S, Yamasaki H, Kitayama Y, et al. Differential effects of two fluorouracil administration regimens for colorectal cancer. Oncol Rep 10:109–113, 2003

McLeod H, Yu J. Cancer pharmacogenomics: SNPs, chips, and the individual patient. Cancer Invest 21:630–640, 2003

McLeod H, Sargent D, Marsh S, Fuchs C, Ramanathan R, Williamson S, Findlay B, Thibodeau S, Petersen G, Goldberg R. Pharmacogenetic analysis of systemic toxicity and response after 5-fluorouracil (5-FU)/CPT-11, 5-FU/oxaliplatin (oxal), or CPT-11/oxal therapy for advanced colorectal cancer (CRC): results from an intergroup trial. Proc Am Soc Clin Oncol 2003. Abstract 1013

McLeod HL, Tan B, Malyapa R, et al. Genotype-guided neoadjuvant therapy for colorectal cancer. Am Soc Clin Oncol Annu Meet 2005. Abstract 3024

McLeod HL, Parodi L, Sargent DJ, et al. UGT1A1*28, toxicity and outcome in advanced colorectal cancer: results from Trial N9741. J Clin Oncol, ASCO Annual Meeting Proceedings. Part I. Vol. 24, No. 18S, 3520, 2006

Nagashima F, Zhang W, Yang D, et al. Polymorphism in sodium-channel alpha 1-subunit (SCN1A) predicts response, TTP, survival, and toxicity in patients with metastatic colorectal cancer treated with 5-FU/ oxaliplatin. J Clin Oncol, ASCO Annual Meeting Proceedings. Part I. Vol. 24, No. 18S (June 20 Supplement), 3533, 2006

Paradiso A, Simone G, Petroni S. et al. Thymidilate synthase and p53 primary tumour expression as predictive factors for advanced colorectal cancer patients. Br J Cancer 82:560–567, 2000

Park DJ, Stoehlmacher J, Zhang W, et al. Xeroderma pigmentosum group D gene polymorphism predicts clinical outcome to platinum-based chemotherapy in patients with advanced colorectal cancer. Cancer Res 61:8654–8658, 2001

Popat S, Matakidou A, Houlston RS, et al. Thymidilate synthase expression and prognosis in colorectal cancer: a systematic review and meta-analysis. J Clin Oncol 22:s29–s36, 2004

Popat S, Hubner R, Houlston RS. Systematic review of microsatellite instability and colorectal cancer prognosis. J Clin Oncol 23:609–617, 2005

Pullarkat ST, Stoehlmacher J, Ghaderi V, et al. Thymidylate synthase gene polymorphism determines response and toxicity of 5-FU based chemotherapy. Pharmacogenomics J 1:65–70, 2001

Rich T, Shepard R, Mosley S. Four decades of continuing innovation with fluorouracil: current and future approaches to fluorouracil chemoradiation therapy. J Clin Oncol 22:2214–2232, 2004

Richman S, Braun MS, Adlard JW, et al. Prognostic value of thymidylate synthase (TS) expression on failure-free survival of fluorouracil-treated metastatic colorectal cancer patients. J Clin Oncol, ASCO Annual Meeting Proceedings. Part I. Vol. 24, No. 18S (June 20 Supplement), 10011, 2006

Rioux P. Clinical trials in pharmacogenetics and pharmacogenomics: methods and applications. Am J Health Syst Pharm 57:887–898, 2000

Rosell R, Cobo M, Isla D, et al. ERCC1 mRNA-based randomized phase III trial of docetaxel (doc) doublets with cisplatin (cis) or gemcitabine (gem) in stage IV non small cell lung cancer (NSCLC) patients (p). J Clin Oncol, ASCO Annual Meeting Proceedings. Part I of II. Vol. 23, No. 16S (June 1 Supplement), 7002, 2005

Salonga D, Danenberg KD, Johnson M, et al. Colorectal tumors responding to 5-fluorouracil have low gene expression levels of dihydropyrimidine dehydrogenase, thymidilate synthase, and thymidine phosphorylase. Clin Cancer Res 6:1322–1327, 2000

Saltz LB, Lenz H, Hochster H, et al. Randomized phase II trial of cetuximab/bevacizumab/irinotecan (CBI) versus cetuximab/bevacizumab (CB) in irinotecan-refractory colorectal cancer. Am Soc Clin Oncol Annu Meet 2005. Abstract 3508

Shirota Y, Stoehlmacher J, Brabender J, et al. ERCC1 and thymidylate synthase mRNA levels predict survival for colorectal cancer patients receiving combination oxaliplatin and fluorouracil chemotherapy. J Clin Oncol 19:4298–4304, 2001

Simpson D, Dunn C, Curran M, et al. Oxaliplatin: a review of its use in combination therapy for advanced colorectal cancer. Drugs 63:2127–2156, 2003

Sobrero AF, Aschele C, Bertino JR. Fluorouracil in colorectal cancer — a tale of two drugs: implications for biochemical modulation. J Clin Oncol 15:368–381, 1997

Srivastava S, Singhal S, Hu X, et al. Differential catalytic efficiency of allelic variants of human glutathione S-transferase Pi in catalyzing the glutathione conjugation of thiotepa. Arch Biochem Biophys 366:89–94, 1999

Stoehlmacher J, Chaderi V, Iqbal S, et al. A polymorphism of the XRCC1 gene predicts for response to platinum based treatment in advanced colorectal cancer. Anticancer Res 21:3075–3079, 2001

Stoehlmacher J, Park DJ, Zhang W, et al. Association between glutathione S-transferase P1, T1, and M1 genetic polymorphism and survival of patients with metastatic colorectal cancer. J Natl Cancer Inst 94:936–942, 2002

Stoehlmacher J, Park DJ, Zhang W, et al. A multivariate analysis of genomic polymorphisms: prediction of clinical outcome to 5FU/oxaliplatin combination chemotherapy in refractory colorectal cancer. Br J Cancer 91:344–354, 2004

Suh KW, Kim JH, Kim YB, et al. Thymidilate synthase gene polymorphism as a prognostic factor for colon cancer. J Gastrointest Surg 9:336–342, 2005
Tournigand C, Andre T, Achille E, et al. FOLFIRI followed by FOLFOX6 or the reverse sequence in advanced colorectal cancer: a randomized GERCOR study. J Clin Oncol 22:229–237, 2004
Van Kuilenburg ABP, Muller EW, Haasjes J, et al. Lethal outcome of a patient with a complete dihydropyrimidine dehydrogenase (DPD) deficiency after administration of 5-fluorouracil: frequency of the common IVS14+1G>A mutation causing DPD deficiency. Clin Cancer Res 7:1149–1153, 2001
Venook A, Niedzwiecki D, Hollis D, et al. Phase III study of irinotecan/5FU/LV (FOLFIRI) or oxaliplatin/5FU/LV (FOLFOX) ± cetuximab for patients (pts) with untreated metastatic adenocarcinoma of the colon or rectum (MCRC): CALGB 80203 preliminary results. J Clin Oncol, ASCO Annual Meeting Proceedings. Part I. Vol. 24, No. 18S (June 20 Supplement), 3509, 2006
Wolmark N, Wieand HS, Kuebler JP, et al. Phase III trial comparing FULV to FULV + oxaliplatin in stage II or III carcinoma of the colon: results of NSABP Protocol C-07. Am Soc Clin Oncol Annu Meet 2005. Abstract LBA 3500
Yang D, Vallbohmer D, Rhodes KE, et al. Molecular prognostic factors of irinotecan efficacy. Am Soc Clin Oncol Annu Meet 2005. Abstract 3621

10 *PIK3CA* Gene Alterations in Human Cancers

Sérgia Velho, Carla Oliveira, and Raquel Seruca

ABSTRACT

Mutations or amplification of the chromosomal region containing the *PIK3CA* gene, coding for the p110α catalytic subunit of class IA of PI3K, were described in diverse tumor models. In this review, we will focus on the localization and type of *PIK3CA* mutations and their association with signaling pathways and cellular effects in different tumor types.

Key Words: PI3K PIK3CA, PI3K signaling, *PIK3CA* mutations, *PIK3CA* amplification, Cancer, AKT, PTEN, BRAF, KRAS, Kinase, Molecular targets,

1 PHOSPHOINOSITOL 3-KINASE (PI3K): OVERVIEW

PI3-kinases were first identified as an 85-kDa phosphoproteins whose appearance correlated with a phosphatidylinositol kinase activity in immunoprecipitates from a number of polyoma middleT mutants (Domin and Waterfield, 1997). They are nowadays defined as lipid kinases responsible for the phosphorylation of inositol lipids at the 3′ position of the inositol ring, and to generate 3-phosphoinositides PtdIns(3)P, PtdIns(3,4)P2 and PtdIns(3,4,5)P3 (Vanhaesebroeck and Waterfield, 1999; Katso et al., 2001; Vivanco and Sawyers, 2002).

The molecular cloning of PI3K demonstrated that PI3-kinases are part of a large and complex family of proteins that contain three classes (I, II, III) and are defined on the basis of their primary structure, regulation and in vitro lipid substrate specificity (Domin and Waterfield, 1997; Wymann and Pirola, 1998; Vanhaesebroeck and Waterfield, 1999; Katso et al., 2001; Vivanco and Sawyers, 2002).

Class I PI3K forms a heterodimeric complex, composed by a catalytic subunit and an adaptor/regulatory subunit which renders the complex responsive to ligand stimulation (Domin and Waterfield, 1997; Wymann and Pirola, 1998; Vanhaesebroeck and Waterfield, 1999; Katso et al., 2001; Vivanco and Sawyers, 2002). In vitro, class I PI3K can use as substrate both PtdIns(4)P and PtdIns(4,5)P_2, but in vivo, they preferentially catalyze the

G.J. Gordon (ed.), *Cancer Drug Discovery and Development: Bioinformatics in Cancer and Cancer Therapy*, DOI: 10.1007/978-1-59745-576-3_10, © Humana Press, a part of Springer Science + Business Media, LLC 2009

Fig. 1. PI3K catalytic subunits domains. *p85 BD* p85-binding domain, *RAS BD* RAS-binding domain, *C2* protein-kinase-C homology-2 (C2) domain, *Helical* helical domain, *Kinase* kinase domain

conversion of PtdIns(4,5)P_2 to PtdIns(3,4,5)P_3 (Domin and Waterfield, 1997; Katso et al., 2001). This class of PI3K can be segregated into two subclasses, IA and IB, based on structural and functional differences (Domin and Waterfield, 1997; Wymann and Pirola, 1998; Katso et al., 2001). Subclass IA, which transmit signals from tyrosine kinase receptors (RTK) and RAS proteins, are heterodimeric proteins having a p110 (α, β, δ) catalytic subunit associated with a p85 (α or β), a p55 (α or γ), or a p50 α adaptor/regulatory subunit (Wymann and Pirola, 1998; Katso et al., 2001). Catalytic subunits are characterized by an N-terminal p85-binding domain, a RAS-binding domain, a protein-kinase-C homology-2 (C2) domain, a helical and a kinase domain in the C-terminus of the protein (Katso et al., 2001) (Fig. 1).

All class IA adaptor/regulatory subunits have two SH2 domains that bind specifically to phosphorylated tyrosine residues in receptor proteins, as well as in other signaling proteins (Wymann and Pirola, 1998; Katso et al., 2001). The main function of the regulatory subunits is to recruit the p110 catalytic subunit to tyrosine phosphorylated proteins at the plasma membrane, where p110 catalytic subunit phosphorylates its lipid substrates (Katso et al., 2001). The interaction between p85 and p110 inhibits the catalytic activity of p110; however, this inhibition is released upon interaction between the SH2 domains of p85 and tyrosine phosphorylated peptides. Thus, recruitment of p110 to the membrane by p85 has two effects on p110 (Wymann and Pirola, 1998; Yu et al., 1998a,b). First, p110 is brought in proximity to the inositol lipid substrates, and second, the catalytic activity of p110 is enhanced as p85 binds to tyrosine kinase receptors or other tyrosine phosphorylated adaptors (Yu et al., 1998a,b). Because p85 affects the activity of p110, as well as its subcellular localization, p85 is more accurately described as a regulatory subunit than as an adaptor subunit (Wymann and Pirola, 1998; Yu et al., 1998a,b).

Upon activation of PI3K, phosphorylated lipids are produced at the cell membrane and contribute to the recruitment and activation of various signaling components (Domin and Waterfield, 1997; Wymann and Pirola, 1998; Vanhaesebroeck and Waterfield, 1999; Katso et al., 2001; Vivanco and Sawyers, 2002).

Of particular interest is the serine/threonine kinase AKT/PKB that is a crucial kinase involved in the PI3K pathway. Once activated, PI3K generates phosphatidylinositol-3,4,5-trisphosphate (PIP(3)), that is essential for the translocation of AKT/PKB to the plasma membrane where it is phosphorylated and activated by phosphoinositide-dependent kinase-1 (PDK-1) and possibly other kinases. Activated AKT/PKB phosphorylates and regulates the function of many cellular proteins responsible for maintenance and regulation of cell homeostasis such as GSK-3, MDM2, mTOR, IKK, proteins of the Forkhead family, BAD, Caspase 9 and AFX (Domin and Waterfield, 1997; Wymann and Pirola, 1998; Cantley and Neel, 1999; Vanhaesebroeck and Waterfield, 1999; Katso et al., 2001; Vivanco and Sawyers, 2002; Osaki et al., 2004b; Samuels and Ericson, 2006).

Additionally, PI3K has been shown to regulate the activity of other cellular targets, such as the serum and glucocorticoid-inducible kinase (SGK), the small GTP-binding proteins RAC1 and CDC42 and protein-kinase-C (PKC) in an AKT-independent manner. However the mechanisms underlying the activation of these targets are poorly characterized (Vivanco and Sawyers, 2002). Figure 2 summarizes the pathways activated by PI3K.

Negative regulation of PI3K—AKT pathway is mainly accomplished by the dual function lipid and protein phosphatase PTEN (phosphatase and tensin homologue deleted

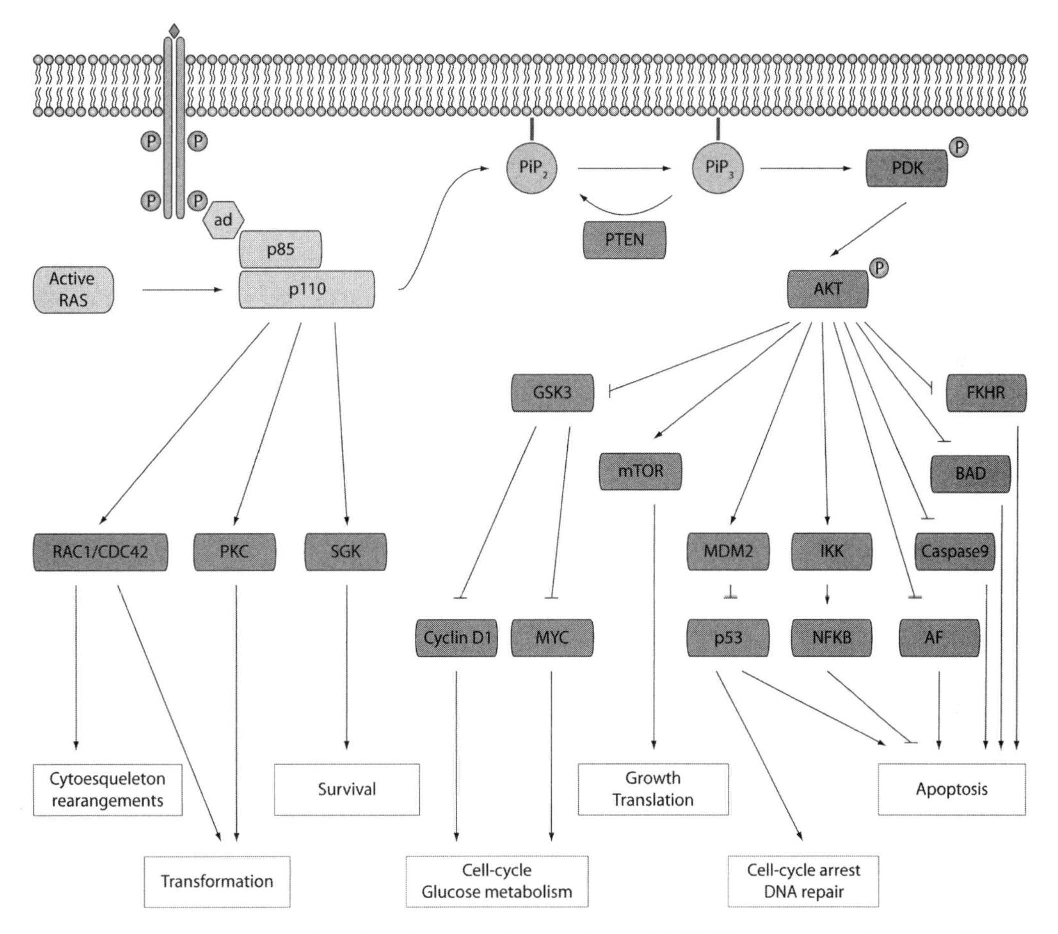

Fig. 2. Class IA PI3K activation, signaling and effects. Subclass IA PI3K is composed by heterodimeric proteins having a p85 regulatory subunit associated with a p110 catalytic subunit. In normal cells, subclass IA PI3K pathway is activated by stimulation of receptor tyrosine kinases (RTK) through the association of p85 subunit with an adaptor protein (schematically represented by *ad*) or active forms of RAS proteins that bind specifically to the RAS-binding domain of p110 subunit. At the cell membrane, active PI3K converts PIP_2 into PIP_3 through phosphorylation of PIP_2. The phosphorylated lipids are essential for the translocation of the serine/threonine kinase AKT to the plasma membrane where it is phosphorylated and activated by phosphoinositide-dependent kinases (PDK). Activated AKT phosphorylates and regulates the function of many cellular proteins responsible for maintenance and regulation of cell homeostasis. Additionally, PI3K regulates the activity of the serum and glucocorticoid-inducible kinase (SGK), the small GTP-binding proteins RAC1 and CDC42 and protein-kinase-C (PKC) in an AKT-independent manner. Negative regulation of PI3K—AKT pathway is mainly accomplished by PTEN (phosphatase and tensin homologue deleted on chromosome 10), which exerts its inhibitory effect through the dephosphorylation of phosphoinositide products of PI3K

on chromosome 10), which was originally identified as a tumor suppressor and is frequently affected by germline and somatic mutations in human cancers (Cantley and Neel, 1999). PTEN exerts its inhibitory effect in the PI3K—AKT pathway through the dephosphorylation of phosphoinositide products of PI3K that hamper the activation of AKT (Maehama and Dixon, 1999).

A disturbed activation of the PI3K—AKT pathway has been associated with the development of diseases such as diabetes mellitus, autoimmunity and cancer. This pathway plays an important role in cell homeostasis as it is involved in the regulation of cell proliferation, adhesion, motility, survival, and differentiation. It has been shown that PI3K—AKT pathway can interfere with cytoskeleton rearrangements, and intracellular trafficking (Domin and Waterfield, 1997; Wymann and Pirola, 1998; Cantley and Neel, 1999; Vanhaesebroeck and Waterfield, 1999; Katso et al., 2001; Hill and Hemmings, 2002; Vivanco and Sawyers, 2002; Osaki et al., 2004b; Samuels and Ericson, 2006).

Genetic aberrations that lead to a gain in PI3K signaling are commonly observed in human cancers. Recently, much attention has been given to the gene encoding for p110α catalytic subunit — *PIK3CA* — due to the increasing evidence of its role in the carcinogenesis process.

2 PI3K SIGNALING AND CANCER

Genetic alterations that lead to a gain in PI3K signaling are (1) activating mutations of p85, (2) mutations in tyrosine kinase receptors, (3) loss of function of PTEN by mutation or transcriptional downregulation, (4) AKT activation by amplification, overexpression or increased phosphorylation, and (5) p110α gain of function due to amplification, and overexpression or somatic mutations (Vivanco and Sawyers, 2002; Osaki et al., 2004b; Bader et al., 2005).

Mutations in the regulatory subunit of PI3K — p85 — have been identified in human tumor samples from the colon and ovary (Philp et al., 2001), which produce deletions in the inter-SH2 region of p85 subunit and lead to PI3K activation, presumably by releasing the p85–p110 complex from negative regulation, bypassing the normal role of RTKs signaling in PI3K activation (Vivanco and Sawyers, 2002).

Activating mutations in RTKs themselves provide additional, although less direct, evidence of the importance of the PI3K—AKT pathway in human cancer. As an example, a truncated variant of the epidermal growth factor receptor (EGFR) lacking the extracellular domain potently activates the PI3K—AKT signaling (Vivanco and Sawyers, 2002).

PTEN, the major downregulator of the PI3K pathway, is inactivated in many tumor types such as endometrium, brain, prostate, ovary, breast, thyroid, head and neck, kidney, lung, melanoma, gastric, lung, lymphomas, hepatocellular carcinomas and renal cell carcinomas (Ali et al., 1999; Vivanco and Sawyers, 2002). Somatic mutations in the *PTEN* gene are frequently found in a variety of sporadic human tumors in which the wild-type allele of the gene is inactivated mostly by deletions, thus conforming to the classical paradigm of tumor suppressor genes (Ali et al., 1999). The absence of functional *PTEN* in cancer cells leads to the constitutive activation of downstream components of the PI3K pathway including the AKT and mTOR kinases (Sansal and Sellers, 2004). In model organisms, inactivation of these kinases can reverse the effects of *PTEN* loss. These data raise the possibility that drugs targeting these kinases, or PI3K itself, may have significant therapeutic activity in *PTEN*-null cancers (Sansal and Sellers, 2004).

Alterations in the activation state of AKT due to amplification, overexpression or overactivation were observed in ovarian, breast, and thyroid cancers and represent an important mechanism of PI3K pathway activation (Vivanco and Sawyers, 2002; Bader et al., 2005).

The p100α catalytic subunit of PI3K is encoded by the *PIK3CA* gene. The *PIK3CA* gene is located in chromosome 3q26.3 (Volinia et al., 1994), and comprises 20 exons predicting a 1,068-amino acid protein. It has been demonstrated by several authors that the *PIK3CA* gene is deregulated in several tumor models. Samuels et al. (2004) performed a large-scale sequence analysis of 117 exons encoding the predicted kinase domains of eight PI3K and eight PI3K-like genes in 35 colorectal cancers, and found that *PIK3CA* gene was the only one harboring somatic mutations. Subsequently, all the coding exons of *PIK3CA* gene were analyzed by Samuels and colleagues in 199 additional colorectal cancers. Mutations were observed in a total of 74 colorectal tumors (32%) (Samuels et al., 2004). Following this work, numerous authors aimed at determining the frequency and role of *PIK3CA* gene mutations in other tumor models.

Mutations of *PIK3CA* in human tumors are somatic, cancer-specific and heterozygous (Bachman et al., 2004; Broderick et al., 2004; Campbell et al., 2004; Samuels and Velculescu, 2004; Samuels et al., 2004, 2005; Lee et al., 2005; Samuels and Ericson, 2006). They are mostly missense mutations; no truncating or nonsense mutations have been identified, but a few cases of in-frame deletions and insertions have been detected (Bader et al., 2005). Rare cases of double mutations, in which two amino acid residues are altered, have also been reported (Lee et al., 2005; Saal et al., 2005). The cancer-specific mutations found in *PIK3CA* are not randomly distributed along the coding sequence (Bader et al., 2005). The majority of mutations (85%) cluster in exons 9 and 20, which encode the catalytic and kinase domains of p110α protein, respectively; these two exons being considered mutational hotspots of *PIK3CA* gene (Broderick et al., 2004; Campbell et al., 2004; Samuels and Velculescu, 2004; Samuels et al., 2004, 2005; Bader et al., 2005; Oda et al., 2005; Samuels and Ericson, 2006). This nonrandomness is indicative of selection for mutations that confer an advantage, and act as dominant oncoproteins (Samuels et al., 2004, 2005; Bader et al., 2005; Samuels and Ericson, 2006), as it is in the case of activating mutations of *KRAS* and *BRAF* genes.

Besides the clustering of mutations in exons 9 and 20, they also associate with specific codons, namely, codons 542 and 545 in exon 9 and codon 1,047 in exon 20 (Ma et al., 2000; Bachman et al., 2004; Broderick et al., 2004; Campbell et al., 2004; Osaki et al., 2004b; Samuels et al., 2004; Buttitta et al., 2005; Garcia-Rostan et al., 2005; Lee et al., 2005; Levine et al., 2005; Oda et al., 2005; Saal et al., 2005; Velho et al., 2005; Wang et al., 2005; Wu et al., 2005b; Li et al., 2006). Regarding the amino acid change within these codons, E542K, E545K and H1047R (Ma et al., 2000; Bachman et al., 2004; Broderick et al., 2004; Campbell et al., 2004; Osaki et al., 2004b; Samuels et al., 2004; Buttitta et al., 2005; Garcia-Rostan et al., 2005; Lee et al., 2005; Levine et al., 2005; Oda et al., 2005; Saal et al., 2005; Velho et al., 2005; Wang et al., 2005; Wu et al., 2005b; Li et al., 2006) are the most frequent alterations found, resulting in a change of charge from negative to strongly positive residues (Levine et al., 2005).

The crystal structure of PIK3CA has not yet been described. Levine et al. (2005) attempted to model the structure of PIK3CA protein based on the well-known structure of the highly homologous PIK3CG protein. Based on this, E542K and E545K seem to be localized on the exposed surface of the molecule and the changes in charge caused by these alterations may affect protein—protein or other intermolecular interactions (Levine et al., 2005). H1047R is located within the kα11 helix, which lies on two sides of the activation loop of *PIK3CA*. Mutations in the activation loop have been shown to affect the specificity of lipid substrates and access to the catalytic core of PIK3CA (Levine et al., 2005).

Amplification of *PIK3CA* gene represents another mechanism of activation of PI3K pathway. Its occurrence was demonstrated in several tumor types such as ovarian cancer (Iwabuchi et al., 1995; Shayesteh et al., 1999; Campbell et al., 2004), cervical cancer (Ma et al., 2000; Zhang et al., 2002; Bertelsen et al., 2005), head and neck squamous cell carcinomas (HNSCCs) (Woenckhaus et al., 2002; Pedrero et al., 2005), nasopharyngeal tumors (Or et al., 2005), lung tumors (Massion et al., 2002, 2004) and gastric carcinomas (Byun et al., 2003), among others. Amplification of *PIK3CA* gene is generally associated with increased *PIK3CA* expression, PI3K activity and activation of AKT, supporting an oncogenic role of *PIK3CA* amplification (Wu et al., 2005a,b).

2.1 PIK3CA Gene Alterations in Distinct Tumor Models

The occurrence of mutations and amplification of *PIK3CA* gene in several types of cancer implies that this pathway may be a valuable target for the development of novel therapies for these cancers. Mutations of *PIK3CA* gene and alternative mechanisms of activating PI3K signaling (namely amplification or PTEN loss) appear to be mutually exclusive (Byun et al., 2003; Saal et al., 2005). We will describe below, the alterations of PI3K pathway in diverse tumor models.

2.1.1 Brain Tumors

Mutations of *PIK3CA* gene were studied in brain tumor types, such as glioblastomas, anaplastic oligodendroglioma, medulloblastoma, anaplastic astrocytoma, low-grade astrocytoma and ependymoma.

The frequency of *PIK3CA* mutations found was not very high. In general, brain tumors are mutated in about 5% of the cases (Broderick et al., 2004). The higher rate of mutations was found in anaplastic oligodendroglioma (14%) (Broderick et al., 2004). Glioblastomas harbor *PIK3CA* mutations in 5–7% of the tumors analyzed (Broderick et al., 2004). This frequency is lower than the one previously found by Samuels et al. (2004) (27%). This difference could be explained by an increased incidence of mutations in nonhotspot exons or it may also depend on the selection of samples. In medulloblastomas *PIK3CA* mutations were found in a frequency ranging from 0 to 5% (Broderick et al., 2004). Three percent of anaplastic astrocytomas were seen to be mutated in *PIK3CA* gene, while no mutations were identified in low-grade astrocytomas and ependymomas, suggesting that *PIK3CA* mutations occur lately in brain carcinogenesis (Broderick et al., 2004). *PIK3CA* mutations in brain tumors occur only in tumors that do not carry *PTEN* mutations (Broderick et al., 2004).

The frequency of *PIK3CA* amplification in brain tumors is still very controversial. While some authors claim to find *PIK3CA* amplification frequently in medulloblastomas and glioblastomas (Hui et al., 2001; Mizoguchi et al., 2004; Tong et al., 2004), others were not able to confirm these results by assessing different series for amplification of the *PIK3CA* gene (Knobbe and Reifenberger, 2003; Broderick et al., 2004).

2.1.2 Head and Neck Squamous Cell Carcinomas

PIK3CA gene amplification is a frequent early event in HNSCCs (Woenckhaus et al., 2002; Pedrero et al., 2005). *PIK3CA* gene amplification was found in premalignant lesions with a frequency (39%) that did not differ significantly from that of advanced tumors (37%).

Association studies between the presence of *PIK3CA* amplification and the clinico-pathological features of HNSCCs did not reach any statistical significance (Pedrero et al., 2005).

In contrast to what is found in other tumor models, in HNSCCs *PIK3CA* gene amplification does not correlate with p110α protein overexpression, pointing to other mechanisms distinct from gene amplification to explain the pattern of p110α expression in HNSCC. A possible explanation is the one postulated by West et al. (2003) who have demonstrated how nicotinic activation of AKT depends upon PI3K and specific nicotinic acetylcholine receptors (nAchRs) in lung carcinoma, a tobacco-related tumor such as HNSCC. However, the mechanism of coupling of nAchRs to the PI3K/AKT pathway is unknown (West et al., 2003; Pedrero et al., 2005).

2.1.3 Barrett's Adenocarcinomas

Barrett's adenocarcinomas develop from a myriad of genetic alterations. These genetic alterations include mutations and allelic losses of tumor suppressor genes such as *p16INK4A*, *p53*, *APC*, *Rb* among others (Blount et al., 1991; Huang et al., 1992; Gonzalez et al., 1997; Wong et al., 2001). In addition, several cellular oncogenes, including the *PIK3CA* gene, were found to be amplified in this tumor model (Miller et al., 2003). *PIK3CA* gene was found to be amplified in 5.7% of Barrett's adenocarcinomas and its amplification was correlated with a specific profile of the tumors. It was significantly correlated with an early tumor stage, small tumor size at the time of resection and the absence of nodal involvement, with all amplifications occurring in stage I and IIA tumors (Miller et al., 2003). These tumors were also at low tumor-node-metastasis (TNM) stage (Miller et al., 2003).

2.1.4 Nasopharyngeal Tumors

Copy number gains/amplification of *PIK3CA* loci were found in 75% of primary nasopharyngeal carcinomas (NPCs), while no *PIK3CA* mutations were detected in this type of primary tumors. These observations suggest that copy number gains, instead of mutations, may be a common mechanism for activation of *PIK3CA* in tumorigenesis in this type of tumors (Or et al., 2005).

2.1.5 Lung Tumors

PIK3CA mutations occur rarely in lung tumors (Samuels et al., 2004; Lee et al., 2005). On the other hand, amplification of the *PIK3CA* gene was found to be a common event in this tumor model. The *PIK3CA* was amplified in 70% of squamous carcinomas, 38% of large cell carcinomas, 19% of adenocarcinomas, and 67% of small cell lung cancers (Massion et al., 2002, 2004). In lung tumors, *PIK3CA* gene copy number did not have prognostic significance (Massion et al., 2002, 2004). In preinvasive lesions, amplification of the *PIK3CA* was associated with severe dysplasia. These observations suggest frequent and early involvement of the PI3-kinase pathway in lung cancer (Massion et al., 2002, 2004).

2.1.6 Breast Cancer

Several studies were aimed at *PIK3CA* mutation screening in breast cancer. The frequency of mutations ranges from 18 to 40% (Bachman et al., 2004; Campbell et al., 2004; Lee et al., 2005; Li et al., 2006). These studies place the *PIK3CA* gene as the most frequently mutated gene in breast cancer (Levine et al., 2005). A lower frequency (8%) was found by Samuels et al. (2004), but only 12 cases were studied. Another possibility to explain this low frequency might be the grade status of the tumors used. In a study where all the breast tumors analyzed were high grade, *PIK3CA* mutations were found in 20.6% of the tumors (Wu et al., 2005b).

The status of estrogen and progesterone receptors (ER/PR), as well as the presence or absence of Her-2/neu amplification, are independent determinants of breast cancer prognosis, as well as predictors of response to targeted therapy (Clemons and Goss, 2001; Ross et al., 2004). Opposing results have been obtained regarding the association between *PIK3CA* mutations and these clinicopathologic markers. While in some studies no associations between *PIK3CA* mutation status and age at diagnosis, estrogen/progesterone, Her-2/neu or nodal status was found (Bachman et al., 2004; Campbell et al., 2004; Levine et al., 2005), in others a strong association between *PIK3CA* mutations, lymph node status, ER and PR positivity and ERBB2 overexpression was obtained and these correlations improved only when stage II tumors were analyzed (Saal et al., 2005). Li et al. (2006) found that mutations at *PIK3CA* were associated with larger tumor size, and positive estrogen and progesterone receptor status, and occur more frequently in tumors with well-differentiated histology. Furthermore, patients with *PIK3CA* mutations showed significantly worse survival, particularly those with positive estrogen receptor expression or nonamplified ERBB2 (Li et al., 2006). *PIK3CA* mutation was an independent factor for worse survival in breast cancer patients with nonamplified ERBB2 (Li et al., 2006).

Regarding the histological type of breast tumors, some authors claim that *PIK3CA* mutations did not correlate with the histological type (ductal/lobular) of breast cancers (Bachman et al., 2004; Campbell et al., 2004; Levine et al., 2005), while others postulate that in primary invasive breast tumors, mutations of *PIK3CA* gene are more frequently found in lobular, less frequent in ductal and uncommon in medullary, mucinous, and papillary tumors (Buttitta et al., 2005).

In the ductal type, the frequency of *PIK3CA* mutations estimated for intraductal breast carcinomas is 13%, while 31% of invasive ductal carcinomas harbor *PIK3CA* mutations (Wu et al., 2005b), suggesting that *PIK3CA* mutations could confer invasive potential to the cancer cells.

PIK3CA mutations in breast cancer were found to be mutually exclusive with the loss of *PTEN* (a repressor of PI3K pathway) (Saal et al., 2005).

Somatic mutations, rather than gain of gene copy number of *PIK3CA*, are the frequent genetic alterations that contribute to human breast cancer progression (Wu et al., 2005b). Although amplification of *PIK3CA* is not a frequent event in breast tumorigenesis (Bachman et al., 2004; Levine et al., 2005; Wu et al., 2005b), the observation of AKT2 amplification and *PIK3CA* mutation in breast cancers implicate the PI3K—AKT pathway among the most important molecular mechanisms underlying sporadic breast cancer, described so far (Levine et al., 2005).

2.1.7 Gastric Cancer

In gastric cancer, *PIK3CA* mutations were identified in 4–11% of the cases (Lee et al., 2005; Li et al., 2005; Velho et al., 2005). A higher frequency was found by Samuels et al. (2004) (25%), but this study was performed in a small series of tumors ($n = 12$). In gastric carcinomas, the mismatch repair status of the tumors is an important parameter to consider when determining the frequency of *PIK3CA* mutations. *PIK3CA* mutations seem to occur preferentially in the context of microsatellite instability (MSI), rather than in microsatellite stable (MSS) gastric carcinomas. In our work (Velho et al., 2005), no mutations of *PIK3CA* gene were found in 21 mismatched repair proficient tumors. According to our data, Li et al. (2005) found a single case from 73 MSS gastric carcinomas analyzed, to be a mutation at *PIK3CA*. The mutational distribution of *PIK3CA* in gastric cancer is similar to the one observed for *KRAS*, which is mutated only in the MSI subset (Brennetot et al., 2003). It is likely that *PIK3CA* mutations do not represent an important oncogenic event for the development/progression of MSS gastric carcinomas, analogous to what is observed for *KRAS* and *BRAF* mutations in mismatched repair proficient gastric carcinomas (Brennetot et al., 2003; Oliveira et al., 2003). In MSI gastric carcinomas we found *PIK3CA* mutations in 19.2% of the cases (Velho et al., 2005). Further, we verified that *PIK3CA* and *KRAS* mutations were mutually exclusive events in MSI gastric carcinogenesis (Velho et al., 2005). These findings are not surprising since PI3K may function as a downstream effector of the RAS pathway (Wymann and Pirola, 1998; Katso et al., 2001). It has been demonstrated that genes involved in the same signaling pathway may manifest mutations in cancer cells in a mutually exclusive manner, presumably due to the lack of selective growth advantage in having a second hit in the already altered pathway (Li et al., 2005). In gastric cancer, *PIK3CA* alterations were found in both early and advanced specimens (Lee et al., 2005), which could indicate an important role in the development of this tumor type.

Genomic amplification of *PIK3CA* was found to occur in 36.4% of gastric tumors and it was strongly associated with increased expression of *PIK3CA* transcript and elevated levels of phosphor-AKT (Byun et al., 2003). Moreover, *PIK3CA* amplification was predominantly detected in tumors with no *PTEN* alterations, suggesting that loss of *PTEN* and activation of *PIK3CA* genes are mutually exclusive events in gastric tumorigenesis (Byun et al., 2003).

Amplification of *PIK3CA* gene was examined in several gastric cancer cell lines (Byun et al., 2003) and showed to be present in 60% of them (Byun et al., 2003). Crossing the result of *PIK3CA* amplification in gastric cancer cell lines with data from the literature concerning the microsatellite status of these cell lines it is interesting to note that some cell lines bearing *PIK3CA* amplification are considered microsatellite stable (Yao et al., 2004). This observation suggests a possible role for the activation of PI3K pathway, through *PIK3CA* amplification, in MSS gastric cancer. Microsatellite instable tumors are more prone to the accumulation of mutations in repetitive and nonrepetitive sequences (Ionov et al., 1993; Philp et al., 2000), thus, *PIK3CA* mutations may represent an event restricted to MSI gastric cancer. These considerations need to be clarified in order to better understand the pattern and role of *PIK3CA* alterations in this type of tumors.

2.1.8 Colon Cancer

Several studies were performed in order to clarify the incidence of *PIK3CA* mutations in colon cancer. The frequency of mutations found in this tumor model ranges from 10 to 32% (Campbell et al., 2004; Samuels et al., 2004; Velho et al., 2005). The higher frequency of mutations (32%) was found when a higher number of exons were analyzed (Samuels et al., 2004). However, this fact may not be the main reason for this discrepancy since 85% of the mutations cluster in the hotspot exons (9 and 20) (Ma et al., 2000; Bachman et al., 2004; Broderick et al., 2004; Campbell et al., 2004; Osaki et al., 2004b; Samuels et al., 2004; Buttitta et al., 2005; Garcia-Rostan et al., 2005; Lee et al., 2005; Levine et al., 2005; Oda et al., 2005; Saal et al., 2005; Velho et al., 2005; Wang et al., 2005; Wu et al., 2005b; Li et al., 2006).

In premalignant colorectal lesions only 2 of 76 cases harbored mutations in the gene indicating that *PIK3CA* mutations arise late in colorectal tumorigenesis (Samuels et al., 2004).

Fifteen percent of sporadic colon carcinoma has microsatellite instability due to mismatched repair deficiency (Ionov et al., 1993; Philp et al., 2000). Two studies have analyzed the relationship between microsatellite instability and mutations in *PIK3CA* gene in colon cancer. *PIK3CA* mutations were identified in a similar frequency in both MSI and MSS colon cancer, suggesting that *PIK3CA* mutations may play a role in the development/progression of both subsets of colorectal carcinomas (Samuels et al., 2004; Velho et al., 2005).

The same studies have correlated the presence of *PIK3CA* mutations with the presence of *KRAS/BRAF* mutations, since it is well known that MAPkinase pathway is involved in sporadic MSI and MSS colorectal carcinoma (Oliveira et al., 2003; Deng et al., 2004; Domingo et al., 2004). We found (Velho et al., 2005), in colon cancer, that *PIK3CA* mutations were significantly more frequent in cases harboring mutations in *KRAS* or *BRAF* than in cases negative for *KRAS* or *BRAF* mutations. There was no difference between the occurrence of concomitant *PIK3CA/KRAS* or *BRAF* mutations between MSI and MSS colorectal cancers (Velho et al., 2005). These findings suggest that these two pathways may operate simultaneously and synergistically in colorectal carcinogenesis and that MSI and MSS colon tumors may trigger the same pathways during their development.

In colorectal cancer, amplification of *PIK3CA* gene seems not to be a common mechanism of activation in this tumor model (Samuels et al., 2004). *PIK3CA* mutations and PTEN protein deregulation were found to be mutually exclusive events in colorectal neoplasias (Frattini et al., 2005).

2.1.9 Hepatocellular Carcinomas

In hepatocellular carcinomas (HCC) *PIK3CA* mutations were identified in 35.6% of 73 samples analyzed (Lee et al., 2005). Fifty percent of the cases displayed the same frameshift mutation (3204_3205insA) (Lee et al., 2005). The 3204_3205insA mutation was never described in any type of cancer analyzed for *PIK3CA* mutation (Lee et al., 2005).

It changes the last C-terminal amino acid (N1068K) in the PIK3CA protein and creates three additional amino acids in the protein (Lee et al., 2005). Whether this mutation has oncogenic activity and how it contributes to tumorigenesis remains to be elucidated. Another frequent *PIK3CA* mutation found in hepatocellular carcinomas in the work of Lee et al. (2005) was the missense mutation A1634C. Campbell et al. (2005) wrote about this mutation as the result of the amplification of pseudogene on chromosome 22 that has an almost exact match with *PIK3CA* exons 9 to 13. In another study aiming at identifying the frequency of *PIK3CA* hotspot mutations in HCC, carried out in Japanese samples, no mutations were observed (Tanaka et al., 2005). In addition, they found abnormally migrating waves near the end of exon 9 in the PCR chromatograms. PCR amplification and subsequent cloning and sequencing revealed that this chromatograms contained two distinct sequences, the wild-type p110α sequence and another sequence corresponding to chromosome 22q11.2 (Tanaka et al., 2005). Therefore, the frequency of *PIK3CA* mutations found by Lee et al. (2005) in hepatocellular carcinomas is not expected to be as high as it was previously found. Lee et al. (2005) analyzed not only hepatocellular carcinomas but also a wider range of common human cancers and found the A1634C mutation only in hepatocellular carcinomas. This result questions whether an unspecific PCR would occur specifically in a fraction of hepatocellular carcinomas, and not in other tumor models.

2.1.10 Thyroid Tumors

In thyroid tumors, *PIK3CA* gene mutations seem not to be a frequent event in benign thyroid adenomas, PTC (papillary thyroid carcinomas), FTC (follicular thyroid carcinomas) and medullary thyroid tumors (Wu et al., 2005a). With respect to the anaplastic variant of thyroid cancer (ATC), opposing results were obtained regarding the frequency of *PIK3CA* mutations. In the work of Wu et al. (2005a) no mutations in the *PIK3CA* gene were observed, while in the work of Garcia-Rostan et al. (2005) mutations were frequently observed in this variant of thyroid cancer.

Amplification on *PIK3CA* gene was found in FTC, PTC, adenomas, MTC (parafollicular C cell-derived medullary thyroid cancer), and in ATC where it appears to be more frequently observed (Liu et al., 2005; Wu et al., 2005a).

2.1.11 Ovarian Cancer

Mutations in the *PIK3CA* gene occur at a low frequency in ovarian tumors. Four to 12% of ovarian cancers harbor mutations in *PIK3CA* gene (Campbell et al., 2004; Levine et al., 2005; Wang et al., 2005).

No association was identified regarding *PIK3CA* mutations and patient age or survival, or stage and grade of the tumors (Levine et al., 2005). The distribution of the mutations between the histological subtypes of ovarian cancer seems not to reach a consensus. They were found in endometrioid, clear cell, serous, mucinous and undifferentiated subtypes (Campbell et al., 2004; Levine et al., 2005; Wang et al., 2005), although, among the literature, mutations of *PIK3CA* gene appear to be more frequent in the endometrioid, clear cell and serous histological subtypes (Campbell et al., 2004; Levine et al., 2005). One study described a strong association between *PIK3CA* mutations in endometrioid and clear cell histological subtypes (Campbell et al., 2004), while other studies do not report such an association (Levine et al., 2005). When advanced ovarian tumors were analyzed, *PIK3CA* mutations were detected more commonly in low-grade and in mucinous or clear cell tumors (Wang et al., 2005).

Amplification of the chromosomal region 3q26 which harbors *PIK3CA* gene has been reported to be a common event in ovarian tumors (Iwabuchi et al., 1995; Shayesteh et al., 1999; Campbell et al., 2004). It is more frequently found in low-grade and low-stage tumors and thus may act as an early event in ovarian cancer development (Iwabuchi et al., 1995; Campbell et al., 2004).

The association between *PIK3CA* gene amplification and the histologic subtype of ovarian cancer (serous, endometrioid, clear cell, mucinous and undifferentiated) was analyzed by Campbell et al. (2004) but no association was found between the two parameters.

In ovarian cancer, mutations of *PIK3CA* and amplification of *PIK3CA* gene are alternative alterations in ovarian cancers (Shayesteh et al., 1999; Campbell et al., 2004; Wang et al., 2005).

2.1.12 Endometrial Carcinoma

Endometrial carcinoma is the most common malignancy of the female genital tract (Kanamori et al., 2001).

The incidence of *PIK3CA* mutations in endometrial cancers is one of the highest among all types of cancers analyzed. About 36% of endometrial carcinomas harbor *PIK3CA* mutations (Oda et al., 2005). Amplification of the *PIK3CA* locus was also detected in endometrial carcinomas (Oda et al., 2005). Mutations in the *PTEN* gene were found to occur in about 30–50% of endometrial carcinomas and is what makes the *PTEN* gene one of the most commonly mutated genes in this type of carcinomas (Risinger et al., 1997; Tashiro et al., 1997; Di Cristofano and Pandolfi, 2000). Furthermore, activation of the PI3K/AKT pathway caused by the loss of *PTEN* was shown to be involved in the mechanism of endometrial carcinogenesis (Kanamori et al., 2001).

Simultaneous mutations in the *PIK3CA* and *PTEN* genes have been thought to be mutually exclusive, as reported in breast carcinoma, glioblastomas and colorectal cancers (Broderick et al., 2004; Frattini et al., 2005; Saal et al., 2005). The same is not true for endometrial carcinomas where the coexistence of the two mutational events was observed in 26% of the tumors (Oda et al., 2005).

In endometrial carcinomas, *PIK3CA* mutations seems not to be associated with the histologic grade, International Federation of Gynecology and Obstetrics (FIGO) stage, lymph node metastasis, and estrogen/progesterone receptor status of the tumors (Oda et al., 2005).

2.1.13 Cervical Tumors

Cervical carcinomas show a recurrent pattern of cytogenetic instability where a gain of the long arm of chromosome 3 is commonly observed (Bertelsen et al., 2005). A possible oncogene in this region is *PIK3CA* at 3q26.3 (Bertelsen et al., 2005). Gain of *PIK3CA* gene copy number is frequently observed in cervical tumors and is clearly accompanied by phosphorylation of AKT (Ma et al., 2000; Zhang et al., 2002; Bertelsen et al., 2005).

In cervical cancer, cell lines harboring amplified *PIK3CA* showed increased expression of *PIK3CA* (p110α) which was subsequently associated with high kinase activity (Ma et al., 2000). In addition, transformation phenotypes in these lines, including increased cell growth and decreased apoptosis, were found to be significantly affected by the treatment with specific PI3-kinase inhibitors, suggesting that increased expression of *PIK3CA* in cervical cancer may result in promoting cell proliferation and reducing apoptosis (Ma et al., 2000). These evidences support that *PIK3CA* is an oncogene in cervical cancer and *PIK3CA* amplification may be linked to cervical tumorigenesis (Ma et al., 2000).

2.2 *Functional Studies*

The high incidence of nonrandom *PIK3CA* mutations detected across different types of tumors strongly suggests a functional significance in tumorigenesis (Kang et al., 2005a). No truncating mutations were observed and 85% of alterations occurred in two small clusters in the helical and kinase domains. This is similar to what is observed for activating mutations in other oncogenes, such as *RAS*, *BRAF*, *CTNNB1*, and members of the tyrosine kinome. The positions of the mutations imply that they are likely to increase kinase activity (Samuels et al., 2004) and the affected residues are highly conserved evolutionarily (Samuels et al., 2004). All these observations suggest an oncogenic role for *PIK3CA* alterations in human cancers.

Functional studies were done by several authors with the purpose of understanding the role of *PIK3CA* mutations in tumorigenesis (Fig. 3). In vitro studies employing transfection of chicken embryo fibroblasts (CEFs) (Kang et al., 2005a), NIH 3T3 cells (Ikenoue et al., 2005), human mammary epithelial (Isakoff et al., 2005; Zhao et al., 2005) or colorectal cancer cell lines (CRC cell lines) (Samuels et al., 2005) were done. In vivo experiments using mice were also selected as experimental systems to test the functional significance of *PIK3CA* mutations in cancer (Samuels et al., 2005; Zhao et al., 2005). These studies demonstrated that the most frequent *PIK3CA* mutants (E542K, E545K and H1047R) and some other mutants that have been reported in the p85-binding domain (R38H and G106V), C2 domain (C420R and E453Q) and in the kinase domain (M1043I) have higher lipid kinase activity than the wild-type protein (Ikenoue et al., 2005; Isakoff et al., 2005; Kang et al., 2005a; Samuels et al., 2005; Zhao et al., 2005).

Furthermore, these mutants were shown to harbor increased transforming potential and morphological changes characteristic of cancer cells such as loss of contact inhibition (Ikenoue et al., 2005), anchorage- and growth factor-independent proliferation (Ikenoue et al., 2005; Isakoff et al., 2005; Kang et al., 2005a; Zhao et al., 2005). The mutations E545K and H1047R were shown to confer in addition to the previously described effects, resistance to apoptosis (Samuels et al., 2005).

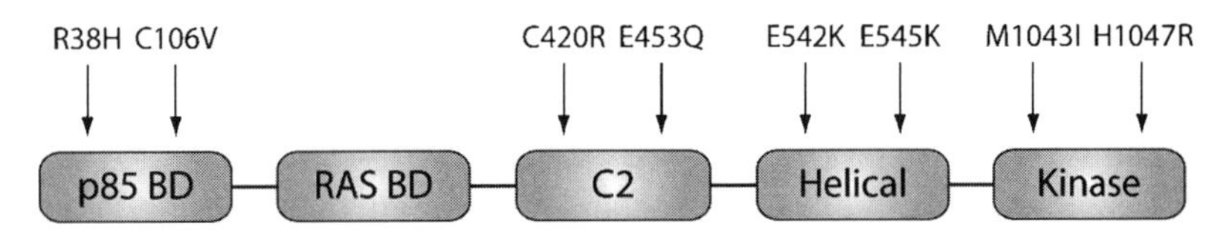

Fig. 3. *PIK3CA* mutations that were submitted to functional studies within the different domains of p110 catalytic subunit of PI3K

Another interesting observation was that CRC cell lines and mammary epithelial cell lines harboring *PIK3CA* mutations are able to induce tumor formation in vivo (Samuels et al., 2005; Zhao et al., 2005). When *PIK3CA* mutant CRC cell lines were injected in mice, tumor cells were able to migrate and invade the adjacent tissue layers, and give rise to metastases (Samuels et al., 2005). This ability of inducing metastases was only found in CRC cell lines and not on mammary epithelial cell lines (Zhao et al., 2005). The different abilities in the formation of metastasis between these distinct tumor cell models may indicate that *PIK3CA* mutations may have different effects regarding metastasis formation depending on the tumor type.

Concerning the downstream pathways activated by mutant *PIK3CA* (R38H, G106V, C420R, E453Q, E542K, E545K, M1043I and H1047R) it was determined that all these mutations highly activate AKT signaling (Ikenoue et al., 2005; Isakoff et al., 2005; Kang et al., 2005a; Samuels et al., 2005; Zhao et al., 2005), with the exception of R38H that exhibit only a slight increase of the phospho-AKT levels (Ikenoue et al., 2005; Zhao et al., 2005). Curiously, the AKT downstream targets that are activated following its phosphorylation in *PIK3CA* mutant cells differ according to the experimental system used. When *PIK3CA* mutant CEFs or NIH 3T3 cells are used as a model, the targets of active AKT seem to be TOR, p70S6 kinase and 4E-binding protein 1, which suggests the involvement of aberrant protein translation control in PI3K-induced oncogenic transformation (Ikenoue et al., 2005; Kang et al., 2005a). On the other hand, when mutant *PIK3CA* CRC cell lines are studied, the downstream effectors of active AKT are FKHR and FKHRL1 proteins (Samuels et al., 2005). The inhibition of forkhead transcription factors by active AKT is thought to confer resistance to apoptosis in CRC cell lines by inhibiting the transcription of proapoptotic genes (Osaki et al., 2004b; Samuels et al., 2005). As it is hypothesized by Samuels et al. (2005), these differences on the targets of AKT activation may depend on the cell type, species, or experimental system analyzed. Additional experiments need to be brought about to address which pathways are targeted upon PIK3CA activation in order to discern the role of oncogenic *PIK3CA* in human tumors. The mechanism by which mutations in *PIK3CA* gene lead to constitutive activation of p110α subunit is not clear.

Oncogenic activation of p110α catalytic subunit of PI3K is not due to an increase of binding affinity to p85 regulatory subunit with concomitant protein stabilization. The evidences to support this assumption are (1) the two p85-binding mutants that were studied (R38H and G106V) displayed modest reductions of binding affinity to p85 (Ikenoue et al., 2005); (2) from all the mutations analyzed, R38H was the less effective in activating the downstream effector of PI3K, AKT (Ikenoue et al., 2005; Zhao et al., 2005); and (3) cells with *PIK3CA* mutants that completely abolish p85-binding domain display increased PI3K activity (Ikenoue et al., 2005; Zhao et al., 2005).

Mutations in *PIK3CA* gene affecting the C2 domain could cause a permanent anchorage of p110α catalytic subunit to the cell membrane. However, membrane anchoring of p110α by myristoylation resulted in a much lower level of PI3K activity compared with the E545K and H1047R mutations (Zhao et al., 2005).

H1047R mutation is located near the activation loop in the catalytic domain of PIK3CA and is likely to affect the specificity or affinity of p110α towards the lipid substrate of PI3K (Bader et al., 2005). Moreover, mutations in the helical domain of

PIK3CA may cause a change in enzyme conformation and possibly disrupt its ability to interact with a still unknown regulator protein (Bader et al., 2005).

2.3 Clinical and Therapeutic Implications of PIK3CA Alterations

The occurrence of mutations and amplification of *PIK3CA* gene in several types of cancer and the frequent clustered mutations within *PIK3CA* implies that this pathway makes an attractive molecular marker for early detection of cancers or for monitoring tumor progression, and a promising therapeutic target for the development of specific inhibitors of p110α subunit.

There are some inhibitors of PI3K pathway that have been extensively used in in vitro and in vivo studies, such as wortmannin and LY294002, which inhibit the catalytic activity of p110 subunits of PI3K (Vivanco and Sawyers, 2002; Osaki et al., 2004a; Samuels and Velculescu, 2004). These compounds were shown to inhibit cell proliferation and/or induce apoptosis in cancer cells via inhibition of PI3K—AKT pathway and also play a role in enhancing the effectiveness of radio- or chemotherapy (Sarkaria et al., 1998; Ng et al., 2000; Osaki et al., 2004a,b; Kim et al., 2006). These inhibitors may also be applied to mutant forms of *PIK3CA*, considering that the transforming potential of *PIK3CA* mutants can still be inhibited by broadly acting PI3K inhibitors such as LY294002 (Samuels et al., 2005). Despite their efficacy towards inhibition of active PI3K pathway, such inhibitors have the disadvantage of nonselectivity (Kim et al., 2006). PI3K is ubiquitously present in many mammalian cells (Kim et al., 2006), thus the use of wortmannin and LY294002 may be toxic to the cells that do not display any alteration of the pathway. The lack of stability of wortmannin and the lack of solubility of LY294002 are additional reasons that hampered further clinical studies of these agents (Kim et al., 2006).

Another possibility of therapy is the use of inhibitors of the downstream targets of PI3K pathway. As it was shown, *PIK3CA* mutations lead to an increase in the phosphorylation of AKT with its concomitant activation (Ikenoue et al., 2005; Isakoff et al., 2005; Kang et al., 2005b; Samuels et al., 2005; Zhao et al., 2005). The repression of the activation of AKT represents an attractive possibility of inhibiting the PI3K pathway, although, no small AKT inhibitors have been established yet (Osaki et al., 2004a).

Further, the inhibition of PI3K—AKT target genes may represent a fighting chance. mTOR is an example of this possibility. Some inhibitors of mTOR are under preclinical and clinical investigations (Kim et al., 2006). It is the case of rapamycin, and its derivatives (Rad001, CCI-779, and AP23573) (Chan, 2004), that were shown to work as potent radiosensitizers of endothelial cells in vitro and led to improved tumor-growth delay of glioma xenografs in vivo (Kim et al., 2006). Moreover, it was demonstrated that in CEFs bearing *PIK3CA* mutations commonly found in human cancers, mTOR was a target for activation, and a key player in the transforming potential of these mutant cells (Kang et al., 2005b). Treatment of these cells with rapamycin interferes with the oncogenicity induced by the cancer-derived mutant *PIK3CA* (Kang et al., 2005b). Together, these results imply that inhibition of the downstream targets may be a valuable approach to indirectly block the oncogenic ability of the active PI3K—AKT pathway. However, it should be taken in account that the downstream effectors of PI3K that are activated upon deregulation of the pathway may depend on the cell type or tumor model. If such inhibitors become a reality then, it would be important to access the active target to be silenced for each tumor.

The best option to prevent the transforming capability of mutant *PIK3CA* is to employ specific inhibitors for the p110α subunit. Patent specifications which described inhibitors of PI3K, including compounds that exhibit some selectivity for individual p110 catalytic isoforms have been published (Ward et al., 2003; Ward and Finan, 2003). Imidazopyridine derivatives are claimed to exhibit excellent PI3K inhibitory activity, especially against p110α, although no isoforms selectivity data are provided (Ward et al., 2003). Quinolone and pyridopyrimidine compounds, closely related to LY294002, are approximately 100-fold more potent against α/β isoforms compared to δ isoforms (Ward et al., 2003).

In the future, the development, if possible, of inhibitors that target specifically the mutant forms of *PIK3CA* would be of particular interest. By this way the toxicity of the already known inhibitors would be overcome since only *PIK3CA* mutant cancer cells would be affected by the action of the inhibitor.

REFERENCES

Ali, I. U., Schriml, L. M. and Dean, M. 1999. Mutational spectra of PTEN/MMAC1 gene: a tumor suppressor with lipid phosphatase activity. J Natl Cancer Inst 91: 1922–1932

Bachman, K. E., Argani, P., Samuels, Y., Silliman, N., Ptak, J., Szabo, S., Konishi, H., Karakas, B., Blair, B. G., Lin, C., Peters, B. A., Velculescu, V. E. and Park, B. H. 2004. The PIK3CA gene is mutated with high frequency in human breast cancers. Cancer Biol Ther 3: 772–775

Bader, A. G., Kang, S., Zhao, L. and Vogt, P. K. 2005. Oncogenic PI3K deregulates transcription and translation. Nat Rev Cancer 5: 921–929

Bertelsen, B. I., Steine, S. J., Sandvei, R., Molven, A. and Laerum, O. D. 2005. Molecular analysis of the PI3K–AKT pathway in uterine cervical neoplasia: frequent PIK3CA amplification and AKT phosphorylation. Int J Cancer, 118: 1877–1883

Blount, P. L., Ramel, S., Raskind, W. H., Haggitt, R. C., Sanchez, C. A., Dean, P. J., Rabinovitch, P. S. and Reid, B. J. 1991. 17p allelic deletions and p53 protein overexpression in Barrett's adenocarcinoma. Cancer Res 51: 5482–5486

Brennetot, C., Duval, A., Hamelin, R., Pinto, M., Oliveira, C., Seruca, R. and Schwartz, S. J. 2003. Frequent Ki-ras mutations in gastric tumors of the MSI phenotype. Gastroenterology 125: 1282

Broderick, D. K., Di, C., Parrett, T. J., Samuels, Y. R., Cummins, J. M., McLendon, R. E., Fults, D. W., Velculescu, V. E., Bigner, D. D. and Yan, H. 2004. Mutations of PIK3CA in anaplastic oligodendrogliomas, high-grade astrocytomas, and medulloblastomas. Cancer Res 64: 5048–5050

Buttitta, F., Felicioni, L., Barassi, F., Martella, C., Paolizzi, D., Fresu, G., Salvatore, S., Cuccurullo, F., Mezzetti, A., Campani, D. and Marchetti, A. 2005. PIK3CA mutation and histological type in breast carcinoma: high frequency of mutations in lobular carcinoma. J Pathol 208: 350–355

Byun, D. S., Cho, K., Ryu, B. K., Lee, M. G., Park, J. I., Chae, K. S., Kim, H. J. and Chi, S. G. 2003. Frequent monoallelic deletion of PTEN and its reciprocal association with PIK3CA amplification in gastric carcinoma. Int J Cancer 104: 318–327

Campbell, I. G., Russell, S. E., Choong, D. Y. H., Montgomery, K. G., Ciavarella, M. L., Hooi, C. S. F., Cristiano, B. E., Pearson, R. B. and Phillips, W. A. 2004. Mutation of the PIK3CA gene in ovarian and breast cancer. Cancer Res 64: 7678–7681

Campbell, I. G., Russell, S. E., Phillips, W. A., Levine, D. A. and Boyd, J. 2005. PIK3CA Mutations in Ovarian Cancer. Clin Cancer Res 11: 7042–7043

Cantley, L. C. and Neel, B. G. 1999. New insights into tumor suppression: PTEN suppresses tumor formation by restraining the phosphoinositide 3-kinase/AKT pathway. Proc Natl Acad Sci USA 96: 4240–4245

Chan, S. 2004. Targeting the mammalian target of rapamycin (mTOR): a new approach to treating cancer. Br J Cancer 91: 1420–1424

Clemons, M. and Goss, P. 2001. Estrogen and the risk of breast cancer. N Engl J Med 344: 276–285

Deng, G., Bell, I., Crawley, S., Gum, J., Terdiman, J. P., Allen, B. A., Truta, B., Sleisenger, M. H. and Kim, Y. S. 2004. BRAF mutation is frequently present in sporadic colorectal cancer with methylated hMLH1, but not in hereditary nonpolyposis colorectal cancer. Clin Cancer Res 10: 191–195

Di Cristofano, A. and Pandolfi, P. P. 2000. The multiple roles of PTEN in tumor suppression. Cell 100: 387–390

Domin, J. and Waterfield, M. D. 1997. Using structure to define the function of phosphoinositide 3-kinase family members. FEBS Lett 410: 91–95

Domingo, E., Espin, E., Armengol, M., Oliveira, C., Pinto, M., Duval, A., Brennetot, C., Seruca, R., Hamelin, R., Yamamoto, H. and Schwartz, S. J. 2004. Activated BRAF targets proximal colon tumors with mismatch repair deficiency and MLH1 inactivation. Genes Chromosomes Cancer 39: 138–142

Frattini, M., Signoroni, S., Pilotti, S., Bertario, L., Benvenuti, S., Zanon, C., Bardelli, A. and Pierotti, M. A. 2005. Phosphatase protein homologue to tensin expression and phosphatidylinositol-3 phosphate kinase mutations in colorectal cancer. Cancer Res 65: 11227

Garcia-Rostan, G., Costa, A. M., Pereira-Castro, I., Salvatore, G., Hernandez, R., Hermsem, M. J. A., Herrero, A., Fusco, A., Cameselle-Teijeiro, J. and Santoro, M. 2005. Mutation of the PIK3CA gene in anaplastic thyroid cancer. Cancer Res 65: 10199–10207

Gonzalez, M. V., Artimez, M. L., Rodrigo, L., Lopez-Larrea, C., Menendez, M. J., Alvarez, V., Perez, R., Fresno, M. F., Perez, M. J., Sampedro, A. and Coto, E. 1997. Mutation analysis of the p53, APC, and p16 genes in the Barrett's oesophagus, dysplasia, and adenocarcinoma. J Clin Pathol 50: 212–217

Hill, M. M. and Hemmings, B. A. 2002. Inhibition of protein kinase B/Akt. implications for cancer therapy. Pharmacol Ther 93: 243–251

Huang, Y., Boynton, R. F., Blount, P. L., Silverstein, R. J., Yin, J., Tong, Y., McDaniel, T. K., Newkirk, C., Resau, J. H., Sridhara, R., et al. 1992. Loss of heterozygosity involves multiple tumor suppressor genes in human esophageal cancers. Cancer Res 52: 6525–6530

Hui, A. B., Lo, K. W., Yin, X. L., Poon, W. S. and Ng, H. K. 2001. Detection of multiple gene amplifications in glioblastoma multiforme using array-based comparative genomic hybridization. Lab Invest 81: 717–723

Ikenoue, T., Kanai, F., Hikiba, Y., Obata, T., Tanaka, Y., Imamura, J., Ohta, M., Jazag, A., Guleng, B., Tateishi, K., Asaoka, Y., Matsumura, M., Kawabe, T. and Omata, M. 2005. Functional analysis of PIK3CA gene mutations in human colorectal cancer. Cancer Res 65: 4562–4567

Ionov, Y., Peinado, M. A., Malkhosyan, S., Shibata, D. and Perucho, M. 1993. Ubiquitous somatic mutations in simple repeated sequences reveal a new mechanism for colonic carcinogenesis. Nature 363: 558–561

Isakoff, S. J., Engelman, J. A., Irie, H. Y., Luo, J., Brachmann, S. M., Pearline, R. V., Cantley, L. C. and Brugge, J. S. 2005. Breast cancer-associated PIK3CA mutations are oncogenic in mammary epithelial cells. Cancer Res 65: 10992–11000

Iwabuchi, H., Sakamoto, M., Sakunaga, H., Ma, Y. Y., Carcangiu, M. L., Pinkel, D., Yang-Feng, T. L. and Gray, J. W. 1995. Genetic analysis of benign, low-grade, and high-grade ovarian tumors. Cancer Res 55: 6172–6180

Kanamori, Y., Kigawa, J., Itamochi, H., Shimada, M., Takahashi, M., Kamazawa, S., Sato, S., Akeshima, R. and Terakawa, N. 2001. Correlation between loss of PTEN expression and Akt phosphorylation in endometrial carcinoma. Clin Cancer Res 7: 892–895

Kang, S., Bader, A. G. and Vogt, P. K. 2005a. Phosphatidylinositol 3-kinase mutations identified in human cancer are oncogenic. Proc Natl Acad Sci USA 102: 802–807

Kang, S., Bader, A. G., Zhao, L. and Vogt, P. K. 2005b. Mutated PI 3-kinases: cancer targets on a silver platter. Cell Cycle 4: 578–581

Katso, R., Okkenhaug, K., Ahmadi, K., White, S., Timms, J. and Waterfield, M. D. 2001. Cellular function of phosphoinositide 3-kinases: implications for development, immunity, homeostasis, and cancer. Annu Rev Cell Dev Biol 17: 615–675

Kim, D. W., Huamani, J., Fu, A. and Hallahan, D. E. 2006. Molecular strategies targeting the host component of cancer to enhance tumor response to radiation therapy. Int J Radiat Oncol Biol Phys 64: 38–46

Knobbe, C. B. and Reifenberger, G. 2003. Genetic alterations and aberrant expression of genes related to the phosphatidyl-inositol-3′-kinase/protein kinase B (Akt) signal transduction pathway in glioblastomas. Brain Pathol 13: 507–518

Lee, J. W., Soung, Y. H., Kim, S. Y., Lee, H. W., Park, W. S., Nam, S. W., Kim, S. H., Lee, J. Y., Yoo, N. J. and Lee, S. H. 2005. PIK3CA gene is frequently mutated in breast carcinomas and hepatocellular carcinomas. Oncogene 24: 1477–1480

Levine, D. A., Bogomolniy, F., Yee, C. J., Lash, A., Barakat, R. R., Borgen, P. I. and Boyd, J. 2005. Frequent mutation of the PIK3CA gene in ovarian and breast cancers. Clin Cancer Res 11: 2875–2878

Li, V., Wong, C., Chan, T., Chan, A., Zhao, W., Chu, K.-M., So, S., Chen, X., Yuen, S. and Leung, S. 2005. Mutations of PIK3CA in gastric adenocarcinoma. BMC Cancer 5: 29

Li, S., Rong, M., Grieu, F. and Iacopetta, B. 2006. PIK3CA mutations in breast cancer are associated with poor outcome. Breast Cancer Res Treat 96: 91–95

Liu, D., Mambo, E., Ladenson, P. W. and Xing, M. 2005. Letter re: uncommon mutation but common amplifications of the PIK3CA gene in thyroid tumors. J Clin Endocrinol Metab 90: 5509

Ma, Y. Y., Wei, S. J., Lin, Y. C., Lung, J. C., Chang, T. C., Whang-Peng, J., Liu, J. M., Yang, D. M., Yang, W. K. and Shen, C. Y. 2000. PIK3CA as an oncogene in cervical cancer. Oncogene 19: 2739–2744

Maehama, T. and Dixon, J. E. 1999. PTEN: a tumour suppressor that functions as a phospholipid phosphatase. Trends Cell Biol 9: 125–128

Massion, P. P., Kuo, W.-L., Stokoe, D., Olshen, A. B., Treseler, P. A., Chin, K., Chen, C., Polikoff, D., Jain, A. N., Pinkel, D., Albertson, D. G., Jablons, D. M. and Gray, J. W. 2002. Genomic copy number analysis of non-small cell lung cancer using array comparative genomic hybridization: implications of the phosphatidylinositol 3-kinase pathway. Cancer Res 62: 3636–3640

Massion, P. P., Taflan, P. M., Shyr, Y., Rahman, S. M. J., Yildiz, P., Shakthour, B., Edgerton, M. E., Ninan, M., Andersen, J. J. and Gonzalez, A. L. 2004. Early involvement of the phosphatidylinositol 3-kinase/Akt pathway in lung cancer progression. Am J Respir Crit Care Med 170: 1088–1094

Miller, C. T., Moy, J. R., Lin, L., Schipper, M., Normolle, D., Brenner, D. E., Iannettoni, M. D., Orringer, M. B. and Beer, D. G. 2003. Gene amplification in esophageal adenocarcinomas and Barrett's with high-grade dysplasia. Clin Cancer Res 9: 4819–4825

Mizoguchi, M., Nutt, C. L., Mohapatra, G. and Louis, D. N. 2004. Genetic alterations of phosphoinositide 3-kinase subunit genes in human glioblastomas. Brain Pathol 14: 372–377

Ng, S. S. W., Tsao, M.-S., Chow, S. and Hedley, D. W. 2000. Inhibition of phosphatidylinositide 3-kinase enhances gemcitabine-induced apoptosis in human pancreatic cancer cells. Cancer Res 60: 5451–5455

Oda, K., Stokoe, D., Taketani, Y. and McCormick, F. 2005. High frequency of coexistent mutations of PIK3CA and PTEN genes in endometrial carcinoma. Cancer Res 65: 10669–10673

Oliveira, C., Pinto, M., Duval, A., Brennetot, C., Domingo, E., Espin, E., Armengol, M., Yamamoto, H., Hamelin, R., Seruca, R. and Schwartz, S. J. 2003. BRAF mutations characterize colon but not gastric cancer with mismatch repair deficiency. Oncogene 22: 9192–9196

Or, Y. Y., Hui, A. B., Tam, K. Y., Huang, D. P. and Lo, K. W. 2005. Characterization of chromosome 3q and 12q amplicons in nasopharyngeal carcinoma cell lines. Int J Oncol 26: 49–56

Osaki, M., Kase, S., Adachi, K., Takeda, A., Hashimoto, K. and Ito, H. 2004a. Inhibition of the PI3K–Akt signaling pathway enhances the sensitivity of Fas-mediated apoptosis in human gastric carcinoma cell line, MKN-45. J Cancer Res Clin Oncol 130: 8–14

Osaki, M., Oshimura, M. and Ito, H. 2004b. PI3K–Akt pathway: its functions and alterations in human cancer. Apoptosis 9: 667–676

Pedrero, J. M., Carracedo, D. G., Pinto, C. M., Zapatero, A. H., Rodrigo, J. P., Nieto, C. S. and Gonzalez, M. V. 2005. Frequent genetic and biochemical alterations of the PI 3-K/AKT/PTEN pathway in head and neck squamous cell carcinoma. Int J Cancer 114: 242–248

Philp, A. J., Phillips, W. A., Rockman, S. P., Vincan, E., Baindur-Hudson, S., Burns, W., Valentine, R. and Thomas, R. J. 2000. Microsatellite instability in gastrointestinal tract tumours. Int J Surg Investig 2: 267–274

Philp, A. J., Campbell, I. G., Leet, C., Vincan, E., Rockman, S. P., Whitehead, R. H., Thomas, R. J. S. and Phillips, W. A. 2001. The phosphatidylinositol 3′-kinase p85{alpha} gene is an oncogene in human ovarian and colon tumors. Cancer Res 61: 7426–7429

Risinger, J. I., Hayes, A. K., Berchuck, A. and Barrett, J. C. 1997. PTEN/MMAC1 mutations in endometrial cancers. Cancer Res 57: 4736–4738

Ross, J. S., Fletcher, J. A., Bloom, K. J., Linette, G. P., Stec, J., Symmans, W. F., Pusztai, L. and Hortobagyi, G. N. 2004. Targeted therapy in breast cancer: the HER-2/neu gene and protein. Mol Cell Proteomics 3: 379–398

Saal, L. H., Holm, K., Maurer, M., Memeo, L., Su, T., Wang, X., Yu, J. S., Malmstrom, P.-O., Mansukhani, M., Enoksson, J., Hibshoosh, H., Borg, A. and Parsons, R. 2005. PIK3CA mutations correlate with hormone receptors, node metastasis, and ERBB2, and are mutually exclusive with PTEN loss in human breast carcinoma. Cancer Res 65: 2554–2559

Samuels, Y. and Ericson, K. 2006. Oncogenic PI3K and its role in cancer. Curr Opin Oncol 18: 77–82

Samuels, Y. and Velculescu, V. E. 2004. Oncogenic mutations of PIK3CA in human cancers. Cell Cycle 3: 1221–1224

Samuels, Y., Wang, Z., Bardelli, A., Silliman, N., Ptak, J., Szabo, S., Yan, H., Gazdar, A., Powell, S. M., Riggins, G. J., Willson, J. K. V., Markowitz, S., Kinzler, K. W., Vogelstein, B. and Velculescu, V. E. 2004. High frequency of mutations of the PIK3CA gene in human cancers. Science 304: 554

Samuels, Y., Diaz, L. A. J., Schmidt-Kittler, O., Cummins, J. M., Delong, L., Cheong, I., Rago, C., Huso, D. L., Lengauer, C., Kinzler, K. W., Vogelstein, B. and Velculescu, V. E. 2005. Mutant PIK3CA promotes cell growth and invasion of human cancer cells. Cancer Cell 7: 561–573

Sansal, I. and Sellers, W. R. 2004. The biology and clinical relevance of the PTEN tumor suppressor pathway. J Clin Oncol 22: 2954–2963

Sarkaria, J. N., Tibbetts, R. S., Busby, E. C., Kennedy, A. P., Hill, D. E. and Abraham, R. T. 1998. Inhibition of phosphoinositide 3-kinase related kinases by the radiosensitizing agent wortmannin. Cancer Res 58: 4375–4382

Shayesteh, L., Lu, Y., Kuo, W. L., Baldocchi, R., Godfrey, T., Collins, C., Pinkel, D., Powell, B., Mills, G. B. and Gray, J. W. 1999. PIK3CA is implicated as an oncogene in ovarian cancer. Nat Genet 21: 99–102

Tanaka, Y., Kanai, F., Tada, M., Asaoka, Y., Guleng, B., Jazag, A., Ohta, M., Ikenoue, T., Tateishi, K., Obi, S., Kawabe, T., Yokosuka, O. and Omata, M. 2005. Absence of PIK3CA hotspot mutations in hepatocellular carcinoma in Japanese patients. Oncogene 25: 2950–2952

Tashiro, H., Blazes, M. S., Wu, R., Cho, K. R., Bose, S., Wang, S. I., Li, J., Parsons, R. and Ellenson, L. H. 1997. Mutations in PTEN are frequent in endometrial carcinoma but rare in other common gynecological malignancies. Cancer Res 57: 3935–3940

Tong, C. Y., Hui, A. B., Yin, X. L., Pang, J. C., Zhu, X. L., Poon, W. S. and Ng, H. K. 2004. Detection of oncogene amplifications in medulloblastomas by comparative genomic hybridization and array-based comparative genomic hybridization. J Neurosurg 100: 187–193

Vanhaesebroeck, B. and Waterfield, M. D. 1999. Signaling by distinct classes of phosphoinositide 3-kinases. Exp Cell Res 253: 239–254

Velho, S., Oliveira, C., Ferreira, A., Ferreira, A. C., Suriano, G., Schwartz, J. S., Duval, A., Carneiro, F. and Machado, J. C. 2005. The prevalence of PIK3CA mutations in gastric and colon cancer. Eur J Cancer 41: 1649–1654

Vivanco, I. and Sawyers, C. L. 2002. The phosphatidylinositol 3-kinase AKT pathway in human cancer. Nat Rev Cancer 2: 489–501

Volinia, S., Hiles, I., Ormondroyd, E., Nizetic, D., Antonacci, R., Rocchi, M. and Waterfield, M. D. 1994. Molecular cloning, cDNA sequence, and chromosomal localization of the human phosphatidylinositol 3-kinase p110 alpha (PIK3CA) gene. Genomics 24: 472–477

Wang, Y., Helland, A., Holm, R., Kristensen, G. B. and Borresen-Dale, A. L. 2005. PIK3CA mutations in advanced ovarian carcinomas. Hum Mutat 25: 322

Ward, S. G. and Finan, P. 2003. Isoform-specific phosphoinositide 3-kinase inhibitors as therapeutic agents. Curr Opin Pharmacol 3: 426–434

Ward, S., Sotsios, Y., Dowden, J., Bruce, I. and Finan, P. 2003. Therapeutic potential of phosphoinositide 3-kinase inhibitors. Chem Biol 10: 207–213

West, K. A., Brognard, J., Clark, A. S., Linnoila, I. R., Yang, X., Swain, S. M., Harris, C., Belinsky, S. and Dennis, P. A. 2003. Rapid Akt activation by nicotine and a tobacco carcinogen modulates the phenotype of normal human airway epithelial cells. J Clin Invest 111: 81–90

Woenckhaus, J., Steger, K., Werner, E., Fenic, I., Gamerdinger, U., Dreyer, T. and Stahl, U. 2002. Genomic gain of PIK3CA and increased expression of p110alpha are associated with progression of dysplasia into invasive squamous cell carcinoma. J Pathol 198: 335–342

Wong, D. J., Paulson, T. G., Prevo, L. J., Galipeau, P. C., Longton, G., Blount, P. L. and Reid, B. J. 2001. p16INK4a lesions are common, early abnormalities that undergo clonal expansion in Barrett's metaplastic epithelium. Cancer Res 61: 8284–8289

Wu, G., Mambo, E., Guo, Z., Hu, S., Huang, X., Gollin, S. M., Trink, B., Ladenson, P. W., Sidransky, D. and Xing, M. 2005a. Uncommon mutation, but common amplifications, of the PIK3CA gene in thyroid tumors. J Clin Endocrinol Metab 90: 4688–4693

Wu, G., Xing, M., Mambo, E., Huang, X., Liu, J., Guo, Z., Chatterjee, A., Goldenberg, D., Gollin, S., Sukumar, S., Trink, B. and Sidransky, D. 2005b. Somatic mutation and gain of copy number of PIK3CA in human breast cancer. Breast Cancer Res 7: R609–R616

Wymann, M. P. and Pirola, L. 1998. Structure and function of phosphoinositide 3-kinases. Biochim Biophys Acta 1436: 127–150

Yao, Y., Tao, H., Kim, J. J., Burkhead, B., Carloni, E., Gasbarrini, A. and Sepulveda, A. R. 2004. Alterations of DNA mismatch repair proteins and microsatellite instability levels in gastric cancer cell lines. Lab Invest 84: 915–922

Yu, J., Wjasow, C. and Backer, J. M. 1998a. Regulation of the p85/p110alpha phosphatidylinositol 3′-kinase. Distinct roles for the N-terminal and C-terminal SH2 domains. J Biol Chem 273: 30199–30203

Yu, J., Zhang, Y., McIlroy, J., Rordorf-Nikolic, T., Orr, G. A. and Backer, J. M. 1998b. Regulation of the p85/p110 phosphatidylinositol 3′-kinase: stabilization and inhibition of the p110alpha catalytic subunit by the p85 regulatory subunit. Mol Cell Biol 18: 1379–1387

Zhang, A., Maner, S., Betz, R., Angstrom, T., Stendahl, U., Bergman, F., Zetterberg, A. and Wallin, K. 2002. Genetic alterations in cervical carcinomas: frequent low-level amplifications of oncogenes are associated with human papillomavirus infection. Int J Cancer 101: 427–433

Zhao, J. J., Liu, Z., Wang, L., Shin, E., Loda, M. F. and Roberts, T. M. 2005. The oncogenic properties of mutant p110{alpha} and p110{beta} phosphatidylinositol 3-kinases in human mammar epithelial cells. Proc Natl Acad Sci USA 102: 18443–18448

Index

Printed in the United States of America